SPIN IN GRAVITY

Is It Possible to Give an Experimental Basis to Torsion?

International School of
Cosmology and Gravitation XV Course

THE SCIENCE AND CULTURE SERIES — PHYSICS

1. Perspectives for New Detectors in Future Supercolliders, 1991
2. Data Structures for Particle Physics Experiments: Evolution or Revolution?, 1991
3. Image Processing for Future High-Energy Physics Detectors, 1992
4. GaAs Detectors and Electronics for High-Energy Physics, 1992
5. Supercolliders and Superdetectors, 1993
6. Properties of SUSY Particles, 1993
7. From Superstrings to Supergravity, 1994
8. Probing the Nuclear Paradigm with Heavy Ion Reactions, 1994
9. Quantum-Like Models and Coherent Effects, 1995
10. Quantum Gravity, 1996
11. Crystalline Beams and Related Issues, 1996
12. The Spin Structure of the Nucleon, 1997
13. Hadron Colliders at the Highest Energy and Luminosity, 1998
14. Universality Features in Multihadron Production and the Leading Effect, 1998
15. Exotic Nuclei, 1998
16. Spin in Gravity: Is It Possible to Give an Experimental Basis to Torsion?, 1998
17. New Detectors, 1999
18. Classical and Quantum Nonlocality, 2000
19. Silicides: Fundamentals and Applications, 2000
20. Superconducting Materials for High Energy Colliders, 2001
21. Deep Inelastic Scattering, 2001
22. Electromagnetic Probes of Fundamental Physics, 2003
23. Epioptics-7, 2004
24. Symmetries in Nuclear Structure, 2004
25. Innovative Detectors for Supercolliders, 2003
26. Complexity, Metastability and Nonextensivity, 2004
27. Epioptics-8, 2004
28. The Physics and Applications of High Brightness Electron Beams, 2005
29. Epioptics-9, 2006
30. Charged and Neutral Particles Channeling Phenomena — Channeling 2008
31. Epioptics-10, 2008
32. Epioptics-11, 2010

THE SCIENCE AND CULTURE SERIES — PHYSICS

SPIN IN GRAVITY

Is It Possible to Give an Experimental Basis to Torsion?

International School of
Cosmology and Gravitation XV Course

Erice, Italy 13 – 20 May 1997

Editors

P. G. Bergmann
New York University

V. de Sabbata
University of Bologna

G. T. Gillies
University of Virginia

P. I. Pronin
University of Moscow

Series Editor
A. Zichichi

World Scientific
Singapore • New Jersey • London • Hong Kong

Published by

World Scientific Publishing Co. Pte. Ltd.
5 Toh Tuck Link, Singapore 596224
USA office: 27 Warren Street, Suite 401-402, Hackensack, NJ 07601
UK office: 57 Shelton Street, Covent Garden, London WC2H 9HE

British Library Cataloguing-in-Publication Data
A catalogue record for this book is available from the British Library.

The Science and Culture Series — Physics
SPIN IN GRAVITY
Is It Possible to Give an Experimental Basis to Torsion?

ISBN-13 978-981-02-3459-1
ISBN-10 981-02-3459-7

PREFACE*

Our get together is about to end. Torsion and spin loom as possible extensions of Albert Einstein's theory of gravity, his general theory of relativity, but their existence and significance have not as yet been established. Why should anyone care?

That is a legitimate question, especially today, when all countries are looking for ways to trim their expenditures. And, programs of experimentation and observation of gravity are getting to be expensive! One major ground for pushing research in the direction of torsion, I believe, is the rapid progress of observational cosmology, along with its remaining major mysteries. Cosmological models have been investigated for the past eighty years. The so-called cosmological principle, which asserts that on a sufficiently large scale the universe is both homogeneous and isotropic, in my opinion never was more than a preliminary working hypothesis; recent observations have cast doubts on its validity.

Another riddle that has not yet been solved is the issue of "dark matter". The shapes of spiral galaxies, including our own, are stable only if their total masses are much greater than the masses of the material visible to us. Possible resolutions might involve torsion fields. This is not to claim that torsion fields are required or established by new observations. Rather, there are facts that we are unable to explain at present with the help of Newton's or Einstein's laws of gravitation. We cannot afford to disregard theories that may possibly provide a way out of these difficulties.

Mankind is about to extend the range of extraterrestial exploration. Understanding the dynamics of large astronomical objects such as our galaxy will then become essential.

---

*This preface has been delivered as 'closing remarks'.

We have been told that among the "universal" constants the Newton's gravitational constant G is least accurately known. To improve that accuracy by earth-bound laboratory experiments appears foreclosed, because of the magnitude of the earth's surface gravitation. Experiments performed in deep space should ameliorate that situation significantly.

Aside from new determination of the value of G, the validity of the inverse-square law of gravity has been questioned recently by several groups. The inverse-square law is supported to a high precision by the orbits of the members of our solar system. Questions arise when we examine our galaxy, or beyond it intergalactic clusters and superclusters. Gravitational interactions between small masses at ranges much smaller than 10^5 km are also being scrutinized.

These are problems calling for the collaboration of experimental and theoretical scientists. A diversity of approaches requires a diversity of talents and experiences. I disagree emphatically with the notion that experimental work is for those individuals who are incapable of doing theoretical research. Rather, the design and the performance of experiments calls for somewhat different aptitudes than theoretical research.

Before adjournment I should like to express our gratitude to our hosts and to all those who have made our get-together pleasurable and productive. Even if we have not resolved our problems, discussing them with each other undoubtedly has clarified them and, we might hope, laid a foundation for future achievements.

Peter Bergmann

Peter Bergmann

Erice, May 14, 1997

Welcome

It is with pleasure that I welcome you all to this fifteenth Course of the International School of Cosmology and Gravitation.

The Director of the Ettore Majorana Centre for Scientific Culture Professor Zichichi has entrusted me with the welcome address on his behalf. We hope that he follow the Course with us for some days.

I am glad to thank the Co-directors of this Course, Prof.Gillies and Prof.Pronin for all the work done to organize this Course and I thank all the lecturers who have accepted our invitation to come in Erice at their own expenses; at this regard I like to thank Prof.Sanders who has authorized the use of Project SEE (Satellite Energy Exchange) funds to pay for the travel of some colleagues.

So, at last, the Course is a reality, thanks to the presence of the lecturers and the numerous participants.

I am glad also to notice that Professor Peter Bergmann, who gives always a strong scientific support to these Courses, is present with his wife.

I think that the introduction of spin in Relativity is a very important point and is necessary if one would try to quantize general relativity.

But the problem is to see if the introduction of spin generalizing the general relativity from a geometric point of view, that is through the concept of torsion, can be experimentally verified. For that reason it is important to have at the same time the presence of theoretical as well experimental phisicists and see if one can give an experimental basis for torsion. In fact, torsion constitutes the more natural and simple way to introduce spin in general relativity. For that reason it is of fundamental importance to see if there are some experiences that indicate, if not directly at least indirectly, the presence of torsion.

We know that until now there is no evidence for torsion but this fact cannot prevent us from considering in some details this implement of research that seems to be important both from a geometrical and a physical point of view.

For this purpose we will discuss the present status of experiments including neutron interferometry, neutron spin rotation induced by torsion in vacuum, anomalous spin-dependent forces with a polarized mass torsion pendulum, space-based searches for spin in gravity, to quote only a few. On the other hand, from the theoretical point of view, different approaches will be considered as, for instance, the nature of torsion and non-metricity, the viability of gravitational theories with spin-torsion and spin-spin interaction, many dimensional gravitational theories with torsion, the spinors in real space-time etc.

First of all I would like to draw the attention of the audience to the fact that Yuval Ne'eman will be present with a series of important lectures on 'World Spinors in Metric and Nonmetric Backgrounds' with spinors on curved space-time, with the interplay between the linear and affine groups, color confinement and other interesting arguments; moreover you will enjoy the presence of theoreticians as Pronin with lectures on the general structure of gravitational theory with torsion and physical effects and measurability of spin-torsion interaction, Borzeszkowski with torsion and Weyl-Cartan geometry, Datta with real spinor fields, Melnikov with D-dimensional cosmological models, Yu Xin (Alfred Yu) with Mach's principle and Torsion in General Relativity, as well as experimentalists like Gillies with recent measurements and searches for variations of Newtonian constant, Ritter with polarized torsion pendulum and searches for anomalous spin coupling to dark matter halo in our galaxy, Sanders with space-based experiments.

We will also enjoy a special lecture by Bergmann on the EIH (Einstein-Infeld-Hoffmann) Theory and the Noether's Theorem.

So we will have a very broad spectrum of subjects that surely will cause many discussions. In fact, as many of you know because they took part in previous courses, after every lecture, that will last not more than fifty minutes, there will be a few minutes' discussion. You can see from the tentative time-table that we will have three lectures in the morning and two lectures in the afternoon with some breaks for coffee (which is free). For lunch and dinner you can choose any of the restaurants approved by the Centre (the list is on the board near the entrance), you sign a list marked "Spin in Gravity" after every meal and you have to pay only for beverages. Now I ask Professor Borzeszkowski to chair the first session. Thank very much.

Venzo de Sabbata

Venzo de Sabbata

CONTENTS

EIH Theory and Noether's Theorem[1]

Peter G. Bergmann

Syracuse University and New York University

Introduction

Classical field theory separates the field, which obeys partial differential equations (the *field equations*), from the *sources*, which satisfy ordinary (*ponderomotive*) differenetial equations, at least when the sources are conceived to be mass points. That the sources of the field are affected by a field's presence is a fact beyond question. But as point-like sources carry a field with them, that tends to infinity at the location of each source, the total field must be split into "incident field" and "self-field" in order to obtain a finite field that determines the behavior of the source.

This procedure is open to criticisms on two scores. First, given the total field, its incident part depends on the choice of self-field, which is far from unique. Second, the whole splitting procedure depends on the linearity of the field laws. As the experimental physicist can only measure the total field, the splitting prescription appears to go against the spirit of a field theory.

General relativity postulates as field equations of the gravitational field partial differential equations that approximate linearity only for weak fields. The equations of motion of binary stars presented a particularly serious conceptual problem, in that each component of the binary system was to travel on geodesics determined by the other only. This choice cannot be justified

---

[1]Dedicated to Engelbert Schucking on the occasion of his seventieth birthday.

as the first step of an approximation. At each component's location the weaker contribution to the total field is adopted as the important part, and the stronger (self-)field discarded.

The Einstein-Infeld-Hoffmann theory avoids these shortcomings. The field is accepted as a whole, and the dynamic laws are given the form of conservation laws. The conserved quantities appear as integrals over domains that do not include the locations of the field sources. The existence of these integrals depends on invariance properties of the theory being examined.

Invariance group and Noether's theorem.

The notions of invariant transformations and of *invariance group* play key roles. An invariant transformation is one that leaves the form of the dynamical law unchanged, whether that is the Lagrangian or the Hamiltonian. The invariant transformations of a dynamical system form a group, the invariance group. The invariance group need not be connected. If it is at least locally connected, there are infinitesimal invariant transformations, and one can construct a canonical (Hamiltonian) formalism.

An infinitesimal invariant transformation has a generator. Noether's theorem states that that generator is a constant of the motion. Loosely, Noether's theorem is a special case of the fact that in a Poisson bracket $[A, B]=C$ either component generates an infinitesimal transformation that changes the other component by an amount C. Noether's theorem results from one component of the Poisson bracket being the Hamiltonian, the other the generator of an infinitesimal invariant transformation, so that C=0.

An example.

Return now to the EIH theory and consider as an example, the gauge properties of the electromagnetic potentials, $\vec{A}$ and Φ, and the invariance of the electromagnetic charge, the source of the electromagnetic field. If the charges present are all within a compact three-dimensional domain, V, then Q, the total charge inside V, equals the two-dimensional integral over the boundary of V,S,

$$Q = \oint \vec{E} \cdot d\vec{S} \,. \tag{1}$$

The canonical momentum densities of the field are $\vec{\pi} = -\frac{1}{c}\frac{\partial \vec{A}}{\partial t}$ and $\pi_0 = 0$.

An infinitesimal gauge transformation leaves the momentum densities unchanged, but affects the configuration variables,

$$\delta \vec{A} = \nabla \psi \;, \; \delta \Phi = -\frac{1}{c}\frac{\partial \psi}{\partial t} \,. \tag{2}$$

ψ is an arbitrary (smooth) function of the space and time coordinates.

The generator of the gauge transformation is

$$\Gamma = \int \left(\vec{\pi} \cdot \nabla \psi - \frac{1}{c}\pi_0 \frac{\partial \psi}{\partial t} \right) d^3x \,. \tag{3}$$

By Noether's theorem the generator Γ is constant. As ψ and $\frac{\partial \psi}{\partial t}$ are arbitrary functions, their coefficients must be zero,

$$\nabla \cdot \vec{E} = 0 \;, \; \pi_0 = 0 \,. \tag{4}$$

If the generators of infinitesimal invariant transformations vanish, Poisson brackets between them must be zero as well. In Dirac's terminology, the generators must not only vanish (they are *constraints*), but they are *first-class constraints.*

That Q does not change in time is proven by way of the Maxwell-Lorentz equations governing the electromagnetic field away from sources. As

$$\frac{dQ}{dt} \equiv \oint \frac{\partial \vec{E}}{\partial t} \cdot d\vec{S} \tag{5}$$

and

$$\frac{\partial \vec{E}}{\partial t} = c \, \nabla \times \vec{B} \,, \tag{6}$$

(6)

Stokes' theorem guarantees the vanishing of the integral (5). Hence the total charge, defined by Eq. (1), is a constant of the motion.

In a somewhat more general situation than in the conservation of electric charge, a two-dimensional integral need not be a constant of the motion, but changes in accordance with an expression that represents a "flux". Consider a field that obeys Euler-Lagrange equations outside a compact domain, the site of the sources. I shall designate the algebraically independent field variables by y^A, where the superscript A runs from 1 to n. The summation convention is to apply to such superscripts and subscripts as well. Coordinate indices are to be identified by lower-case Greek letters.

The field equations outside the source domain are

$$\frac{\delta L}{\delta y^A} \equiv \frac{\partial L}{\partial y^A} - \left(\frac{\partial L}{\partial y^A{}_{,\rho}}\right)_{,\rho} = 0. \qquad (7)$$

A "symmetry" of the field equations is indicated where a mapping of the field on itself leaves the form of the Lagrangian, i.e., the function

$$L(y^A, y^A{}_{,\rho})$$

unchanged. These are the invariant transformations defined earlier. I assume that there are symmetries that include the neighborhood of the identity mapping. The existence of other symmetries is not excluded. An infinitesimal transformation of the field variable y^A will be written as ψ^A. The change in the form of L resulting from the infinitesimal transformation will be

$$\begin{aligned} \delta L &= -\left(\frac{\partial L}{\partial y^A}\psi^A + \frac{\partial L}{\partial y^A{}_{,\rho}}\psi^A{}_{,\rho} + Q^\rho{}_{,\rho}\right) \\ &= -\left(\psi^A \frac{\delta L}{\delta y^A} + C^\mu{}_{,\mu}\right), \\ C^\mu &\equiv \psi^A \frac{\partial L}{\partial y^A{}_{,\mu}} + Q^\mu . \end{aligned} \qquad (8)$$

The generalized EIH theory.

Q^ρ (and therefore also C^ρ) is a function to be chosen arbitrarily, as it makes no contribution to the Euler-Lagrange equations. The transformation ψ^A will be invariant if C^μ can be chosen so that δL vanishes.

$$\Psi^A \frac{\delta L}{\delta y^A} + C^\mu{}_{,\mu} = 0. \tag{9}$$

Such a choice is

$$C^\mu = \Psi^A \frac{\partial L}{\partial y^A{}_{,\mu}}. \tag{10}$$

Other choices are obtained by adding to C^μ expressions of the form $U^{\mu\nu}{}_{,\nu}$ where $U^{\mu\nu}$ must be skewsymmetric,

$$U^{\mu\nu} + U^{\nu\mu} = 0, \tag{11}$$

but is otherwise arbitrary.

The conservation of electric charge provides a hint how one may formulate dynamical laws for the sources of a field without integrating across the sources themselves. A volume integral can be converted into a surface integral, whose dimension may be reduced once more by the applications of Stokes' theorem. By Gauss's theorem a four-dimensional integral turns into a three-dimensional integral taken over the boundary of the original domain of integration. If the original integrand, $C^\mu{}_{,\mu}$, vanishes, then the corresponding three-dimensional integral,

$$\mathcal{H}_B = \oint C^\mu \, d^3\Sigma_\mu = 0, \quad d^3\Sigma_\mu \equiv \frac{1}{6}\,\delta_{\mu\rho\sigma\tau}\, dx^\rho \wedge dx^\sigma \wedge dx^\tau, \tag{12}$$

vanishes as well.

As the next step we cut the integral domain in two. The cutting surface will be two-dimensional, and the two three-dimensional pieces have their boundaries in common. Obviously, as the two resulting three-dimensional pieces, after adjusting their orientations, are equal,

$$\mathcal{H} = \int_M C^\mu d^3\Sigma_\mu = \int_N C^\mu d^3\Sigma_\mu \,, \tag{13}$$

their individual magnitudes depend only on the common two-dimensional boundary,

$$\mathcal{H} = \frac{1}{2}\oint W^{\mu\nu}\, d^2\Sigma_{\mu\nu} \,, \quad d^2\Sigma_{\mu\nu} \equiv \frac{1}{2}\delta_{\mu\nu\rho\sigma}\, dx^\rho \wedge dx^\sigma ,$$
$$W^{\mu\nu} + W^{\nu\mu} = 0,$$
$$W^{\mu\rho}{}_{,\rho} = C^\mu . \tag{14}$$

The integrand $W^{\mu\nu}$ is not uniquely determined: Given one solution compatible with (14), another solution is obtained by adding a term

$$W^{\mu\nu\prime} = W^{\mu\nu} + T^{\mu\nu\sigma}{}_{,\sigma} \,, \tag{15}$$

with

$$T^{\mu\nu\sigma} = \delta^{\mu\nu\sigma\tau} V_\tau \,. \tag{16}$$

V_τ is arbitrary.

The EIH theory is based on the fact that the two-dimensional integral (14) equals the three-dimensional integral (13). Neither domain of integration includes the surroundings of the sources.

Figure 1 shows an arrangement, which underlies the interpretation that follows. The three-dimensional integral (13) has the topology of a sphere S^2 multiplied by a finite curve segment. It surrounds the domain containing the sources, $S^2 \times R$. Its boundary, the domain of integrations (14), will consist of two spheres, S^2.

Fig. 1 . The slender long tube encloses a source of the field. It is surrounded by a three-dimensional surface that intersects the source region nowhere and which is bounded by two disjoint two-dimensional domains.

After adjusting the orientation of the two pieces of the boundary, the relationship between Eqs. (13) and (14) can be given the form

$$\mathcal{H}_{II} = H_{I} + \int C^{\mu} \, d^3 \Sigma_{\mu} \, . \tag{17}$$

The change in the value of the two-dimensional integral going from M to N is determined by the "flux", C^{μ} . If that flux vanishes, then $H_{II} = H_{I}$.

Concluding Remarks

In this presentation I have avoided introducing the components of the metric tensor as the field variables, without ruling out that choice. To this extent the sources of the field need not be components of a tensor. Directions of line elements may, but need not be labeled as space-like and time-like.

I have not used symbols for covariant differentiation. All symbols for differentiation indicate partial derivatives.

References:

A. Einstein, L. Infeld, and B. Hoffmann, Ann. Math. 39, 65-100 (1938)

E. Noether, Goett. Nachr. 37, 235 (1918).

May 6, 1997

Torsion and the Weyl-Cartan Space Problem in Purely Affine Theory

HORST-HEINO V. BORZESZKOWSKI
Technische Universität Berlin, Institut für Theoretische Physik,
Hardenbergstr. 36, D-10623 Berlin

AND

HANS-JÜRGEN TREDER
Rosa-Luxemburg-Str. 17a, D-14482 Potsdam

ABSTRACT

According to Poincaré, only the epistemological sum of geometry and physics is accessible to experience. This does not however mean that the geometric part can be chosen arbitrarily. On the contrary, there are requirements of measurement to be imposed on geometry because, otherwise, the theory resting on this geometry cannot be physically interpreted. In particular, the Weyl-Cartan space problem must be solved. This means that it must be guaranteed that the comparison of distances is compatible with the Levi-Civita transport. In this lecture, we discuss these requirements of measurement in case of purely affine theories.

In purely affine theories, the affine field Γ^i_{kl} represents Poincaré's sum, and, therefore, the solution of the space problem has to be done so that it implies a physically meaningful splitting of the connection in a metrical part and in matter fields. This separation can be performed differently. A necessary condition for this procedure is the vanishing of the homothetic curvature.

In Schrödinger's separation, where this condition is satisfied, the connection Γ^i_{kl} is split into the asterisk connection Γ^{*i}_{kl} and the torsion vector Γ_l which takes the part of a vector potential. The latter satisfies Maxwell-like equations in a dielectric medium described by the tensor g_{ik} which is proportional to the symmetric part of the Ricci tensor formed from the asterisk connection. This means, the tensor g_{ik} describes both the space-time metric and the dielectric medium. In this case of separation the covariant derivative of g_{ik} performed with respect to Γ^{*i}_{kl} is not equal to zero so that, from the point of view of the Weyl-Cartan space problem, this separation is not complete. However, we shall show that there is a separation solving the Weyl-Cartan space problem, where this solution simultaneously determines the matter field via Einstein's equations. Now, the matter fields are given by the *torsion tensor* $\Gamma^i_{[kl]}$.

1. INTRODUCTION

As was argued by Poincaré(1), only the sum of geometrical and physical properties is accessible to experience.[1] If one adopts this point of view then the problem to be solved is to isolate, as far as possible, the "items" of this sum from each other, i.e., to isolate geometrical structures by conceiving arrangements that test them by the motion of physical bodies and then to formulate the (dynamical) laws of physics with respect to this space-time structure. The former part of this task first requires us to define a metric that allows us to determine infinitesimal distances measurable by rigid bodies (Helmholtz, Lie) or by scaled bars (Riemann). This definition implies the conception of the comparison of distances along a given curve. Therefore, a second problem arises if one defines an independent parallel transport by gyroscopes (Eddington, Levi-Civita) leading to the rule of Levi-Civita's absolute parallelism. Then one must solve the *Weyl-Cartan space problem*(3-5), i.e., one must guarantee that the comparison of distances given by the metric is compatible with the Levi-Civita transport.

In Einstein's purely metric theory of general relativity with Riemann's definition of a metric, this compatibility requires us to satisfy the Ricci lemma [see Eq.(1)].[2] This situation does not either change in so-called mixed theories of general relativity assuming a metric and a connection as independent basic variables; the requirement of compatibility again leads to Ricci's lemma.

However, in purely affine theories assuming the components of the connection as the only basic variables the situation changes drastically. Here neither metric nor matter are *a priori* given, both have to be introduced simultaneously in order to arrive at measurable statements in such a unified field theory . Therefore, in this case the problem mentioned above has to be solved entirely. The connection field and the corresponding field equations describe only the Poincaré sum, and one has to split it into metric and matter such that (i) the space-time structure introduced this way satisfies the condition of compatibility between metric and connection and (ii) the geometrical remainder can be interpreted as matter.

In the following we shall show that in the case of the Einstein-Schrödinger theory(6, 7) this splitting can really be performed, where the requirements arising from the solution of the Weyl-

---

[1] Einstein(2) commented on Poincaré's statement as follows: „Geometry (*G*) says nothing about the behavior of real things, but only geometry together with the quintessence (*P*) of the physical laws. Symbolically speaking, we can state that only the sum (*G*) +(*P*) is subject to the control of experience.

[2] In Riemannian geometry, this condition of compatibility is automatically valid: There no independent connection is defined, and Eq. (1) with the covariant derivative referred to the Christoffel connection is identically satisfied."

Cartan space problem just provide a guideline for the definition of matter and thus for the physical interpretation of this theory.

To this end, first we shall consider the Einstein-Schrödinger equations in the Schrödinger gauge because then the necessary condition for the solution of the Weyl-Cartan space problem, namely the vanishing of the homothetic curvature, is satisfied (Sec. 3). After summarizing Schrödinger's separation of geometry and matter (Sec. 4), it will be shown that the solution of the Weyl-Cartan problem also leads to a splitting of Poincaré's sum (Sec. 5). Finally, in Sec. 6 some remarks on differential identities and conservation laws will be made. To prepare for this discussion, in Sec. 2, we shall repeat requirements to impose on space-time for reasons of measurement. (Although we shall only deal with the case of tensorial matter, we shall discuss the requirements of measurement more generally.[3])

Before starting, it should still be mentioned that we do not reconsider the Einstein-Schrödinger theory here in order to test its physical truth, but in order to discuss the mathematically and physically interesting relation between the solution of the Weyl-Cartan space problem and the determination of matter.[4]

2. REQUIREMENTS OF MEASUREMENT TO BE IMPOSED ON SPACE-TIME

(i) *The Ricci Lemma (or metricity condition)*

$$g_{ik;l} = 0. \tag{1}$$

It specifies the relation between metric g and connection Γ so that Γ is given as (cf. Ref. 10)

$$\Gamma^i_{kl} = \{ {}^{\,i}_{kl} \} + g^{ir} g_{km} \Gamma^m_{[rl]} + g^{ir} g_{lm} \Gamma^m_{[rk]} + \Gamma^i_{[kl]}, \tag{2}$$

where

$$\{ {}^{\,i}_{kl} \} = \frac{1}{2} g^{im} (g_{mk,l} - g_{kl,m} + g_{lm,k}). \tag{3}$$

Here the skew-symmetric part $\Gamma^i_{[kl]}$ can be chosen arbitrarily, and the symmetric part consisting of the first three terms on the right is then uniquely determined by g_{ik} and $\Gamma^i_{[kl]}$ [10]. In GRT, where $\Gamma^i_{[kl]} = 0$, Eq.(2) leads to $\Gamma^i_{kl} = \{ {}^{\,i}_{kl} \}$. As was argued by Schrödinger [7], Eq.(1) is a sufficient condition for $ds^2 = g_{ik} dx^i dx^k$ to be in accordance with the affine measure

[3] See for this also Ref. 8.

[4] This remark, however, should not conceal the fact that, at present, the Einstein-Schrödinger theory is the only example of a physically meaningful affine field theory (for this cf, Ref. 9).

of distances along every geodesic. It guarantees that metric and distance measurement, on the one hand, and connection and direction comparison, on the other, are compatible.

Moreover, Eq.(1) is also a necessary condition. Indeed, from the physical point of view, the major meaning of Eq.(1) consists in the fact that it ensures the telecomparison of lengths. Only if (1) is satisfied does one get

$$(A^i A^k g_{ik})_{;l} \frac{dx^l}{d\tau} = 0 , \tag{4}$$

where A^i denotes an arbitrary vector parallel transported along the curve $x^i = x^i(\tau)$. The physical meaning of Eq.(4) and thus of Eq.(1) was clarified by Einstein's criticism [11] of Weyl's 1918 paper [12], wherein Weyl suggested giving up the relation (1).

(ii) *The Weyl Lemma* [13]

$$h^{\;A}_{i,l} + \Gamma^A_{Bl} h^B_i - \Gamma^k_{il} h^A_k = 0 . \tag{5}$$

This specifies the relation between the tetrads h^A_i, replacing g_{ik} via $g_{ik} = \eta_{AB} h^A_i h^B_k$, and the connection. Otherwise, if one does not look at the tetrads as the "root" of the metric g but as anholonomic coordinates, replacing the coordinates x^i, then Eq.(5) has to be interpreted as a rule for the transformation of the connection from holonomic coordinates x^i to anholonomic coordinates h^A_i, and vice versa.

From the first point of view, Eq.(5) can be considered as an alternative form of (1): It states the covariant constancy of the basic quantity h, just as (1) states that one of g, and it can be derived from the Einstein-Hilbert Lagrangian $\mathbf{L}_{EH}$ (for pure vacuum) by an independent variation of Γ, similarly as (1) can be obtained from $\mathbf{L}_{EH}$ by an independent variation of Γ. Viewed from the second standpoint, (5) means that genuine tensorial matter and tensorial matter occurring as the result of the fusion of spinorial matter are parallely transported in the same manner (for this cf. Ref. 14).

(iii) *The Levi-Civita condition* [15]

$$[A_l A^k_{;l} - A^k A_{l;l}] \frac{dx^l}{d\tau} = 0 \tag{6}$$

Unlike the conditions (1) and (5), this is a requirement on Γ only. It defines a parallel transport along the curve $x^i = x^i(\tau)$. Because in GRT the metric is symmetric, the general solution of (6) reads (cf. Ref. 16)

$$\Gamma^i_{kl} = \{ {}^{\,i}_{kl} \} + \delta^i_k \Phi_l , \tag{7}$$

where Φ_l is an arbitrary vector. This means that the definition of the Levi-Civita transport allows a non-symmetric connection Γ which can be considered as resulting from the Christoffel connection $\{ {}^{i}_{kl} \}$ by regauging it. To make the condition of telecomparison of lengths given by (1) and of parallel transport of vectors defined by Eq.(7) compatible one has, however, to assume the gauge $\Phi_l = 0$.

Physically speaking, Eq.(6) says that there are gyroscopes which, when transported along a world line $x^i = x^i(\tau)$, only measure the rotation caused by $u^i = dx^i/d\tau$ (which means that there are no "cosmic" Coriolis forces).

In Einstein's GRT with minimally coupled matter based on Riemann space-times (with vanishing torsion and nonmetricity) these three requirements are satisfied. If one turns to more general spaces, however, then one arrives at theories where not all of these conditions hold true.

3. PURELY AFFINE SPACES AND THE EINSTEIN-SCHRÖDINGER EQUATIONS

Assuming a nonsymmetric connection $\Gamma^i_{kl} = \{ {}^{i}_{kl} \} + \Omega^i_{kl}$ as basic geometrical field then the curvature tensor,

$$P^i_{klm} = -\Gamma^i_{kl,m} + \Gamma^i_{km,l} - \Gamma^r_{kl}\Gamma^i_{rm} + \Gamma^r_{km}\Gamma^i_{rl}, \tag{8}$$

formed from Γ and its first derivatives is skew-symmetric in the last two indices but not in the first two indices. Therefore, one can construct two second-rank tensors from it by index contraction, namely the Ricci tensor,

$$P_{kl} = P^m_{klm}, \tag{9}$$

and the tensor of homothetic curvature,

$$S_{kl} = P^m_{mkl}. \tag{10}$$

The Einstein-Schrödinger equations stem from the Lagrangian

$$\mathfrak{H} = \frac{2}{\lambda}(-\det P_{kl})^{1/2} \tag{11}$$

and read,[3, 4]

$$P_{ik,l} - P_{rk}\Gamma^r_{il} - P_{ir}\Gamma^r_{lk} - \frac{2}{3}(P_{ik}\Gamma_l + P_{il}\Gamma_k) = 0, \tag{11a}$$

where the conditions of integrability

$$\left\{\sqrt{-\det P_{ik}}\,P^{[ik]}\right\}_{,k} = 0 \tag{11b}$$

have to be satisfied ($\Gamma_l = \Gamma^r{}_{[lr]}$ is the torsion vector). Equations (11a, b) are not only coordinate-covariant but also invariant under the Einstein A-transformation,

$$\overline{\Gamma}^i{}_{kl} = \Gamma^i{}_{kl} + \delta^i_k A_l, \tag{12}$$

where A_l is an arbitrary gauge vector. This additional invariance means that the field equations (11a) determine only 60 of the 64 components of $\Gamma^i{}_{kl}$ (the Γ_l remain undetermined).[5]

In the Schrödinger gauge, where $A_k = \frac{2}{3}\Gamma_k$, one has

$$\Gamma^{*l}_{ik} = \Gamma^l_{ik} + \frac{2}{3}\delta^l_i\Gamma_k \tag{13}$$

leading to $\Gamma^*_l = 0$, and the field equations take on the form[(4)]

$$P_{ik;\underset{*}{l}} + 2P_{ir}\Gamma^{*r}{}_{[kl]} = 0 \tag{14}$$

(the asterisk under the differentiation index denotes the covariant derivative with respect to the connection Γ^*). In this gauge, the curvature tensor is given by

$$P^{*i}{}_{klm} = P^i{}_{klm} - \frac{2}{3}\delta^i_k(\Gamma_{l,m} - \Gamma_{m,l}) \tag{15}$$

so that the Ricci tensor can be written as

$$P^*_{kl} = P_{kl} + \frac{2}{3}(\Gamma_{l,k} - \Gamma_{k,l}), \tag{16}$$

while the homothetic curvature vanishes:

$$S^*_{kl} = 0. \tag{17}$$

4. SCHRÖDINGER'S SEPARATION: TORSION VECTOR AND HOMOTHETIC CURVATURE

In the Schrödinger gauge[(4)] the relation $\Gamma^*_l = \Gamma^{*n}_{ln} = 0$ is satisfied and equation (11a) can be written as the "plus-minus lemma"[6]

$$P_{ik,l} - P_{rk}\Gamma^{*r}_{il} - P_{ir}\Gamma^{*r}_{lk} = \lambda(g_{ik,l} - g_{rk}\Gamma^{*r}_{il} - g_{ir}\Gamma^{*r}_{lk}) = 0, \tag{18}$$

---

[5] For details, see Appendix II.

[6] In contrast to the following, in this section we call the nonsymmetric metric introduced by the Schrödinger assumption by g instead of γ.

where

$$P_{ik} = P^*_{ik} - \frac{2}{3}(\Gamma_{i,k} - \Gamma_{k,i}). \tag{19}$$

Therefore, this separation implies a vanishing homothetic curvature. (For a separation, where the plus-minus relation is satisfied but the torsion vector is unequal to zero, one finds that the homothetic curvature is equal to the to the rotation of the torsion vector.)

The tensor $F_{ik}(=-F_{ki})$,

$$F_{ik} = P^*_{ik} - P_{ik} = P^*_{[ik]} - P_{[ik]} = \frac{2}{3}(\Gamma_{i,k} - \Gamma_{k,i}), \tag{20}$$

is connected with the Hermitian part S^H_{ik} of the tensor of homothetic curvature via the relation,

$$S^H_{ik} := \frac{1}{2}(S_{ik} + \bar{S}_{ik}) = \frac{3}{2}F_{ik} \tag{21}$$

$$S_{ik} = \delta^m_r P^r_{mik} = -\Gamma^{\ \ r}_{(ir),k} + \Gamma^r_{(kr),i} + \Gamma_{i,k} - \Gamma_{k,i}. \tag{22}$$

As a consequence, the tensor F_{ik} satisfies the first group of the Maxwell equations identically,

$$F_{ik,l} + F_{kl,i} + F_{li,k} = 0. \tag{23}$$

Furthermore, according to (11b) the following relations are valid:

$$\lambda(\sqrt{-g}\, g^{[ik]})_{,k} = (\sqrt{-g}\{P^{*[ik]} + F^{ik}\})_{,k} = 0. \tag{24}$$

These relations define a current density $\sqrt{-g}\, J^i$ with $(\sqrt{-g}\, J^i)_{,i} = 0$ by the second group of the Maxwell-like equations

$$(\sqrt{-g}\, F^{ik})_{,k} = \sqrt{-g}\, J^i \quad \text{with} \quad \sqrt{-g}\, J^i := (\sqrt{-g}\, P^{*[ki]})_{,k}. \tag{25}$$

The current is independent of the functions Γ_l (i.e., of the vector potential). Since however the g_{ik} are given by the P_{ik} one finds "vacuum polarization".

Due to the covariance of $\mathbf{H}$, for each variation δx^l of the coordinates x^l the relation

$$\delta\mathbf{H} = N^{ik}_l \delta U^l_{ik} - (\sqrt{-g}\, g^{ik} \delta U^r_{ik})_{,r} = 0 \tag{26}$$

is satisfied, where

$$U^l_{ik} = \Gamma^l_{ik} - \delta^l_k \Gamma^r_{ir} \tag{27}$$

is Einstein's affinity. In virtue of the field equations, this identity provides Einstein's conservation law,

$$(\sqrt{-g}\, g^{ik} \delta U^r_{ik})_{,r} = 0. \tag{28}$$

In the case of translations, where $(\delta x^i)_{,l} = 0$, the conservation law (28) can be written as

$$-\lambda\frac{\delta \mathbf{H}}{\delta x^k} = \{\sqrt{-g}\,P^{mn}(\Gamma^r_{mn} - \delta^r_n\Gamma^s_{ms})_{,k}\}_{,r} = 0\,. \tag{29}$$

The affine tensor density

$$\vartheta^i_k = \sqrt{-g}\,P^{mn}(\Gamma^i_{mn} - \delta^i_n\Gamma^s_{ms})_{,k} = \lambda\sqrt{-g}\,g^{mn}U^i_{mn,k} \tag{30}$$

satisfying the conservation law (29) is a canonical energy-momentum complex ($\sqrt{-g}\,g^{mn}U^i_{mn,k}$ is the Einstein affine complex).

As was shown by P. G. Bergmann[17], the principles of general relativity (i.e., of coordinate covariance) and of equivalence (i.e., of minimal coupling) always imply the existence of a covariant vector density

$$\vartheta^i = \mathbf{L}\xi^i - \frac{\delta\mathbf{L}}{\delta\Gamma^l_{mn,i}}\bar{\delta}\Gamma^l_{mn} - \frac{\delta\mathbf{L}}{\delta\psi_{A,i}}\bar{\delta}\psi_A \tag{31}$$

satisfying the equation

$$\vartheta^i_{,i} = 0 \tag{32}$$

where the divergence is a scalar density ($\bar{\delta}K$ denotes the change of a quantity K with fixed values of the coordinates). As was also mentioned in Ref. 17, in the case of Einstein's 1915 theory starting from the covariant Lagrangian $\mathbf{L} = \mathbf{L}_{EH} + \mathbf{L}_{mat}[g,\psi]$ one is led to Møller's expression reading for pure gravity

$$\vartheta^i_k = \{\sqrt{-g}(g_{kn,m} - g_{km,n})g^{im}g^{ln}\}_{,l} \qquad (\vartheta^k = \vartheta^k_i\alpha^i,\ \ \xi^i = \alpha^i = const.) \tag{33}$$

In the present case, where one starts from the Lagrangian (11), one obtains (28).

Returning now to the discussion of the affine tensor density (30), with the asterisk affinities the expression $\sqrt{-g}\,\vartheta^i_k$ splits up into four terms,

$$\frac{1}{\sqrt{-g}}\vartheta^i_k = P^{*mn}U^{*i}_{mn,k} - \frac{1}{3}S^{mn}U^{*i}_{mn,k} + \frac{4}{3}P^{*[in]}\Gamma_{n,k} - \frac{2}{3}S^{in}\Gamma_{n,k}\,. \tag{34}$$

Seeing that, according to (23) and (25), Γ^{*i}_{kl} can be considered as a background field and $\Phi_l = (2/3)\Gamma_l$ as gauge field (with the field strength $(1/2)S_{ik}$) the relation (46) can be interpreted as follows. The second and the third terms of the right-hand side describe the interaction between gauge field and background, while the fourth term demonstrates that the homothetic curvature S^{in} and the torsion vector (of the unified field Γ^i_{kl}) behave like canonically conjugate quantities.

Thus, in the gauge of Schrödinger one arrives at Maxwell-like equations (23) and (25) in a polarized and charged medium. The torsion vector $\Gamma_l = \Gamma^n_{[ln]}$ can be considered as vector potential $A_i = (2/3)\Gamma_i$ and the curvature tensor $F_{ik} = P^*_{ik} - P_{ik}$ plays the role of the field-strength, and the torsion vector and the homothetic curvature are canonically conjugate variables in the same sense as the vector potential and the field strength are such variables in the electrodynamics of charged and polarized media.

If one now regards that

$$P^i_{klm} = P^{*i}_{klm} - \frac{2}{3}\delta^i_k(\Gamma_{l,m} - \Gamma_{m,l}) \tag{35}$$

such that

$$P_{(kl)} = P^*_{(kl)} = R_{kl} + \Omega_{(kl)} \tag{36}$$

then one can rewrite relation (11a) or, equivalently,

$$P_{kl} = \lambda g_{kl} \tag{37}$$

in the equation

$$R_{kl} = -\Omega_{(kl)} + \lambda g_{(kl)}\,, \tag{38}$$

where $\Omega_{(kl)}$ can be interpreted as the energy-momentum tensor of the electromagnetic field F_{kl} moving in a medium.

Therefore, in the gauging of Schrödinger the unified field Γ^i_{kl} is splitted into a field Γ_l satisfying Maxwell's equations and a Riemannian part which, via (38), defines the matter tensor

$$\tilde{T}_{kl} = -\frac{1}{\kappa}\Omega_{(kl)}\,. \tag{39}$$

This version of a unified theory is thus an Einstein-Cartan theory.

5. THE DETERMINATION OF METRIC AND MATTER IN THE EINSTEIN-SCHRÖDINGER THEORY

Now let us separate metric and matter in a manner that simultaneously provides a solution of the Weyl-Cartan space problem[18]. To this end, we split Eq.(14) into its symmetric and antisymmetric parts. This provides the equations

$$2P_{(ik);\underset{*}{m}} = -(P_{kr}\Gamma^{*r}_{[im]} + P_{ir}\Gamma^{*r}_{[km]}) \tag{40}$$

and, as a preliminary (see below),

$$P_{[ik];\underset{*}{m}} + 2P_{ir}\Gamma^{*r}_{km} = 0\,. \tag{41}$$

Considering now the class of solutions satisfying the relations

$$P_{kr}\Gamma^{*r}{}_{[lm]} = - P_{lr}\Gamma^{*r}{}_{[km]} \tag{42}$$

(which represent 40 equations for the P_{kl} and $\Gamma^{*r}{}_{kl}$) then one has, instead of (40) and (41), the equations

$$P_{(ik);\underset{\bullet}{m}} = P^{*}{}_{(ik);\underset{\bullet}{m}} = 0 \tag{43}$$

and

$$P_{[ik];\underset{\bullet}{m}} + 2P_{mr}\Gamma^{*r}{}_{[ik]} = 0. \tag{44}$$

According to the definition of P_{ik}, $\Gamma^{i}{}_{kl}$, and $\Gamma^{*i}{}_{kl}$, the equations (43) and (44) are $40 + 24 = 64$ equations for the 60 components of $\Gamma^{*i}{}_{kl}$ and the four components Γ_{l}. This means that with respect to Einstein's A-transformation (12), our *ansatz* implies a special gauge.

As already mentioned, following Schrödinger one can define a covariant metric via the relation

$$P_{(ik)} = P^{*}{}_{(ik)} = \lambda g_{(ik)}, \tag{45}$$

where λ is introduced for dimensional reasons. (The contravariant metric is defined by $g_{ir}g^{lr} = \delta^{l}_{i}$.) Therefore, Eq.(43) can be interpreted as the lemma of Ricci (40 conditions), and (44) represents 24 equations for $\Gamma^{*r}{}_{ik}$ and Γ_{i}.

The solution of (43) is[7]

$$\Gamma^{*i}{}_{kl} = \{{}^{i}{}_{kl}\} + (-S^{*}{}_{k}{}^{i}{}_{l} - S^{*}{}_{l}{}^{i}{}_{k} + S^{*i}{}_{kl}) := \{{}^{i}{}_{kl}\} + \Omega^{i}{}_{kl} \tag{46}$$

with $S^{*i}{}_{kl} = \Gamma^{*i}{}_{[kl]}$. ($\Omega$ satisfies the conditions $\Omega_{ikl} + \Omega_{kil} = 0$, $\Omega^{m}{}_{mi} = \Omega^{m}{}_{im} = 0$.) Accordingly, curvature and Ricci tensor, respectively, can be written as follows:

$$P^{*i}{}_{klm} = R^{i}{}_{klm} - \Omega^{i}{}_{kl\perp m} + \Omega^{i}{}_{km\perp l} - \Omega^{r}{}_{kl}\Omega^{i}{}_{mr} + \Omega^{r}{}_{km}\Omega^{i}{}_{lr},$$

$$\begin{aligned} P^{*}{}_{kl} &= R_{kl} - \Omega^{r}{}_{kl\perp r} + \Omega^{r}{}_{kr\perp l} - \Omega^{r}{}_{kl}\Omega^{m}{}_{mr} + \Omega^{r}{}_{km}\Omega^{m}{}_{lr} \\ &= R_{kl} - \Omega^{r}{}_{kl\perp r} + \Omega^{r}{}_{km}\Omega^{m}{}_{lr} := R_{kl} + \Omega_{kl} \end{aligned} \tag{47}$$

where $R^{i}{}_{klm}$ and R_{kl} are the curvature and the Ricci tensor formed from the Christoffel connection, and $\perp$ denotes the covariant derivative with respect to this connection.

From the latter relation in (47) one obtains (cf. Eq.(36))

$$P^*_{(kl)} - R_{kl} = P_{(kl)} - R_{kl} = \frac{1}{2}\,(\Omega_{kl} + \Omega_{lk}) \tag{48}$$

or

$$R_{kl} = \lambda g_{kl} - \frac{1}{2}\,(\Omega_{kl} + \Omega_{lk}). \tag{49}$$

This provides the following Einstein equations

$$R_{kl} - \frac{1}{2} g_{kl} R = -\lambda g_{kl} - \kappa\, T_{kl} \tag{50}$$

with

$$\kappa\, T_{kl} = \frac{1}{2}\,(\Omega_{kl} + \Omega_{lk} - g_{kl} g^{mn} \Omega_{mn}), \tag{51}$$

where

$$T_k{}^l{}_{\perp l} = 0. \tag{52}$$

The dynamical equations (52) are automatically satisfied when the geometrical structure is determined in the manner described above.

6. THE HELMHOLTZ-LIE PROBLEM

The Einstein-Schrödinger equations result from the Lagrange density (11) which, by use of the reduced minor

$$\frac{2}{\lambda \mathbf{H}} \frac{\partial \mathbf{H}}{\partial P_{ik}} = \gamma^{ik} \tag{53}$$

or

$$\frac{\partial \mathbf{H}}{\partial P_{ik}} = \sqrt{-\gamma}\,\gamma^{ik} \quad \text{(with } \gamma := \det \gamma_{jk}\text{)}, \tag{54}$$

can be written as follows,

$$\mathbf{H} = 2\sqrt{-\gamma}, \tag{55}$$

where

$$\begin{aligned} P_{ik} &= \lambda \gamma_{ik}, \\ P^{ik} &= \frac{1}{\lambda}\gamma^{ik} = \gamma^{im}\gamma^{kn}\gamma_{mn} \\ (\gamma_{il}\gamma^{kl} &= \gamma_{li}\gamma^{lk} = \delta_i^k) \end{aligned} \tag{56}$$

Introducing now again the variables (27) the Ricci tensor is given as,

$$P_{ik} = -U^{s}_{ik,s} + U^{s}_{il}U^{l}_{sk} - \frac{1}{3}U^{s}_{is}U^{l}_{lk} \tag{57}$$

and the variation of $\mathbf{H}$ by the field variables U^{l}_{ik} provides[6]

$$\delta\mathbf{H} = N^{ik}_{l}U^{l}_{ik} - (\sqrt{-\gamma}\gamma^{ik}\delta U^{r}_{ik})_{,r}\,, \tag{58}$$

where

$$\begin{aligned} N^{ik}_{l} = (\sqrt{-\gamma}\gamma^{ik})_{,l} + (\sqrt{-\gamma}\gamma^{mk})(U^{i}_{ml} - \frac{1}{2}\delta^{i}_{l}U^{r}_{mr}) \\ +\sqrt{-\gamma}\gamma^{im}(U^{k}_{lm} - \frac{1}{2}\delta^{k}_{l}U^{r}_{rm}) \end{aligned} \tag{59}$$

and

$$N^{[ik]}{}_{k} = \left(\sqrt{-\gamma}\ {}^{[ik]}\right)_{,k}. \tag{60}$$

The condition $N^{ik}{}_{l} = 0$ is identical with the Einstein-Schrödinger equations (11a).

The Lie variation of $\mathbf{H}$ induced by infinitesimal coordinate transformations is given by the expression

$$-\delta\mathbf{H} = \frac{\partial\mathbf{H}}{\partial(\sqrt{-\gamma}\gamma^{mn})}\delta(\sqrt{-\gamma}\gamma^{mn}) = R_{mn}\delta(\sqrt{-\gamma}\gamma^{mn}), \tag{61}$$

with

$$\delta(\sqrt{-\gamma}\gamma^{ik}) = \sqrt{-\gamma}(\gamma^{sk}\delta x^{i}{}_{,s} - \gamma^{si}\delta x^{k}{}_{,s}) - \delta(\sqrt{-\gamma}\gamma^{ik}\delta x^{s})_{,s}, \tag{62}$$

such that

$$-\delta\mathbf{H} = \{\lambda(\sqrt{-\gamma}(\gamma^{ir}\gamma_{kr} + \sqrt{-\gamma}\gamma^{ri}\gamma_{rk})_{,i} - \sqrt{-\gamma}\gamma^{mn}\gamma_{mn,k}\}\ \delta x^{k} \tag{63}$$

Due to the properties of a determinant and its minors, the latter expression is equal to zero. Therefore, here the (Bianchi-like) conservation laws are valid. Physically non-trivial identities (dynamical equations), however, follow only by the separation of the affine field into metrical and matter fields.

Because of $N^{ik}{}_{l} = 0$, one finds also. for the 00-component of the Einstein energy-momentum complex

$$d t^{0}_{0} = \frac{\delta t^{0}_{0}}{\delta P^{mn}}\delta P^{mn} + \frac{\delta t^{0}_{0}}{\delta U^{l}{}_{ik}}\delta U^{l}{}_{ik} = U^{l}{}_{mn,0}\delta^{0}_{l}R^{mn} + R^{mn}{}_{,0}\delta^{0}_{l}U^{l}{}_{mn}\,, \tag{64}$$

such that the Hamilton equations are fulfilled, where the $R^{mn}\delta^{0}_{k}$ are the momenta and the $U^{l}{}_{mn}$ the coordinates.[9]

In the purely affine theory, the Hamiltonian density belonging to the Lagrangian (11) does not contain a separate interaction potential, but only some type of kinetic energy density (see Appendix I). If one interprets the equivalence class $\{P_{\mu\nu}\}$ of the $P_{\mu\nu}$ as the points of a superspace Σ [19], then the Hamiltonian metric is given by

$$d\sigma^2 = \frac{1}{2}(P^{\mu\nu}P^{\alpha\beta} + P^{\mu\beta}P^{\alpha\nu} - P^{\mu\alpha}P^{\alpha\nu})dP_{\mu\alpha}dP_{\beta\nu}. \qquad (65)$$

Now, the Hertz principle of the straightest path, which is complementary to the Jacobi principle of the shortest path, could be supposed as

$$\delta \int d\sigma = 0 \qquad (66)$$

From this point of view, space-time should be interpreted as the straightest in the momentum superspace of $\{P_{\mu\nu}\}$.

According to the definition of the Ricci tensor (see Appendix I, (A.27)), the Hamiltonian superspace Σ is a subspace of the superspace Σ^*: $\{U^{\lambda}{}_{\mu\nu}\}$ with the $U^{\lambda}{}_{\mu\nu}$ as coordinates. The subspace Σ is defined by the conditions

$$R_{\mu\nu} = -U^{r}{}_{\mu\nu,r} + U^{m}{}_{\mu r}U^{r}{}_{m\nu} - \frac{1}{3}U^{m}{}_{\mu m}U^{r}{}_{r\nu} = 0. \qquad (67)$$

The straightest path in Σ is the straightest path in Σ^* under the constraint (67). In the spirit of Hertz' dynamics one finds no interaction potential, but only constraints.

7. CONCLUSIONS

In the affine theory, one has to perform the separation of the affine unified field so that one gets a measurement-theoretically reasonable space-time structure and terms physically interpretable as matter. This separation can be performed differently. A necessary condition is the vanishing of the homothetic curvature[10].

In Schrödinger's separation summarized in Sec. 4 this necessary condition is also satisfied. There the connection Γ is split into the asterisk connection Γ^* and a vector Γ_l which takes the part of a vector potential satisfying Maxwell-like equations in a dielectric medium described by the material tensor g_{ik}. In this case one has $g_{ik;l} \neq 0$ and g_{ik} embraces both the space-time metric and the dielectric medium. In other words, the electromagnetic field is isolated from the connection Γ such that the affine field is separated into the two items "electromagnetism" and "metric plus dielectric medium". This does not, however, provide a solution of the Weyl-Cartan space problem.

However, one can find a solution of this problem that simultaneously determines the matter via Einstein's equations. That the definition of a background geometry satisfying the requirement of measurement and the determination of matter defined by Eqs. (50)-(52) are the same problems is physically plausible because both determinations concern the coupling of gravity and matter, in the first case the interaction of gravitational fields with measurement devices and in the second case the coupling of gravity with its matter source.

The fact that the quantity γ_{ik} introduced by Eqs. (56) can be separated into a metric g and a remainder F which can be interpreted as matter field,

$$\gamma_{ik} = g_{ik} + F_{ik}, \text{ with } \quad \gamma_{(ik)} = g_{ik}, \quad \text{and} \quad \gamma_{[ik]} = F_{ik}, \tag{68}$$

rests on the "pythagoricity of the measure" (Weyl[4]). Due to the fact that the differentials of the coordinates are contravariant vectors, one has,

$$ds^2 = \gamma_{ik} dx^i dx^k = \gamma_{(ik)} dx^i dx^k = g_{ik} dx^i dx^k = \lambda^{-1} P_{(ik)} dx^i dx^k, \tag{69}$$

and this index symmetry enables us to isolate a Einstein-Riemann space-time.

The purely affine space -time V_4 is reduced to Levi-Civita's absolute parallelism given by a nonsymmetric connection $\Gamma^i_{kl} \neq \Gamma^i_{lk}$. The metrization of this V_4 performed by the solution of the Weyl-Cartan space problem determines a Riemann-Einstein V_4 which contains the torsion field $\Gamma^i_{[kl]}$, as the unified matter field in locally Lorentz-covaraint terms. Therefore, Einstein's general principle of relativity is satisfied, and the field $\Gamma^i_{[kl]}$ describes usual matter. The cosmic system is then given by the straightest path in the superspace $\{p_{\mu\nu}\}$ $(\mu, \nu = 1, 2, 3)$ defined by the Einstein tensor P_{ik} formed from the affine connection Γ^i_{kl} (see Appendix I).

On the other hand, the Machian "space-time without space and time" reduces the space V_3 to the conception of the totality of the simultaneous distances (r_{AB}) of all matter points of the universe (Einstein 1956) and makes the time to the conception of the mean value of these distances. The solution of this problem, in analogy to the above situation let us call it the solution of the "Mach-Einstein space problem", leads to a Riemannian space-time with an Einstein-Cartan teleparallelism of the reference tetrads $h^A{}_i$ (where $g_{ik} = h^A{}_i h_{kA}$). Here a torsion field is defined by the (Einstein-Cartan) anholonomy objects $h^i{}_A h^A{}_{[k,l]}$ which are no Lorentz-covariant objects. The field equations are of Einstein-Mayer type and define the anholonomy object as a matter field that does not show local Lorentz symmetries.[7] This matter field has features that are typical for "dark matter" whose existence is supposed by astronomers in order to explain the motion of cosmic systems (cf. Ref. .20).

Finally, a remark on the-- physical comparison between the purely affine approach here under consideration and the two mixed (or metric-affine) approaches cited above. In contradiction to Hehl[(21)], Moffat[(22)] allows an antisymmetric part in the metric tensor. Therefore, it is not surprising that the purely affine theory is more similar to the latter approach. Indeed, the purely affine field equations are equivalent to the following form of the Einstein-Schrödinger equations:

$$\gamma_{ik,l} - \gamma_{rk}\Gamma^r_{il} - \gamma_{ir}\Gamma^r_{lk} - \frac{2}{3}(\gamma_{ik}\Gamma_l + \gamma_{il}\Gamma_k) = 0, \tag{70}$$

$$\left(\sqrt{-g}\,\gamma^{[ik]}\right)_{,k} = 0 \tag{71}$$

$$P_{ik} = \lambda\gamma_{ik}. \tag{72}$$

Formally, they mainly differ from the equations in Ref. 16 by Eq.(72), and this makes the physical difference: In the purely affine field theory it is impossible to introduce an additional matter field since Eq.(72) defining the metric leaves no room for a matter term. Therefore, to specify usual matter fields here one has to apply separation methods of the kind described above.

APPENDIX I: CANONICAL EQUATIONS AND GENERAL RELATIVITY[8]

Due to general covariance, the Bianchi identify holds. It has the form

$$\delta\mathbf{H} = \mathbf{N}^{ik}{}_l\delta U^l{}_{ik} - \partial_l(\frac{\partial\mathbf{H}}{\partial R_{mn}}U^l{}_{mn}) = \mathbf{N}^{ik}{}_l\delta U^l{}_{ik} - \partial_l(\mathbf{G}^{mn}U^l{}_{mn}) = 0 \tag{A.1}$$

where the variation of U is given by

$$\delta U^l{}_{ik} = U^r{}_{ik}\delta x^l{}_{,r} - U^l{}_{rk}\delta x^r{}_{,i} - U^l{}_{ir}\delta x^r{}_{,k} + \delta x^l{}_{,ik} - \delta x^r{}_{,ir}\delta^l_k - U^l{}_{ik,r}\delta x^r \tag{A.2}$$

and

$$\mathbf{N}^{ik}_l := \frac{\delta\mathbf{H}}{\delta U^l_{ik}}. \tag{A.3}$$

Using the explicit function $\mathbf{H}$, given by Eq: (11), and

[7] Instead of general relativity, we find the "relativity of inertia".
[8] In the Appendices we mainly follow two sections of Ref. 9.

Using the explicit function $\mathbf{H}$, given by Eq: (11), and

$$\delta \mathfrak{G}^{ik} = \mathfrak{G}^{rk} \delta x^i_{,r} + \mathfrak{G}^{ir} \delta x^k_{,r} - (\mathfrak{G}^{ik} \delta x^r)_{,r} \tag{A.4}$$

we get another form of the Bianchi identity (A1) [6, 7, 23]

$$\frac{\delta \mathbf{H}}{\delta x^l} = - \frac{\delta \mathbf{H}}{\delta \mathfrak{G}^{ik}} \frac{\delta \mathfrak{G}^{ik}}{\delta x^l} = (\mathfrak{G}^{ik} R_{lk} + \mathfrak{G}^{ki} R_{kl})_{,i} - \mathfrak{G}^{ik} R_{ik,l} = 0 . \tag{A.5}$$

This is Einstein's dynamical equation in his "generalized theory of gravitation". With the field equations $\mathbf{N}^l_{ik} = 0$ we get the conservation law

$$\delta \mathbf{H} = (\frac{\partial \mathbf{H}}{\partial R_{mn}} \delta U^l_{mn})_{,l} = \vartheta^l_{,l} = 0 \tag{A.6}$$

for the contravariant vector density

$$\vartheta^l = \mathfrak{G}^{mn} \delta U^l_{mn} = \frac{1}{2\lambda^2} \mathbf{H} \mathbf{R}^{MN} \delta^l_r \delta U^l_{mn} = - \frac{\partial \mathbf{H}}{\partial U_{mn,l}} \delta U^r_{mn} . \tag{A.7}$$

In the case of an infinitesimal translation, $\delta x^i_{,l} = 0$, we find $\delta U^l_{mn} = -U^l_{mn,r} \delta x^r$, and the conservation law

$$\vartheta^i_{k,i} = 0 \tag{A.8}$$

for Einstein's affine energy-momentum complex

$$\vartheta^i_k = - \mathfrak{G}^{mn} \delta^i_r \delta U^r_{mn,k} = - \frac{1}{2\lambda^2} \mathbf{H} R^{mn} \delta^i_r \delta U^r_{mn,k} = \frac{\partial \mathbf{H}}{\partial U_{mn,i}} \delta U^r_{mn,k} . \tag{A.9}$$

The Hamiltonian density $\chi = \vartheta^0_0$ can now be calculated to be

$$\chi = \frac{\partial \mathbf{H}}{\partial U^l_{mn,0}} U^l_{mn,0} = - \frac{1}{2\lambda^2} \mathbf{H} R^{mn} \delta^0_r U^0_{mn,0} = \frac{1}{2} \mathbf{H} \Theta , \tag{A.10}$$

where

$$\Theta = \sqrt{-g} \chi . \tag{A.11}$$

Here, the canonical coordinates are Einstein's affine tensor U^l_{mn}, and the canonical momenta are the tensor components $\Pi^{mn}{}_l$, given by

$$\begin{aligned} \mathbf{P}^{mn}{}_l &= \sqrt{-g} \Pi^{mn}{}_l = \frac{\partial \mathbf{H}}{\partial U^l_{mn,0}} \\ &= - \frac{1}{2\lambda^2} \mathbf{H} R^{mn} \delta^0_l = - \mathfrak{G}^{mn} \delta^0_l = - \frac{1}{\lambda} \sqrt{-g} R^{mn} \delta^0_l \end{aligned} \tag{A.12}$$

The canonical field equations imply that, up to a total derivative, the equation

$$\delta \chi = \frac{\delta \chi}{\delta \mathbf{P}^{mn}{}_l} \delta \mathbf{P}^{mn}{}_l + \frac{\delta \chi}{\delta U^l_{mn}} \delta U^l_{mn} = U^l_{mn} \delta \mathbf{P}^{mn}{}_l - \mathbf{P}^{mn}{}_l \delta U^l_{mn} \tag{A.13}$$

is valid. Here, χ is the Hamiltonian given by Eq.(A.10), with the Lagrangian $\mathbf{H}$. - We have

$$\frac{\delta\chi}{\delta \mathbf{P}^{mn}{}_l} = \frac{\partial\chi}{\partial \mathbf{P}^{mn}{}_l} = \partial_0 U^l{}_{mn} = \dot{U}^l{}_{mn} \tag{A.14}$$

and

$$\begin{aligned}\frac{\delta\chi}{\delta U^l{}_{mn}}\delta U^l{}_{mn} = & - \partial_0 \mathbf{P}^{mn}{}_l \delta U^l{}_{mn} + \partial_0(\mathbf{P}^{mn}{}_l \delta U^l{}_{mn}) \\ & - \frac{1}{2}\mathbf{N}^{mn}{}_l \Theta \delta U^l{}_{mn} + \frac{1}{2}\partial_l(\mathbf{G}^{mn}\Theta\delta U^l{}_{mn})\end{aligned} \tag{A.15}$$

Hence,

$$\frac{\delta\chi}{\delta U^l{}_{mn}} + \dot{\mathbf{P}}^{mn}{}_l - \frac{1}{2}\mathbf{N}^{mn}{}_l\Theta = 0 . \tag{A.16}$$

If $\Theta \neq 0$ (and therefore $\chi \neq 0$) the canonical equations (16) are the field equations $\mathbf{N}^l{}_{ik} = 0$. The Hamiltonian, Eq.(A.10), vanishes when the $U^0{}_{mn}$ do not depend on time.

The postulate of symmetry in the lower indices

$$\Gamma^i{}_{kl} = \Gamma^i{}_{lk} = \Gamma^i{}_{(kl)} \tag{A.17}$$

which breaks Einstein' s gauge invariance) makes the connections equal to the Christoffel symbols. We then get

$$R_{[lr]}\Gamma^r{}_{(ik)} = \frac{1}{2}(- R_{[ik],l} + R_{(kl),i} + R_{[li],k}) \tag{A.18}$$

and in addition $\Gamma^r{}_{ri,k} - \Gamma^r{}_{rk,i} = 0$. Consequently, the R_{ik} are symmetric, and the derived metric g_{ik} as well. The space becomes a Riemannian-Einstein space with cosmological constant, $R_{ik} = \lambda g_{ik}$. In this case, Einstein's energy-momentum complex is equivalent to Møller's general-relativistic complex

$$\vartheta_i{}^k = (\sqrt{-g}(g_{in,m} - g_{im,n})g^{km}g^{ln})_{,l} = S_i{}^{kl}{}_{,l} \tag{A.19}$$

with the antisymmetric superpotential $S_i{}^{kl} = - S_i{}^{lk}$. In the mixed representation, the Lagrange density of such spaces is $\mathbf{L} = \sqrt{-g}(R - 2\lambda)$, and the Bianchi identity reads[6, 7, 17, 24]

$$\delta\mathbf{L} = \frac{\partial \mathbf{L}}{\partial g^{ik}}\delta g^{ik} + \mathbf{N}^{mn}{}_l \delta U^l{}_{mn} - \partial_l(\mathbf{G}^{mn}\delta U^l{}_{mn}) = 0 \tag{A.20}$$

with the field equations

$$\frac{\delta\mathbf{L}}{\delta g^{ik}} = \sqrt{-g}(R_{ik} - (1/2)g_{ik} + \lambda g_{ik}) \text{ and } \mathbf{N}^{mn}{}_l = 0 . \tag{A.21}$$

Now we may substitute g_{ik} for $\lambda^{-1}R_{ik}$ to get the so-called Eddington -Einstein theory[25, 26].

In such Riemann-Einstein spaces the field equations are valid in the form[30]

$$\sqrt{-g}(R_{ik} - (1/4)g_{ik}) = 0\,,\ \ \mathbf{N}^{mn}{}_{,l} = 0\,,\ R = 4\lambda\,. \tag{A.22}$$

The Hamiltonian of such a space - the Hamiltonian of the Eddington-Einstein space - can be written as

$$\chi = \vartheta_0{}^0 = (\sqrt{-g}(g_{0n,m} - g_{m,n})g^{0m}g^{ln})_{,l} = \sqrt{-g}\,g_{0m,n}(g^{0m}g^{ln} - g^{0n}g^{lm})_{,l}\,. \tag{A.23}$$

The spaces of maximal mobility with a lie group of 10 parameters are the de Sitter universes. Their Riemann curvature tensor is

$$R_{iklm} = -\frac{\lambda}{3}(g_{il}g_{km} - g_{im}g_{kl}) \tag{A.24}$$

their Weyl tensor of conformal curvature vanishes. The de Sitter metric is a conformally flat metric and can be written (with $x^0 = ct$) in the form

$$g_{ik} = \frac{3}{\lambda c^2 t^2}\eta_{ik} = \frac{1}{\lambda}R_{ik}\,. \tag{A.25}$$

Hence, this metric defines the signal velocity c and in the case of $\lambda > 0$ an event horizon $\sqrt{3/\lambda}$ connected with the constant Hubble parameter $\sqrt{3/\lambda}\,c$. According to the cosmological point of view, only a positive cosmological constant is meaningful.[27]

With symmetric metric and Ricci tensor, the Bianchi identity (A.5) gets the known form

$$(\sqrt{-g}\,E^i{}_k)_{,i} - \frac{1}{2}\sqrt{-g}\,E^{mn}\,g_{mn,k} = \sqrt{-g}\,E^i{}_k)_{;i} = 0\,. \tag{A.26}$$

The tensor $E_{ik} := R_{ik} - (1/2)g_{ik}$ is the Einstein tensor with Ricci tensor R_{ik} given by

$$\begin{aligned} R_{ik} &= -\Gamma^r{}_{ik,r} + \Gamma^r{}_{ir,k} - \Gamma^r{}_{ik}\Gamma^m{}_{rm} + \Gamma^r{}_{im}\Gamma^m{}_{rk} \\ &= -U^r{}_{ik,r} + U^m{}_{ir}U^r{}_{mk} - \frac{1}{3}U^m{}_{im}U^r{}_{rk} \end{aligned} \tag{A.27}$$

and the connection is given by the Christoffel symbols. The lemma of Ricci indicates the metricity of the connection,

$$g_{ik;l} = g_{ik,l} - g_{rk}\{{}^{r}_{il}\} - g_{ir}\{{}^{r}_{kl}\} = 0\,. \tag{A.28}$$

According to the Einstein-Infeld- Hoffmann method, a point-like testparticle is a monopole-like singularity of the Einstein equations endowed with a very small mass.[28] Therefore, in the case of a symmetric connection, the path of such a particle in a purely affine field is a geodesic world-line with respect to the metric defined by the partial derivative $\mathfrak{G}^{ik} = \partial\mathbf{H}/\partial R_{ik}$.

In the Einstein identity,

$$(\mathfrak{G}^{mn}\delta U^{k}{}_{mn})_{,k} = 0, \tag{A.29}$$

the conserved quantity is a covariant vector density because $\delta U^{k}{}_{mn}$ is an infinitesimal variation of the affine tensor $U^{k}{}_{mn}$. However, the restriction of the infinitesimal variation to the affine transformation with $x^{r}{}_{,i} = 0$ gives the conservation law

$$(\mathfrak{G}^{mn}\delta U^{k}{}_{mn,i})_{,k} = 0 \tag{A.30}$$

in which the energy-momentum complex is an affine-tensor density only.

APPENDIX II: SUPER-GAUGE AND PSEUDO-HERMITICITY

The fundamental concept of the affine geometry is the Levi-Civita definition of an "absolute parallelism" defined by the equation

$$A_i(A_{k,l} - \Gamma^{r}{}_{kl}A_r)\frac{dx^l}{ds} = A_k(A_{i,l} - \Gamma^{r}{}_{il}A_r)\frac{dx^l}{ds}. \tag{A.31}$$

These equations are covariant with respect to coordinate transformations $\bar{x}^l = \bar{x}^l(x^r)$ (Einstein's T group), with

$$\bar{A}_i = \frac{\partial x^r}{\partial \bar{x}^i}A_r \tag{A.32}$$

and

$$\begin{aligned} \bar{\Gamma}^{i}{}_{kl} &= \frac{\partial \bar{x}^i}{\partial x^a}\frac{\partial x^b}{\partial \bar{x}^k}\frac{\partial x^c}{\partial \bar{x}^l}\Gamma^{a}{}_{bc} + \frac{\partial \bar{x}^i}{\partial x^r}\frac{\partial^2 x^r}{\partial \bar{x}^k \partial \bar{x}^l} \\ \bar{U}^{i}{}_{kl} &= \frac{\partial \bar{x}^i}{\partial x^a}\frac{\partial x^b}{\partial \bar{x}^k}\frac{\partial x^c}{\partial \bar{x}^l}U^{a}{}_{bc} + \frac{\partial \bar{x}^i}{\partial x^r}\frac{\partial^2 x^r}{\partial \bar{x}^k \partial \bar{x}^l} - \delta^i_l\frac{\partial \bar{x}^m}{\partial x^r}\frac{\partial^2 x^r}{\partial \bar{x}^k \partial \bar{x}^m} \end{aligned} \tag{A.33}$$

The equations (A.14) are invariant also with respect to the super-gauge transformations of the connections $\Gamma^{i}{}_{kl}$ (Einstein's A group)

$$\begin{aligned} \Gamma'^{i}{}_{kl} &= \Gamma^{i}{}_{kl} + \delta^i_k\Phi_l \\ \Gamma'_k &= \Gamma_k - \frac{2}{3}\Phi_k \\ U'^{i}{}_{kl} &= U^{i}{}_{kl} + \delta^i_k\Phi_l - \delta^i_l\Phi_k \end{aligned} \tag{A.34}$$

with an arbitrary gauge vector Φ_l. The A group implies

$$\begin{aligned} R'^{i}{}_{klm} &= R^{i}{}_{klm} - \delta^i_k(\Phi_{l,m} - \Phi_{m,l}) \\ R'_{kl} &= R_{kl} + \Phi_{k,l} - \Phi_{l,k} \end{aligned} \tag{A.35}$$

According to the definition of the tensor g_{ik} via the expression

$$\frac{g^{ik}}{\sqrt{-det\, g^{ab}}} = \mathfrak{G}^{ik} = \frac{\partial \mathfrak{H}}{R_{ik}[\Gamma]} \tag{A.36}$$

we get

$$g'_{ik} = g_{ik} + \frac{1}{\lambda}(\Phi_{i,k} - \Phi_{k,i}). \tag{A.37}$$

The T group and the A group built Einstein's super-group with $TA = AT$. Hence, the T-A transformations destroy any index symmetry of the metric tensor g_{ik} and the affine connection Γ'^{i}_{kl} as well. In contrast to that, the Levi-Civita connection is preserved.

The Einstein-Schrödinger purely affine theory is super-gauge invariant: With respect to the variations $\delta\Phi_l$ of the connections

$$\delta\Gamma'^{i}_{kl} = \delta^{i}_{k}\Phi_l \tag{A.38}$$

we obtain

$$\frac{\delta\mathfrak{H}}{\delta\Phi_l} = -2\mathfrak{G}^{[lk]}{}_{,k} = \frac{1}{\lambda^2}(\mathfrak{H}R^{[lk]})_{,k} \tag{A.39}$$

and this vector density vanishes according to the condition of integrability of the Einstein-Schrödinger equations (11a). Indeed, these equations are only 60 conditions on the 64 Γ'^{i}_{kl}, and we can choose the torsion vector Γ_l by choice of the gauge vector

$$g_{ik,l} - g_{rk}\Gamma'^{r}_{il} - g_{ir}\Gamma'^{r}_{lk} - \frac{2}{3}(g_{ik}\Gamma_l + g_{il}\Gamma_k)$$

$$= g_{ik,l} - g_{rk}\Gamma''^{r}_{il} - g_{ir}\Gamma''^{r}_{lk} - \frac{2}{3}(g_{ik}\Gamma'_l + g_{il}\Gamma'_k). \tag{A.40}$$

The choice $\Phi_i = (2/3)\Gamma_i$ of the gauge function yields Schrödinger's asterisk affinity

$$\Gamma^{*i}_{kl} = \Gamma'^{i}_{kl} + \frac{2}{3}\delta^{i}_{k}\Gamma_l. \tag{A.41}$$

Therefore we get

$$R_{ik,l} - R_{rk}\Gamma^{*r}_{il} - R_{ir}\Gamma^{*r}_{lk} = 0$$
$$R_{ik} = R^{*}_{ik} - \frac{2}{3}(\Gamma_{i,k} - \Gamma_{k,i}) = \lambda g_{ik} \tag{A.42}$$

These equations represent an alternative formulation of the Einstein-Schrödinger theory. We may deduce them also from a "mixed" Lagrangian[29]

$$\mathbf{L} = \sqrt{-g}(R - 2\lambda) = \mathfrak{G}^{ik} R_{ik} - 2\lambda\sqrt{-g} \tag{A.43}$$

in which $\mathfrak{G}^{kl}$ and $\Gamma^{i}{}_{kl}$ (and $U^{i}{}_{kl}$ respectively) are independent field variables. The Euler variations are then immediately given by

$$\frac{\delta \mathbf{L}}{\delta \mathfrak{G}^{ik}} = R_{ik} - \lambda g_{ik} \quad \text{and} \quad \frac{\delta \mathbf{L}}{\delta U^{l}{}_{ik}} = \mathbf{N}^{ik}{}_{l}. \tag{A.44}$$

Now, the equation $R_{ik} - \lambda g_{ik} = 0$ is no identity but a field equation, and the Bianchi identities then read

$$\delta\mathbf{L} = R_{ik}\delta\,\mathfrak{G}^{ik} + \mathbf{N}^{ik}{}_{l}\delta U^{i}{}_{kl} - (\mathfrak{G}^{ik}\delta U^{l}{}_{ik})_{,l} = 0\,. \tag{A.45}$$

With Schrödinger's asterisk connection, the dynamical identities can be written in the form[6, 7]

$$(\mathfrak{G}^{(ik)}R^{*}{}_{(ik)})_{,i} - \frac{1}{2}\mathfrak{G}^{(mn)}R^{*}{}_{(mn),l} = \frac{1}{2}\mathfrak{G}^{[mn]}R^{*}{}_{[mn],l]} = -\frac{1}{\lambda^{2}}\mathbf{H}R^{[mn]}R^{*}{}_{[mn],l]}. \tag{A.46}$$

The non-gauge-invariant condition $\Gamma^{i}{}_{kl} = \Gamma^{i}{}_{lk}$ yields the general-relativistic Einstein spaces with $\Gamma^{i}{}_{lk} = \{^{i}_{kl}\}$. - The weaker conditions

$$R_{ik} = R_{ki} \quad (\text{equivalent to } g_{ik} = g_{ki}) \tag{A.47}$$

lead to

$$\Gamma^{i}{}_{kl} = \{^{i}_{kl}\} + \delta^{i}_{k}\Phi_{,l} \quad \text{and} \quad \Gamma_{i} = -\frac{2}{3}\Phi_{,l}. \tag{A.48}$$

We obtain the relation

$$g_{\underset{+}{i}\underset{-}{k};l} = g_{ik,l} - g_{rk}\{^{r}_{il}\} - g_{rk}\delta^{r}_{i}\Phi_{,l} - g_{ir}\{^{r}_{lk}\} - g_{ir}\delta^{r}_{k}\Phi_{,l} = -2g_{ik}\Phi_{,l}\,. \tag{A.49}$$

The Bianchi identity takes the form

$$(\sqrt{-g}\,E^{i}{}_{k})_{,i} - \frac{1}{2}\sqrt{-g}\,E^{mn}g_{mn,k} = (\sqrt{-g}\,E^{i}{}_{k})_{;i} - 4\sqrt{-g}\,E^{ik}\Phi_{,i} = 0\,. \tag{A.50}$$

This leads to the following equation of motion for a test-particle

$$\left(\frac{dx^{i}}{ds}\right)_{;l}\frac{dx^{l}}{ds} = 4\frac{d\Phi}{ds}\frac{dx^{i}}{ds}. \tag{A.51}$$

The path of a test-particle is an auto-parallel world line measured with a variable rule (scalar-tensor theory of Weyl's general relativity).

The general affine field theory breaks any index symmetry. However, Einstein's "transposition invariance" (or pseudo-hermiticity) holds with respect to the field variables $U^{i}{}_{kl}$

(which are canonicel variables). The substitution of $U^i{}_{kl}$ by $\tilde{U}^i{}_{kl} = U^i{}_{lk}$ yields $\tilde{R}_{ik} = R_{ik}$, $\tilde{g}_{ik} = g_{ik}$ and therefore

$$\tilde{\mathbf{N}}^i{}_{kl} = \mathbf{N}^i{}_{kl}. \tag{A.52}$$

In this sense, the Einstein-Schrödinger field equations are pseudo-hermitian.[6, 26, 29] Einstein believed that the pseudo-hermiticity reflects the symmetry of matter and antimatter. If one accepts this view, then the index symmetry of Riemann spaces corresponds to the identity of "gravitation" and "antigravitation".

REFERENCES

1. H. Poincaré (1902), *La Science et l'Hypothèse* (Flamarion, Paris), Chap. 3-5.
2. A. Einstein (1922), *Geometrie und Erfahrung* [Geometry and Experience] (Springer, Berlin).
3. H. Weyl (1921), "Das Raumproblem" [The Problem of Space], *Jahresber. Dt. Math. Ges.* **30**. (Cf. also: Hermann Weyl, Gesammelte Abhandlungen, ed. by K. Shandrasekharan, Vol. II, No. 45 (Springer, Berlin etc., 1968).
4. H. Weyl (1923), "Die Einzigartigkeit der Pythagoreischen Maßbestimmung" [The Uniqueness of the ‚Pythagorean Measure], *Math. Zeitschr.* **12**, 114. (Cf. also: Hermann Weyl, Gesammelte Abhandlungen, *op. cit.*, No. 49).
5. E. Cartan (1923), "Sur un théorème fundamental de H. Weyl", Jour. d. math. p. et a. 2, 167. (Cf. also: Cartan, Œuvres complétes, p. 633 (Gauthier-Villars, Paris, 1955).
6. A. Einstein (1955), *The Meaning of Relativity*, 5th ed., App. II (Princeton University Press).
7. E. Schrödinger (1950), *Space-Time Structure* (Cambridge University Press).
8. H.-H. v. Borzeszkowski and H.-J. Treder (1996), "Mixed field theory and Einstein-Cartan theory", in *Quantum Gravity*, ed. by P G. Bergmann, H.-J. Treder, and V. de Sabbata, (World Scientific, Singapore).
9. H.-J. Treder (1994), "Hamiltonian Dynamics of Purely Affine Fields", *Astron. Nachr.* **315**, 9.
10. J. A. Schouten (1954), *Ricci-Calculus* (Springer, Berlin etc.).
11. A. Einstein (1921), "Über eine naheliegende Ergänzung des Fundaments der allgemeinen Relativitätstheorie" [On an Itself-Suggesting Supplementation of the Foundation of General Relativity], *Ber. Preuss. Akad. Wiss.* 1921, p. 261.

12. H. Weyl (1918), "Gravitation und Elektrizität" [Gravitation and Electricity], *Ber. Preuss. Akad. Wiss.* 1918, p. 465. (Cf. also: Hermann Weyl, Gesammelte Abhandlungen, *op. cit.*, No. 31.

13. H. Weyl(1929), "Elektron und Gravitation" [Electron and Gravitation], *Zeitschr. Phys.* **56**, 330-352. (Hermann Weyl, (Cf. also: Gesammelte Abhandlungen, *op. cit.*, No. 85).

14. H.-J. Treder (1971), *Gen. Rel. Grav.* **2**, 313; H.-J. Treder and H.-H. v. Borzeszkowski (1973), *Int. J. Theor. Phys.* **8**, 319.

15. T. Levi-Civita (1926), *The Absolute Differential Calculus* (Blackie, London and Glasgow).

16. L. P. Eisenhart, *Non-Riemannian Geometry* (Amer. Math. Soc. Publ., New York, 1926).

17. P. G. Bergmann (1962), In *Encyclopedia of Physics*, Vol. IV, ed. S. Flügge (Springer, Berlin etc.).

18. H.-H. v. Borzeszkowski and H.-J. Treder (1997), *Gen. Rel. Grav.* **29**, 455.

19. H.-J. Treder, U. Bleyer, and D.-E. Liebscher (1996), Jacobi's Principle and Hertz' Definition of Time" in Gravity, *Particles and Space-Time*, ed. by P. Pronin and G. Sardanashvily (World Scientific, Singapore).

20. H.-H. v. Borzeszkowski and H.-J. Treder (1997), *Found. Phys.* **27**, No. 4.

21. F. W. Hehl, J. D. McCrea, E. W. Mielke, Y. Ne'eman (1995), "Metric-Affine Gauge Theory of Gravity: Field Equations, Noether Identities, World Spinors, and Breaking of Dilation Invariance", *Physics Reports* **258**, pp. 1 - 171 (see also the early Hehl papers on this subject cited therein).

22. J. W. Moffat (1991), in *Gravitation (Proc. Banff Summer Institute)*, R. D. Mann and P. Wesson, eds. (World Scientific, Singapore). For recent discussions, cf. also: N. J. Cornish, J. W. Moffat, and D. C. Tatarski (1995), *Gen. Rel. Grav.* **27** , 933, and the literature cited therein.

23. H.-J. Treder (1983), *Ann. Physik (Leipzig)* **40**, 378.

24. E. Schrödinger (1948), "Final affine field laws", *Proc. Roy. Irish Acad. A* **52**, 1-9.

25. A. S. Eddington (1948), Fundamental Theory (Cambridge University Press).

26. W. Pauli (1951), *Theory of Relativity* (Pergamon Press, London).

27. A. S. Eddington (1925), *Relativitätstheorie in mathematischer Behandlung* (Springer, Berlin). This German edition differs from the English edition (*Mathematical Theory of Relativity*, Cambridge, 1923) by Einstein's appendix "Eddingtons Theorie und Hamiltonsches Prinzip" and by Eddington's new section "Die neue Einsteinsche Theorie" in the chapter "Die Weltgeometrie".

28. P. G. Bergmann (1942), *Introduction to the Theory of Relativity* (Prentice Hall, New York).
29. E. Schrödinger (1946), "The general affine field laws", *Proc. Roy. Irish Acad. A* **51**, 41-
30. C. Møller (1966), "Survey of investigations on energy-momentum complex", in *Entstehung, Entwicklung und Perspektiven der Einsteinschen Gravitationstheorie*, ed. by H.-J. Treder (Akademie-Verlag, Berlin).

HESTENES' GEOMETRIC ALGEBRA AND REAL SPINOR FIELDS

Bidyut Kumar Datta*
and
Venzo de Sabbata

Istituto di Fisica dell' Universita
Via Irnerio 46, 40126 Bologna, ITALY
and
World Laboratory, Lausanne, SWITZERLAND

Abstract

We give a brief review of Hestenes' geometric algebra as applied to Minkowski space-time and transformations within it. The connection between the algebra of Dirac matrices and the real Dirac algebra, a Clifford algebra based on Minkowski space-time, is discussed giving the full geometric significance of the Dirac γ_μ. Real spinor fields expressed in terms of the real Dirac algebra are shown to have one-to-one correspondence with Dirac spinors. The "Observables" appearing in the Dirac theory are expressed in terms of a real spinor field and a covariant C transformation, known as charge conjugation, is constructed within it.

---

* **Permanent address :**
ICSC - World Laboratory, Calcutta Branch,
108/8 Maniktala Main Road, Suite-J/77,
Calcutta-700 054, INDIA.

1. INTRODUCTION

(A) OBJECTIVES

This note is prepared with the following objectives in mind before introducing the real spinor fields.

(1) To clarify the geometrical and physical interpretation of the Dirac theory,

(2) To provide some understanding of the mathematical structure of the theory, i.e. of Dirac algebra,

(3) To show that in the language of space-time algebra the Dirac theory can be given a completely geometrical formulation, involving neither matrices nor complex numbers,

(4) To show that the algebra of gamma matrices has the same geometric significance as tensor algebra,

(5) To show that the unit imaginary $\sqrt{-1}$ in the formulation of the Dirac theory has a definite geometrical and physical significance; it can be replaced by a spacelike bivector, a geometric entity.

(B) THE ALGEBRA OF DIRAC MATRICES

The four gamma matrices of the Dirac theory over the complex numbers generate the complete algebra of 4×4 matrices. The 4×4 matrix algebra with the geometric interpretation induced by the conditions :

$$\frac{1}{2}(\gamma_\mu\gamma_\nu + \gamma_\nu\gamma_\mu) = g_{\mu\nu} I \;, \qquad \ldots(1.1a)$$

$$\frac{1}{4}\,\mathrm{Tr}\,\gamma_\mu\,\gamma_\nu = g_{\mu\nu} \;, \qquad \ldots(1.1b)$$

where the $g_{\mu\nu}$ $(\mu, \nu = 0,1,2,3)$ are the components of space-time metric tensor and I is the unit matrix, is called the algebra of Dirac matrices. Condition (1.1b) implies that the matrices γ_μ are irreducible. Conditions (1.1) show that the matrices are traceless :

$$\frac{1}{4}\,\mathrm{Tr}\,\gamma_\mu = 0 \,. \qquad \ldots(1.2)$$

(C) GEOMETRIC SIGNIFICANCE OF THE γ_μ

In the conventional approach the full geometric significance of the γ_μ can be understood only with the specification of their relation to a Dirac spinor. Essentially the relation (1.1a) is independent on the assumption that the γ_μ are matrices. This can be realized by considering that the γ_μ belong to an associative non-commutative algebra. This is achieved by Hestenes[1-3,5,6] by interpreting the γ_μ as vectors of a space-time frame with γ_0 the reference frame's 4-velocity, instead of as matrices. So the conditions (1.1) for matrices correspond to the single equation

$$\frac{1}{2}(\gamma_\mu\gamma_\nu + \gamma_\nu\gamma_\mu) = \gamma_\mu \cdot \gamma_\nu = g_{\mu\nu} \qquad \ldots(1.3)$$

for vectors.

The vectors γ_μ generate an associative algebra over the reals, which provides a direct and complete algebraic characterization of the geometric properties of Minkowski space-time. This associative algebra has been dubbed by Hestenes[1] as "space-time algebra".

As the geometric interpretation of the γ_μ of the space-time algebra is independent of the notion of spinor, the γ_μ assumes a central position in the mathematical description of all physical systems in space-time including relativistic quantum theory.

2. SPACE-TIME ALGEBRA (STA)

We give a brief review of Hestenes' geometric algebra[1-10] as applied to Minkowski space-time and transformations within it.

The geometric product ab, denoted by simple juxtaposition, of two proper vectors a and b in space-time can be understood geometrically by separating it into symmetric and antisymmetric parts :

$$ab = a.b + a \;\; b \,, \qquad \ldots(2.1a)$$

where

$$a.b = \frac{1}{2}(ab + ba) = a.b \,, \qquad \ldots(2.1b)$$

$$a \wedge b = \frac{1}{2}(ab - ba) = -b \wedge a . \qquad \ldots(2.1c)$$

(2.1b) tells that (a.b) is a scalar quantity, the usual inner product of space-time vectors. Here "scalar" means "real number". The quantity $a \wedge b$, called the outer product of a and b, is a (proper) bivector and is the same one used with differential forms[11]. The geometric product is defined to be associative. So the outer product with its antisymmetric character implies that

$$a \wedge b \wedge c \wedge d \wedge e = 0 \qquad \ldots(2.2)$$

for any five or more vectors in space-time.

The Clifford algebra of real four-dimensional space-time is generated by four orthonormal vectors $\{\gamma_\mu\}$ and is spanned by

$$\underset{\text{(scalar)}}{1}, \quad \underset{\text{(vectors)}}{\{\gamma_\mu\}}, \quad \underset{\text{(bivectors)}}{\{\sigma_k, \gamma_5 \sigma_k\}}, \quad \underset{\text{(trivectors or pseudovectors)}}{\{\gamma_5 \gamma_\mu\}}, \quad \underset{\text{(pseudoscalar or quadrivector)}}{\gamma_5} \qquad (2.3)$$

$$(\mu = 0,1,2,3 \ ; \quad k = 1,2,3) ,$$

where γ_5 is the unit pseudoscalar or quadrivector for space-time :

$$\gamma_5 = \gamma_0\gamma_1\gamma_2\gamma_3 = \sigma_1\sigma_2\sigma_3 = i, \qquad \ldots(2.4a)$$

$$\sigma_k = \gamma_k \gamma_0 , \qquad \ldots(2.4b)$$

$$\gamma_5^2 = i^2 = -1 . \qquad \ldots(2.4c)$$

i is the unit pseudoscalar for three-dimensional Euclidean space. The significant result (2.4a) connects two distinct geometric entities γ_5 and i algebraically.

The algebra (2.3) is the space-time algebra or real Dirac algebra $\mathbb{D}$ having sixteen components and so is a sixteen - dimensional linear space. In Dirac algebra

$$\gamma_5 \gamma_\mu = - \gamma_\mu \gamma_5 \quad . \qquad \ldots(2.5)$$

The even elements of the basis (2.3) for the space-time

$$1 \, , \left\{ \sigma_k \, , i \, \sigma_k \right\} \, , \, i \qquad \ldots(2.6)$$

coincide with Pauli algebra $\mathbb{P}$ because of the relation (2.4b) which satisfies

$$\sigma_k \cdot \sigma_j = \frac{1}{2} (\sigma_k \sigma_j + \sigma_j \sigma_k) = \delta_{kj} \quad , \qquad \ldots(2.7)$$

the requirement of a basis in three-dimensional Euclidean space. Because of the important relation (2.4a) between γ_5 and i, we replace γ_5 by i in $\mathbb{P}$. Moreover, in $\mathbb{P}$, i commutes with σ_k :

$$i \, \sigma_k = \sigma_k \, i \quad , \qquad \ldots(2.8)$$

indicating its similarity in character with the unit imaginary.

(2.4b) shows that the vectors σ_k in $\mathbb{P}$ become bivectors as viewed from $\mathbb{D}$. In STA the four Dirac γ_μ are no longer viewed as four matrix-valued components of a single isospace vector, but as four linearly independent basis vectors for "real" space-time. Stated more explicitly, γ_0 is a unit vector in the forward light cone and the γ_i, $(i = 1,2,3)$, are a dextral set of spacelike vectors. The algebra $\mathbb{P}$ (2.6) is an even subalgebra of $\mathbb{D}$ with respect to the selection of the timelike vector γ_0, and is an eight-dimensional linear space of spinors.

3. CONJUGATIONS

Following Hestenes[2], we define four types of conjugation operators in real Dirac algebra $\mathbb{D}$. Any multivector M in $\mathbb{D}$ can be written as

$$M = M_S + M_V + M_B + M_T + M_P \quad , \qquad \ldots(3.1)$$

where the subscripts S,V,B,T,P mean, respectively, scalar, vector, bivector, trivector (pseudovector) and pseudoscalar parts of the multivector M.

(i) Conjugate Multivectors (Reversion)

We define conjugate multivector $\widetilde{M}$ of M of $\mathbb{D}$ by reversing the order of the products of all vectors of M. This takes M into $\widetilde{M}$ as

$$\widetilde{M} = M_S + M_V - M_B - M_T + M_P \; . \qquad \ldots(3.2)$$

As it is independent of any basis in the algebra, it is an invariant type of conjugation.

(ii) **Space-time Conjugation**

By using the unit pseudoscalar γ_5, we define the space-time conjugation $\overline{M}$ of M of $\mathbb{D}$ by the operation which reverses the direction of all vectors in space-time. This maps M into $\overline{M}$ as

$$\overline{M} = -\gamma_5 M \gamma_5$$

$$= M_S - M_V + M_B - M_T + M_P \quad . \qquad \ldots(3.3)$$

It is to be noted that M is even if $\overline{M} = M$ and odd if $\overline{M} = -M$.

(iii) **Space Conjugation**

We introduce an operation, called space conjugation, which takes M of $\mathbb{D}$ into M^* such that

$$M^* = \gamma_0 M \gamma_0 \quad . \qquad \ldots(3.4)$$

This operation depends on the choice of γ_0 and changes $\{\sigma_i\}$ and $\{\gamma_i\}$ into left-handed frames without affecting γ_0 :

$$\sigma_i^* = \gamma_0 \sigma_i \gamma_0 = -\sigma_i \quad , \qquad \ldots(3.5a)$$

$$\gamma_\mu^* = \gamma_0 \gamma_\mu \gamma_0 = (\gamma_0 \, , -\gamma_i) \quad . \qquad \ldots(3.5b)$$

(iv) Hermitian Conjugation

We introduce another operation, called Hermitian conjugation, which takes M of $\mathbb{D}$ into $M^\dagger$ such that

$$M^\dagger = \gamma_0 \widetilde{M} \gamma_0 . \qquad \ldots(3.6)$$

$M^\dagger$ depends on the choice of γ_0 and corresponds to the Hermitian conjugate of M in the usual matrix representations of the Dirac algebra.

4. LORENTZ TRANSFORMATIONS : LORENTZ ROTATIONS

In the algebra of Dirac matrices the conditions (1.1) do not determine the Dirac matrices uniquely. Any two sets of Dirac matrices $\{\gamma_\mu\}$ and $\{\gamma'_\mu\}$ are related by a similarity transformation

$$\gamma'_\mu = L \gamma_\mu L^{-1} , \qquad \ldots(4.1)$$

where L is a nonsingular matrix. This, in fact, gives a change in representation of the Dirac matrices.

What does eq. (4.1) mean in the space-time algebra, where the γ_μ are vectors ? The geometrical requirement that the γ_μ in (4.1) must be vectors implies that they can be expressed as

$$\gamma'_\mu = a_\mu{}^\nu \gamma_\nu . \qquad \ldots(4.1a)$$

This means that (4.1) must be invariant under reversion

$$\tilde{L}\gamma_\mu = \gamma_\mu \tilde{L}$$

or $$\tilde{L} L \gamma_\mu L^{-1} = \gamma_\mu \tilde{L} .$$

So, one may choose L such that

$$\tilde{L} L = 1 \quad \text{or} \quad L^{-1} = \tilde{L} . \qquad \ldots(4.2)$$

Then (4.1) assumes the form

$$\gamma'_\mu = a_\mu{}^\nu \gamma_\nu = L \gamma_\mu \tilde{L} \qquad \ldots(4.3)$$

describing a Lorentz transformation from a frame of vectors $\{\gamma_\mu\}$ into a frame $\{\gamma'_\mu\}$. Furthermore, one can solve eq. (4.3) for L as a function of the γ_μ and the γ'_μ only. This implies that L is a multivector and that every Lorentz transformation can be expressed in that form.

Hestenes[1] has shown that (4.3) is a proper Lorentz transformation (i.e., transformations continuously connected to the identity) if and only if L is an even multivector satisfying

$$L\tilde{L} = 1 \qquad \ldots(4.4)$$

from which one can write

$$L = \pm \exp(-\Phi/2) , \qquad \ldots(4.5)$$

where Φ is a bivector. L is sometimes referred to as a Lorentz rotation.

Following Hestenes[2], we call (4.3) a spatial rotation if $L = L^*$, whence we write

$$L = \exp(-\tfrac{1}{2}\gamma_5 \lambda) \quad , \qquad \ldots(4.6a)$$

and a special timelike rotation if $L = L^\dagger$ for which

$$L = \exp(-\tfrac{1}{2}\mu) \quad , \qquad \ldots(4.6b)$$

where λ and μ are bivectors satisfying

$$\lambda^* = -\lambda \, , \qquad \mu^* = -\mu \quad . \qquad \ldots(4.7)$$

Any Lorentz rotation can be expressed as a spatial rotation followed by a special timelike rotation, i.e.

$$L = \exp(-\tfrac{1}{2}\mu - \tfrac{1}{2}\gamma_5 \lambda) \quad . \qquad \ldots(4.8)$$

5. REAL SPINOR FIELDS

A real spinor field $\psi(x)$ can be expressed in terms of real Dirac algebra by[2]

$$\psi(x) = \exp(\tfrac{1}{2} v(x)) \quad , \qquad \ldots(5.1)$$

where $v(x)$ is an even multivector in $\mathbb{D}$:

$$v(x) = \alpha(x) + \gamma_5\beta + \Phi \, . \qquad \ldots(5.2)$$

Here α and β are scalars and Φ is a bivector.

By using (4.5), (5.1) and (5.2), the spinor field $\psi(x)$ can be put in the form :

$$\psi = (\varrho e^{\gamma_5 \beta})^{1/2} L , \qquad ...(5.3)$$

where

$$\varrho = e^{\alpha(x)} > 0 . \qquad ...(5.4)$$

(5.2) shows that the even multivector $v(x)$ has eight linearly independent components which correspond to the column matrix with four complex components of a Dirac spinor. Thus real spinors as defined above are equivalent to Dirac spinors, because the two can be put into one-to-one correspondence[2].

We define four vectors J_μ by

$$J_\mu = \psi \gamma_\mu \tilde{\psi} = \varrho e_\mu , \qquad ...(5.5)$$

where the e_μ are the orthonormal vectors defined by

$$e_\mu = L \gamma_\mu \tilde{L} . \qquad ...(5.6)$$

since $\varrho_0 > 0$ (see eq. (5.4)), J_0 is a timelike vector in the forward light cone and is equivalent to the probability current density of the Dirac theory. So, $\varrho(x)$ may be interpreted as the proper probability density.

A restriction on the wave equation for ψ is imposed by the requirement of the conservation of probability :

$$\Box \cdot J_0 = 0 . \qquad ...(5.7)$$

The vector J_3 represents the orientation of the electron spin. The plane containing J_1 and J_2 described by the space-like bivector $J_1 \wedge J_2$ corresponds to the abstract complex plane in which the phase factor of the Dirac wave function is described. This plane may quite plausibly be called "phase plane" (or "spin plane") of the wave function[2]. Now the unit imaginary $(-1)^{1/2}$ in the electron wave function can be interpreted geometrically as the generator $\gamma_2\gamma_1$ of rotations in the phase plane.

The relation

$$\gamma_5 \sigma_3 = \gamma_5 \gamma_3 \gamma_0 = \gamma_2 \gamma_1 \qquad \ldots(5.8)$$

shows that the phase giving the magnitude of a rotation in the phase plane and the spin describing the orientation of the phase plane are inextricably unified.

Given the observables J_μ in (5.5), one can uniquely determine ψ by considering the bilinear function of ψ :

$$\psi\tilde{\psi} = \rho\, e^{\gamma_5\beta} \quad . \qquad \ldots(5.9)$$

One may relate the factor $e^{\gamma_5\beta}$ to an "observable" of the Dirac theory by considering it[2] to be the relative admixture of positive and negative energy components of ψ. In view of the above, the scalar part of $\psi\tilde{\psi}$:

$$(\psi\tilde{\psi})_0 = \rho \cos \beta \qquad \ldots(5.10)$$

can plausibly be interpreted as the proper particle density of the spinor field ψ. Then a negative value given by (5.10) would mean the likelihood of observing an antiparticle.

6. CHARGE CONJUGATION

Dirac equation for an electron can be written in Hestenes' algebra as[2,6]

$$\Box \psi = (m \psi \gamma_0 + e A \psi) \gamma_2 \gamma_1 , \qquad \ldots(6.1)$$

$$(\hbar = c = 1)$$

where $\psi(x)$ is the wave function, m is the mass, e is the charge of the electron, and A is the electromagnetic vector potential. We discuss here certain symmetries of the spinor field $\psi(x)$ that map the field $\psi(x)$ onto itself preserving the wave equation (6.1) or changing it in a definite and physically meaningful way.

The wave function ψ uniquely determines a frame of tangent vectors

$$J_\mu(x) = \psi \gamma_\mu \tilde{\psi} = \rho e_\mu \qquad \ldots(6.2)$$

at each point of space-time, and inversely, except for a factor $e^{\gamma_5 \beta}$, the tangent vectors J_μ determine ψ. In view of the above, one can give a geometric interpretation to transformations of ψ, because any transformation of the tangent vectors can be interpreted geometrically.

Symmetries of a spinor field can be interpreted geometrically as some combination of two distinct types of geometrical transformations : (i) a transformation of the tangent vectors $J_\mu(x)$ at a point x of space-time into a new set of tangent vectors $J'_\mu(x)$ at the same point x, and (ii) a point transformation

$$x = x^\mu \gamma_\mu \rightarrow x' = x'^\mu \gamma_\mu , \qquad \ldots(6.3)$$

wherein the tangent vectors $J_\mu(x)$ at a point x of space-time are mapped into equivalent vectors at a different point x'.

For obvious reason we restrict our discussion to only one type of symmetry, viz. C transformation, leaving the other for later discussion.

The transformation C, known as the charge conjugation, changes the sign of the electromagnetic coupling leaving the rest of the Dirac equation invariant. This can be achieved if we take the charge conjugation as

$$C : \psi \rightarrow \psi^C = \psi \gamma_2 \gamma_0 \qquad \ldots(6.4a)$$

or,

$$C : \psi \rightarrow \psi^C = \psi \gamma_1 \gamma_0 . \qquad \ldots(6.4b)$$

It is easy to see that under charge conjugation (6.4a) or (6.4b) the Dirac equation (6.1) becomes

$$\Box \psi = (m \psi \gamma_0 - e A \psi) \gamma_2 \gamma_1 . \qquad \ldots(6.5)$$

The charge conjugation (6.4a) induces a rotation of π of the J_i about J_2 axis :

$$J_0 \to J_0\,,\quad J_1 \to -J_1\,,\quad J_2 \to J_2\,,\quad J_3 \to -J_3\,; \qquad \ldots(6.6a)$$

while the charge conjugation (6.4b) induces a rotation of π of the J_i about J_1 axis :

$$J_0 \to J_0\,,\quad J_1 \to J_1\,,\quad J_2 \to -J_2\,,\quad J_3 \to -J_3\,. \qquad \ldots(6.6b)$$

Physically this means that in both the cases the spin remains invariant.

Moreover, under charge conjugation (6.4a) or (6.4b) the bilinear function of ψ changes sign :

$$\psi^C \tilde{\psi}^C = -\psi\tilde{\psi} = -\rho\, e^{\gamma_5 \beta}\,. \qquad \ldots(6.7)$$

So, the scalar part of the bilinear function of the wave function also changes sign :

$$(\psi^C \tilde{\psi}^C)_0 = -(\psi\tilde{\psi})_0 = -\rho \cos\beta\,. \qquad \ldots(6.8)$$

In view of the interpretation endowed to eq.(5.10), we note that the charge conjugation considered above changes the sign of the proper particle density of the spinor field ψ, indicating the likelihood of observing an antiparticle.

7. REMARKS

The vectors $\gamma_0, \gamma_1, \gamma_2$ appearing in the Dirac equation (6.1) belong to a set of arbitrarily chosen orthonormal vectors γ_μ (μ = 0,1,2,3). The choice of a coordinate frame $\{\gamma_\mu\}$ with γ_0 the reference frame's 4-velocity corresponds to the standard matrix representation of the Dirac theory for which γ_0 is Hermitian and the γ_k (k = 1,2,3) are anti-Hermitian. Furthermore, the standard matrix representation

$$\gamma_0 = \begin{pmatrix} I & 0 \\ 0 & -I \end{pmatrix} , \qquad \gamma_k = \begin{pmatrix} 0 & -\hat{\sigma}_k \\ \hat{\sigma}_k & 0 \end{pmatrix}$$

with

$$\hat{\sigma}_1 \hat{\sigma}_2 \hat{\sigma}_3 = \sqrt{-1}\; I ,$$

(I is the 2x2 unit matrix and the $\hat{\sigma}_k$ are the usual 2x2 Pauli matrices.) associates the unit imaginary $\sqrt{-1}$ of the matrix representation with the bivector $\gamma_2\gamma_1$ in eq. (6.1). So, the Hermitian conjugation and the complex numbers of the standard matrix representation can be related to some intrinsic features of the Dirac equation.

Acknowledgements

The authors places on record their grateful indebtedness to Professor A. Zichichi, President, ICSC - World Laboratory, LAUSANNE (SWITZERLAND), for financial support under E-8 GRAVI PROJECT.

References

1. D. Hestenes, "Space-Time Algebra" (Gordon and Breach, New York, 1966).

2. D. Hestenes, J. Math. Phys., **8**, 798 (1967).

3. D. Hestenes, Am. J. Phys., **39**, 1013 (1971).

4. D. Hestenes and R. Gurtler, Am. J. Phys., **39**, 1028 (1971).

5. D. Hestenes, J. Math. Phys., **14**, 893 (1973).

6. D. Hestenes, J. Math. Phys., **16**, 556 (1975).

7. D. Hestenes and R. Gurtler, J. Math. Phys., **16**, 573 (1975).

8. D. Hestenes, Am. J. Phys., **47**, 399 (1979).

9. J.D. Hamilton, J. Math. Phys., **25**, 1823 (1984).

10. B.K. Datta, Physical theories in space-time algebra, in "Quantum Gravity", eds. P.G. Bergmann, V. de Sabbata and H. -J. Treder (World Scientific, Singapore, 1996), pp. 54-79.

11. H. Flander, "Differential Forms" (Academic, New York, 1963).

Erice 1997
International School of Cosmology and Gravitation

EVIDENCE FOR TORSION IN GRAVITY ?

Venzo de Sabbata

World Laboratory, Lausanne, Switzerland
Dept. of Physics, University of Bologna and Ferrara, Italy
Istituto di Fisica Nucleare, sezione di Ferrara, Italy

Abstract

We consider general relativity with spin in the sense of Einstein-Cartan theory that is introducing torsion. After an outline of the theory, we consider very briefly some consequences as for instance the connection between torsion and magnetism, photon-torsion interaction which leads to rotation of polarization plane and dispersion, the neutrino oscillations and neutron interferometry. At last we make some considerations to the possibility that torsion constitute a way to go toward quantization of gravity. At this regard we introduce the concept of a spinorial space-time using the Hestenes space-time algebra because we think that both tensors and spinors are to be described in a unique manifold, the real space-time.

EVIDENCE FOR TORSION IN GRAVITY?

Venzo de Sabbata
World Laboratory, Lausanne, Switzerland
Dept. of Physics, University of Bologna and Ferrara, Italy
Istituto di Fisica Nucleare, sezione di Ferrara, Italy

1. Introduction

We think that a problem of primary importance is that of introducing the spin in the Einstein theory of gravity. Spin is in fact a fundamental property of all elementary particles as the mass: mass and spin are two elementary and independent original concepts. Then when we consider the early universe, we are faced with the cosmological problem which is strictly connected with elementary particle physics. So we must taken into account also the spin; as the mass is connected with the curvature of space-time, the spin will be connected with another geometrical property of the space-time which, according to Cartan [1 - 4] is represented by torsion. For that reason we will consider the Einstein-Cartan theory where the affine connection is no more symmetric and where the torsion $Q_{\alpha\beta.}{}^{\mu}$ is given by the antisymmetric part of the affine connection $\Gamma_{\alpha\beta}{}^{\mu}$ i.e.

$$Q_{\alpha\beta.}{}^{\mu} = (1/2)(\Gamma_{\alpha\beta}{}^{\mu} - \Gamma_{\beta\alpha}{}^{\mu}) = \Gamma_{[\alpha\beta]}{}^{\mu} \qquad (1)$$

We will see also that through torsion we have a way to go toward the quantization of space-time.

One important geometric meaning of torsion that is of the antisymmetric part of the connection (which is easy to show to be a tensor), is that the closed paths are open in the tangent space; while in the case of the curvature a vector which goes along a closed path changes direction when returns to the starting point (Fig.1a), in the case of torsion the vector is suffering a translation (Fig.1b); obviously if both curvature and torsion are present, the vector will have a rotation and a translation (Fig.1c).

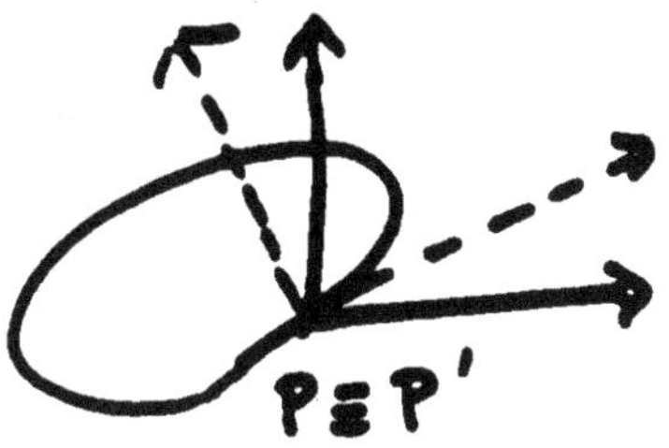

Fig.1a - curvature

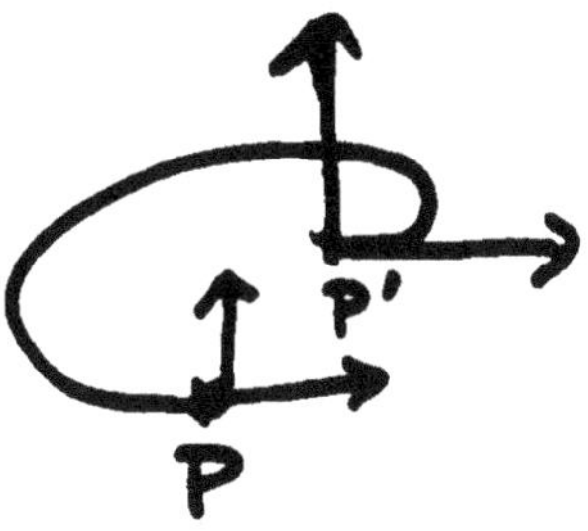

Fig.1b - torsion

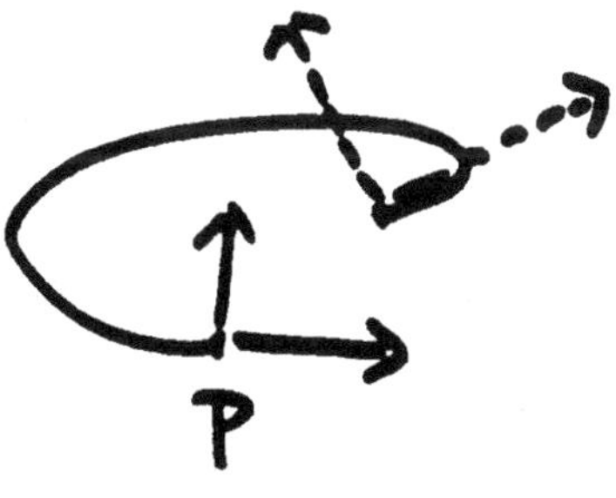

Fig.1c - curvature and torsion

Fig.1a,b,c, - when a vector describes an infinitesimal closed path, and when this path is developed in the flat space tangent to the manifold, we have a rotation (if there is only curvature), or a translation (if there is only torsion) or both (if there is curvature and torsion)

This leads to important physical consequences; although the modification of the starting point is minimal (a asymmetric connection instead of the symmetric one of the Einstein theory: as said by Trautman "the Einstein-Cartan theory is the simplest and the most natural modification of the original Einstein theory of gravitation" [5]), the physical consequence are enormous. To quote only some of them we will find

a) connection betwen torsion and magnetism
b) interaction photon-torsion which will lead to rotation of the polarization plane and dispersion
c) neutrino oscillations
d) neutron interferometry

The important thing is to see if it is possible to perform some experiments which will indicate the presence of torsion. Until now we have not any direct evidence for torsion, although one can interpret some physical facts as an indirect evidence of the presence of torsion.

Before that we will briefly illustrate the theoretical aspect of the introduction of torsion. First of all remember that it is usual to denote the Riemann-Cartan space-time of the Einstein-Cartan theory (when an affine asymmetric connection is introduced) as U_4 to distinguish it from the Riemann space-time (with a symmetric affine connection) which is denoted by V_4.

We know that in general relativity the dynamical relation between the stress-energy-momentum tensor and curvature is expressed by the Einstein equations so one may think that it must be an analogous dynamical relation including spin tensor, a relation that would connect the spin tensor with another geometric property of the space-time. This obviously is not possible in the ambit of general relativity so that one is necessitated to modify the theory in order to introduce torsion and related it to the spin tensor (see for instance Hehl et al.[6]).

Now we have said that the closed paths are open in the tangent space. To see better this point, consider an infinitesimal parallelogram in the tangent space at a point P; in general they do not close (Fig.2), the failure of closing being proportional to torsion.

In other words a closed contour (P, A_1 , A_2 , A_3) in an U_4 manifold becomes in general a non-closed contour (P, B_1,B_2,B_3) in the flat space E tangent to U_4 in P (Fig.2)

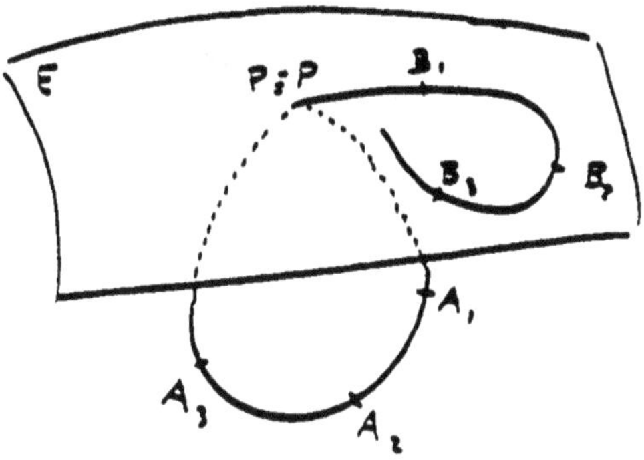

Fig.2

Consider in fact the infinitesimal parallelogram PACB in the tangent space E, and put $\overline{PA}$ = dx', $\overline{PB}$ = dx'' (Fig.3)

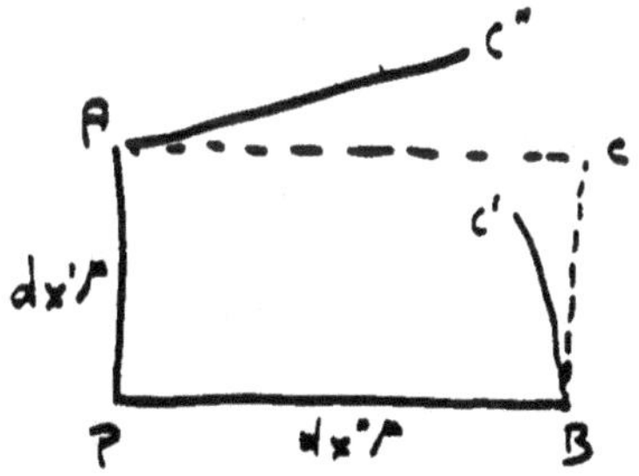

Fig.3

Displacing dx' along dx'' in the corresponding U_4 manifold we have

$$\delta(dx'^{\mu}) = - \Gamma_{\alpha\beta}{}^{\mu}\, dx'^{\beta} dx''^{\alpha} \tag{2}$$

while for the displacement of dx'' along dx'

$$\delta(dx''^{\mu}) = - \Gamma_{\alpha\beta}{}^{\mu} dx''^{\beta} dx'^{\alpha} \qquad (3)$$

Therefore

$$\delta(dx''^{\mu}) - \delta(dx'^{\mu}) = - \Gamma_{\alpha\beta}{}^{\mu}(dx'^{\beta}dx''^{\alpha} - dx'^{\alpha}dx''^{\beta})$$

which we can write as

$$= - (\Gamma_{\alpha\beta}{}^{\mu} - \Gamma_{\beta\alpha}{}^{\mu}) dx'^{\beta}dx''^{\alpha} = 2\, Q_{\alpha\beta}{}^{\mu}\, dx'^{\beta}dx''^{\alpha} \qquad (4)$$

and as in the presence of torsion $Q_{\alpha\beta}{}^{\mu} \neq 0$, we have in general that the points C', C'' and C do not coincide so that the parallelogram is not closed. In the Riemannian space the Γ are symmetric (the Christoffel symbols) and then $Q = 0$.

This property that we have in the Riemannian space is then clearly a consequence of the symmetry of the affine connection. This is no more valid if the connection is asymmetric: in this last case we have two different points and, as we have seen, the parallelogram is not closed; this is just the geometric characteristic of torsion.

The closure failure behaves as dislocations (or defects) in crystals and we can say that torsion plays the role of defect density and the defect in space-time topology leads naturally to a possible quantization of space-time. We think in fact that when one likes to quantize gravity, one cannot quantize gravity in analogy to the other interactions in nature because, according to Einstein, there is no gravity force but curvature of space-time; so if one like to quantize gravity one must go towards the quantization of space-time itself. The torsion seems to represent a way to go through the quantization of space-time.

But before that, some words on the mathematical side: in order to find the field equations one proceeds as in the Einstein case; the only difference is that the affine connection Γ is not symmetric. First of all it is important to note that in Riemann-Cartan manifold, denoted as U_4, the geodesic equation [7]

$$d^2x^\mu/ds^2 + \left\{ {\mu \atop \alpha\beta} \right\} (dx^\alpha/ds)(dx^\beta/ds) = 0 \qquad (5)$$

can be obtained, as in Riemann manifold V_4, by the variational principle

$$\delta \int (g_{\mu\nu} dx^\mu dx^\nu)^{1/2} = 0 \qquad (6)$$

It defines curves of extremal length with respect to the metric, and it is different, in general, from the autoparallel equation

$$d^2x^\mu/ds^2 + \Gamma_{\alpha\beta}{}^\mu (dx^\alpha/ds)(dx^\beta/ds) = 0 \qquad (7)$$

which defines curves along which a vector is displaced parallelely to itself, according to the affine connection of the manifold. In fact only the symmetric part of the affine connection enters in eq.(7) (because of the symmetry of the product $dx^\alpha dx^\beta = dx^\beta dx^\alpha$) and the symmetric part do not coincide with the Christoffel symbols; we have:

$$d^2x^\mu/ds^2 + \Gamma_{(\alpha\beta)}{}^\mu (dx^\alpha/ds)(dx^\beta/ds) =$$
$$= d^2x^\mu/ds^2 + \left[\left\{ {\mu \atop \alpha\beta} \right\} + 2\, Q^\mu{}_{(\alpha\beta)} \right] (dx^\alpha/ds)(ds^\beta/ds) = 0 \qquad (8)$$

This equation coincides with a geodesic in V_4 (where $Q^\mu{}_{\alpha\beta} = 0$) or in a U_4 manifold in which torsion is a totally antisymmetric tensor

$$Q_{\mu\alpha\beta} = Q_{[\mu\alpha\beta]} \qquad (9)$$

as in this case we have $Q_{\mu(\alpha\beta)} = 0$.

We will see that neither the geodesic equation (5) nor the autoparallel equation (7) represent the motion equation.

Continuing along these lines, one introduces the curvature tensor which is defined as in a Riemannian space, using however the U_4 connection instead of Christoffel symbols. Therefore:

$$R_{\alpha\beta\mu}{}^\nu = \partial_\alpha \Gamma_{\beta\mu}{}^\nu - \partial_\beta \Gamma_{\alpha\mu}{}^\nu + \Gamma_{\alpha\lambda}{}^\nu \Gamma_{\beta\mu}{}^\lambda - \Gamma_{\beta\lambda}{}^\nu \Gamma_{\alpha\mu}{}^\lambda \qquad (10)$$

and using this explicit expression one may verify that the following identities are satisfied:

$$R_{(\alpha\beta)\mu\nu} = 0 = R_{\alpha\beta(\mu\nu)} \qquad (11)$$

$$R_{[\alpha\beta\mu]}{}^{\nu} = 2\,\nabla_{[\alpha}Q_{\beta\mu]}{}^{\nu} - 4\,Q_{[\alpha\beta}{}^{\lambda}\,Q_{\mu]\lambda}{}^{\nu} \qquad (12)$$

The Bianchi identities are generalized as follows:

$$\nabla_{[\alpha}R_{\beta\mu]\nu}{}^{\lambda} = 2\,Q_{[\alpha\beta}{}^{\rho}R_{\mu]\rho\nu}{}^{\lambda} \qquad (13)$$

and the Einstein tensor is defined, as usual:

$$G_{\mu\nu} = R_{\mu\nu} - (1/2)g_{\mu\nu}R \qquad (14)$$

where $R_{\mu\nu} = R_{\alpha\mu\nu}{}^{\alpha}$ is the Ricci tensor and $R = R_{\mu}{}^{\mu}$. So formally the Einstein tensor is the same as in Riemann space-time but now is important to stress that, in the Riemann-Cartan space-time $G_{\mu\nu}$ and $R_{\mu\nu}$ are no longer symmetric tensors.

Remember now that if $\mathfrak{L}$ denotes the material Lagrangian density in general relativity we have for the dynamical stress-energy-momentum tensor:

$$T^{\mu\nu} = \frac{2}{\sqrt{-g}}\,\frac{\delta\,\mathfrak{L}}{\delta g_{\mu\nu}} \qquad (15)$$

($\sqrt{-g} = (-\det g_{\mu\nu})^{1/2}$ is the root of the determinant of the metric).

Now if $\mathfrak{L}$ depends not only on the tensor $g_{\mu\nu}$ and its derivatives but also on a <u>new</u> geometrical quantity $K_{\mu\nu}{}^{\alpha}$, we have the following dynamical definition of spin density tensor:

$$S_{\alpha}^{\cdot\mu\nu} = \frac{1}{\sqrt{-g}}\,\frac{\delta\,\mathfrak{L}}{\delta K_{\mu\nu\cdot}{}^{\alpha}} \qquad (16)$$

$K_{\mu\nu\cdot}{}^{\alpha}$ is called the contorsion tensor and is connected with torsion tensor

$$K_{\mu\nu\cdot}{}^{\alpha} = -\,Q_{\mu\nu\cdot}{}^{\alpha} - Q^{\alpha}{}_{\cdot\nu\mu} + Q_{\mu\cdot\nu}{}^{\alpha} = -\,K_{\nu\cdot\mu}{}^{\alpha} \qquad (17)$$

so that we can write for the affine connection

$$\Gamma_{\mu\nu}{}^{\alpha} = \left\{ {\alpha \atop \mu\nu} \right\} - K_{\mu\nu\cdot}{}^{\alpha} \qquad (18)$$

We like to give the explicit expression for the Einstein tensor $G_{\mu\nu}$ and the scalar curvature R of an U_4 manifold in which the usual riemannian parts $G_{\mu\nu}\left(\{\ \}\right)$ and $R\left(\{\ \}\right)$ are explicity separated in order to put into evidence the torsion contributions. Starting with the definition (10), inserting the explicit expression (18) of the connection, we obtain

$$G^{\mu\nu}(\Gamma) = G^{\mu\nu}\left(\{\ \}\right) + \overset{*}{\nabla}(T^{\mu\nu\alpha} - T^{\nu\alpha\mu} + T^{\alpha\mu\nu}) +$$

$$+ 4\, T^{\mu\alpha}{}_{[\beta}T^{\nu\beta}{}_{\alpha]} + 2\, T^{\mu\alpha\beta}\, T^{\nu}{}_{\alpha\beta} - T^{\alpha\beta\mu}\, T_{\alpha\beta}{}^{\nu} -$$

$$- (1/2)g^{\mu\nu}(4\, T_{\lambda}{}^{\alpha}{}_{[\beta}\, T^{\lambda\beta}{}_{\alpha]} + T^{\alpha\beta\lambda}\, T_{\alpha\beta\lambda}) \qquad (19)$$

and

$$(-g)^{1/2}R(\Gamma) = (-g)^{1/2}R\left(\{\ \}\right) + (-g)^{1/2}T^{\alpha\mu\nu}\, K_{\alpha\mu\nu} +$$

$$+ \partial_{\mu}(2(-g^{1/2})K_{\nu}{}^{\mu\nu}) \qquad (20)$$

where we have defined for simplicity

$$\overset{*}{\nabla}_{\alpha} = \nabla_{\alpha} + 2\, Q_{\alpha} \qquad (21)$$

and where $G^{\mu\nu}\left(\{\ \}\right) = G^{\nu\mu}\left(\{\ \}\right)$ and $R\left(\{\ \}\right)$ denote the Einstein tensor and the scalar curvature constructed with the Christoffel symbols. Obviously when $Q_{\mu\nu}{}^{\alpha} = 0$ then $T_{\mu\nu\alpha} = 0$ and $G_{\mu\nu} = G_{\mu\nu}\left(\{\ \}\right)$, $R = R\left(\{\ \}\right)$, i.e. in the absence of torsion all the U_4 geometric objects are reduced to the usual Riemannian ones.

This new geometrical property of the Einstein tensor (i.e.the fact that $G_{\mu\nu}$ is no longer a symmetric tensor) is related to the conservation of the total angular momentum in the presence of a non zero intrinsic spin density. At this point, starting from a Lagrangian which will be not only function of the metric tensor and its derivatives but also of the torsion tensor (and, if that is the case, that is if we consider a propagating torsion, also of its derivatives) and applying the usual variational method, we arrive at the field equations [7].

The field equations are then

$$G^{\mu\nu} = \chi\,\theta^{\mu\nu}$$
$$T^{\alpha\mu\nu} = \chi\,S^{\alpha\mu\nu} \qquad (22)$$

where the first equation is similar to Einstein equation but $\theta^{\mu\nu}$ contains also the spin density tensor, while the second one is the new relation which connects the geometrical quantity $T^{\alpha\mu\nu}$ (defined as $T_{\alpha}{}^{\mu\nu} = -(1/2)(-g)^{1/2}\delta R/\delta K_{\nu\mu}{}^{\alpha}$ being R the scalar curvature and K the contorsion tensor defined in (17)) with the spin density tensor $S^{\alpha\mu\nu}$ as defined in (16) (see [7]).

The tensor $\theta^{\mu\nu}$ can be shown to coincide with the canonical energy-momentum tensor (see [7]) which in curvilinear coordinates is defined as

$$\sqrt{-g}\,\theta_{\mu}{}^{\nu} = \delta_{\mu}{}^{\nu}\mathcal{L}_m - \nabla_{\mu}\psi\left[\partial\mathcal{L}_m/\partial(\partial_{\nu}\psi)\right] \qquad (23)$$

which is related to $T_{\mu\nu}$ by a symmetrization procedure.

The equation of motion for the Einstein-Cartan theory is

$$(dp^{\mu}/ds) + \left\{ {\mu \atop \nu\alpha} \right\} p^{\alpha}u^{\nu} + F^{\mu} = 0 \qquad (24)$$

where

$$F^{\mu} = (dt/ds)\int d^3x'(-g)^{1/2}\Phi^{\mu} \qquad (25)$$

represents a force term arising from the interaction of the spin with the geometry of the U_4 manifold.
Φ^{μ} is defined as (see for details [7])

$$\Phi^{\mu} = (-g)^{-1/2}\partial_{\nu}(\sqrt{-g}\,\theta^{[\mu\nu]}) + K^{\mu}{}_{\nu\alpha}\theta^{[\alpha\nu]} - S_{\alpha\beta\nu}R^{\mu\nu\alpha\beta} \qquad (26)$$

If the test particle is spinless (or if the macroscopic test body is unpolarized) then $S_{\mu\nu\alpha} = 0$, $\Phi^{\mu} = 0$, $\theta^{\mu\nu} = T^{\mu\nu}$ and the equation of motion reduces to the usual geodesic equation

$$(du^{\mu}/ds) + \left\{ {\mu \atop \nu\alpha} \right\} u^{\nu}u^{\alpha} = 0 \qquad (27).$$

The important thing that we will stress is that torsion may constitute a way to go towards the quantization of gravity.

In fact the property of non-closure leads to the relation

$$l^{\alpha} = \oint Q_{\beta\gamma}{}^{\alpha}\, dA^{\beta\gamma} \qquad (28)$$

but before considering that argument we will give some physical consequences.

2. Spin-Torsion Lagrangian; Torsion and Magnetism

Now we will explore the connection that torsion seems to have with magnetism. For that reason we start from spin-torsion interaction Lagrangian in a Riemann-Cartan space-time:

$$\mathcal{L} = -(1/2)\, S^{kij}\, K_{ijk} \qquad (29)$$

where S^{kij} is the spin tensor density (see 16) and K_{ijk} is the contorsion tensor defined in (17).

Eq.(29) may be rewritten as (since $S_{ijk} = -S_{jik}$)

$$\mathcal{L} = (1/2)\, S^{kij}[Q_{kij} - 2Q_{j(ki)}] = (1/2)S^{kij}Q_{kij} \qquad (30)$$

In the case of Dirac particle (the spin is a totally antisymmetric tensor) we have (see ref.[7])

$$S^{ijk} = \eta^{ijkl}\, S_l \qquad (31)$$

(where η^{ijkl} is the completely antisymmetric symbol) , and the interaction Lagrangian contains only the axial-vector part of the torsion tensor

$$\mathcal{L} = S_k\, Q^k \qquad (32)$$

(being $Q_k = (1/2)\eta_{kijl}Q^{ijl}$) and going to the particle rest system, where $S^k = (0,\bar{S})$, the interaction energy between a spin $\bar{S}$ and a torsion $\bar{Q}$ may be written as

$$E = -\mathcal{L} = -\bar{S}\cdot\bar{Q} \qquad (33)$$

We can observe that this formula is formally analogous to the interaction energy of a magnetic moment $\bar{\mu}$ in a constant magnetic field $\bar{H}$ namely:

$$E = - \bar{\mu}\cdot\bar{H} \tag{34}$$

so that we can understand, without going through the calculation (see ref.[8] and [9]) that torsion induces spin alignement as well as a magnetic field and if to each spin is associated a magnetic moment, torsion can also produce a magnetic field inside the matter.

The formal analogy of this expression (33) with the interaction energy of a magnetic dipole μ in a constant magnetic field H, (34), shows that a spin in a space with torsion behaves like a dipole in a magnetic field, that is it will be predominantly oriented along the torsion direction.

Therefore torsion, inducing the alignement of the spins (and then of the intrinsic magnetic moments) of the particles inside matter, can produce a magnetic field.

Whithout going into calculation (see ref.[8]) and applying these consideration to the early universe we find that to each portion of hadronic matter containing a total angular momentum density J and then a torsion $Q = 4\pi GJ/c^2$, it was associated a magnetic field (produced by the nucleons orientation)

$$B = 8\pi/3c\ (2\ \alpha\ G_f)^{1/2} J \tag{35}$$

where $\alpha = e^2/\hbar c$ is the fine structure constant (that comes through the magnetic moment) and G_f the strong gravity constant (according to the Dirac hypothesis [9,10] and according to Isham, Salam and Strathdee theory [11]) (for the necessity of the introduction of strong gravity constant see [12,13,14]), so that we have a dipole magnetic moment

$$U \sim (\alpha\ G_f)^{1/2}\ S/c \tag{36}$$

At this point we can observe that the same relation between angular and magnetic moment holds also in the case of a particle.

In fact:

$$\mu/\hbar \simeq e/mc \simeq (\alpha G_f)^{1/2}/c \qquad (37)$$

This result fits nicely with Dirac's two metric hypothesis [11] that is the atomic metric operative at a microscopic level and the usual macroscopic metric connected with gravitation. These two metrics coincide at the beginning of the hadron era, but the atomic metric has not changed in time, while the macroscopic metric developed with the universe.

Therefore the ratio U/S for a particle was frozen to the value (37) it had at that epoch ($t_0 = 10^{-23}$sec when, according to Dirac large number hypothesis [10] for which the gravitational constant decreases in time as $G \propto t^{-1}$, the gravitational constant coincides with the strong gravity constant G_f), while the magnetic field of a celestial body has developed in time up to its present value

$$U \sim (\alpha G)^{1/2} S/c \qquad (38)$$

This explain also why Blackett's experiment [15], which measured the magnetic field of a rapidly rotating body, provided a null result. In fact this laws holds only for astronomical bodies (and not for all rotating matter) bacause it is a consequence of a very hot and collapsed initial state of the universe. So we find the Blackett relation

$$S = qU \qquad (39)$$

where

$$q = c/(\alpha G)^{1/2} \qquad (40)$$

which is in remarkable agreement with astrophysical data (that is with $q = 1.3 \cdot 10^{15} g^{1/2} cm^{-1/2}$ (see also [16]).

Thus the spin-torsion interaction naturally accounts for the Blackett's empirical relation eq.(39).

On these lines we can provide also a primordial magnetic field in the Universe. In fact if the universe is rotating and posesses a torsion field and a non vanishing total angular momentum density [17], then it must have a universal magnetic field B_U (produced by the orientation of the nucleons by the torsion-spin interaction):

$$B_U = (4/c)(2\alpha G)^{1/2} S_U / R_U^3 \qquad (41)$$

where, if we refer to the value of today, we have $R_U \sim 10^{28}$cm (the Hubble radius) and $S_U \sim 10^{120}\hbar$, the total angular momentum of the Universe, being related to the spin of elementary particle by the condition (see [13] and [17]) $S_U = \hbar\,(R_U/r)^3$ (being r the Compton wave-length of a hadron $r = \hbar/mc$) so that we find

$$B_U \sim 10^{-6}\text{Gauss} \qquad (42)$$

Summarizing we can say that considering torsion-spin interaction, we have found for the magnetic field in terms of the spin density σ [8]:

$$B \approx (8\pi/3c)(2\alpha G)^{1/2}\,\sigma \qquad (43)$$

Now in the case of early universe very hot and dense, under the hypothesis of tensor-meson dominance, hot hadronic matter undergoes strong gravitational interactions [11, 13] (by exchange of f-mesons) with a strong coupling constant $G_f \approx \hbar c/m^2 \approx 10^{39}$G and so G must be replaced by G_f in eq.(43) (see eq.(35)).

Also inside a neutron star which can be regarded, in many respects, as a sort of single big nucleon, the distance between two neutrons is of the order of the nuclear radius; at such distance, the hadronic gravitational interactions take place with the exchange of the f-mesons, and Newtonian gravity must be replaced by the strong gravity.

Thus for a neutron star with all the nucleon spins aligned by the strong gravity spin-torsion interaction, eq.(43) (with $G = G_f$) would give a magnetic field $B \approx 10^{15}$Gauss [18]. This is equivalent to the alignement of the magnetic moments $\mu_n \approx e\hbar/2mc \approx 10^{-23}$ erg/Gauss of all the nucleons in the neutron star [19].

3. Rotation and dispersion of polarized e.m. waves propagating in a medium with torsion

In general one say that the electromagnetic field cannot be coupled to torsion in order to preserve local gauge invariance; in other words one cannot apply the minimal coupling principle

$$\eta_{\mu\nu} \Longrightarrow g_{\mu\nu} \quad , \quad \partial_\mu \Longrightarrow \nabla_\mu \qquad (44)$$

that is one cannot replace the flat with curved metric and the derivative with covariant derivative in the presence of torsion to the special relativistic photon lagrangian without breaking gauge invariance of the theory.

This result finds a natural justification in the framework of the Poincaré gauge field theory of gravitation (see ref.[20] for an extensive bibliography on this subject).

Several attempts have been made in order to apply minimal coupling without breaking gauge invariance: they are successful, however, only at the cost of imposing arbitrary geometrical constraints upon torsion [21], or introducing a modified definition of gauge transformation [22].

Therefore these attempts lie beyond the limits of the Einstein-Cartan theory.

In order to preserve the gauge invariance in the Einstein theory, the most simple hypothesis is to assume that [23,24,25]

"photons neither produce nor feel torsion" (45)

i.e.that Maxwell equations in a Riemann-Cartan space U_4 are identical to the same equations written in the Riemann space V_4 obtained from U_4 putting torsion to zero.

We like to emphasize that the statement (45) can be assumed to be strictly valid only as long as the electromagnetic (e.m.) field is treated as a classical, not quantized field. When a quantum point of view is adopted, however, (45) is recovered only as a first approximation, since, in general, according to a quantum description of the e.m. field in a space with torsion, one may always expect an interaction between the photons and the torsionic background.

In fact a photon, with a process of the second order in the perturbative development of the e.m. interaction, can virtually disintegrate into an electron-positron pair (vacuum polarization effect); since these particles are massive fermions, which couple to torsion, they feel the presence of a torsionic background; as a consequence, also the e.m. field is affected by torsion (see

Fig.4); even if torsion does not directly interact with the photon field, it does interact with the virtual pairs produced in vacuum by a "physical" photon. This interaction preserves the gauge invariance and the Maxwell equations are modified by a quantum contribution of the second order, so that, to the zeroth order, i.e. in the classical field approximation, the coupling with torsion disappears and we recover the usual form as the Maxwell equations.

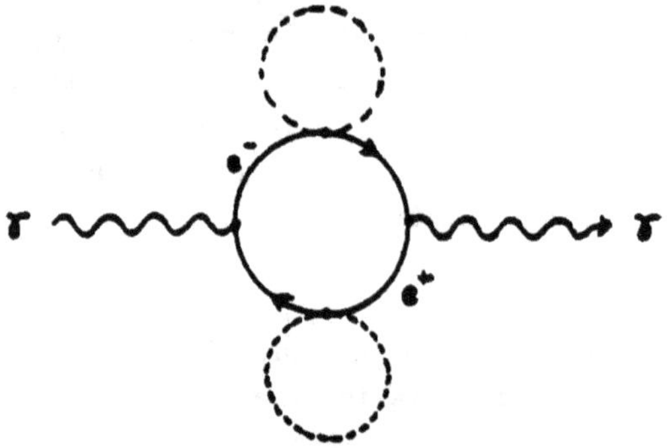

Fig.(4) The dashed lines represent the interaction of the virtual pair with the external torsionic background

We will not go through the calculation (see [26 - 32]); we will notice only that the virtual pair production induces in vacuum a current density proportional to field intensity. The resulting Maxwell equations are then, in the presence of torsion:

$$\left.\begin{aligned} \partial_\mu F^{\nu\mu} &= 4\pi J^\nu + (2\alpha/3\pi)\ \eta^{\nu\mu\rho\sigma} F_{\mu\rho} Q_\sigma \\ \partial_{[\nu} F_{\sigma\mu]} &= 0 \end{aligned}\right\} \qquad (46)$$

These eqs.(46) can be derived from a Lagrangian where the photon-torsion term interaction term $\mathcal{L}_I$ is given by [30,31,32]

$$\mathcal{L}_I = (-g)^{1/2} \alpha \eta^{\mu\nu\alpha\beta} A_\mu F_{\nu\alpha} Q_\beta \qquad (47)$$

with $F_{\nu\mu} = 2\partial_{[\nu} A_{\mu]}$.

α is the fine-structure constant $\alpha = e^2/\hbar c$ (remember that the fine structure constant enter in the photon-torsion interaction lagrangian when an explicit computation is carried out which takes into account the interaction between torsion and the virtual pairs e^+- e^- associated to a propagating electromagnetic field; see for instance ref.[29,30,31], and [7] at page 268) and Q_σ is an axial vector related to the torsion tensor $Q_{\mu\nu}{}^\rho = \Gamma_{[\mu\nu]}{}^\rho$ by

$$Q^\sigma = (1/16)\eta^{\mu\nu\rho\sigma}K_{\mu\nu\rho} = (1/16)\eta^{\mu\nu\rho\sigma}(Q_{\nu\rho\mu} - Q_{\mu\nu\rho} - Q_{\rho\mu\nu})$$

where $\eta^{\mu\nu\rho\sigma}$ is the totally antisymmetric tensor.

This lowest order photon-torsion interaction has been obtained through an explicit perturbative calculation.

1) <u>rotation of linear polarization</u>

In the weak field approximation, the Maxwell eqs. can be written as [26]

$$\bar\nabla\cdot\bar E = -(\bar E/\varepsilon)\cdot\bar\nabla\varepsilon \qquad \bar\nabla\cdot\bar H = -(\bar H/\varepsilon)\cdot\bar\nabla\varepsilon \tag{48}$$

$$\bar\nabla\wedge\bar E = -(\varepsilon/c)(\partial\bar H/\partial t) \qquad \bar\nabla\wedge\bar H = (\varepsilon/c)(\partial\bar E/\partial t) - \eta\varepsilon^2\bar H \tag{49}$$

where $\varepsilon = (-g)^{1/2}$ is connected with the metric and $\eta=(4\alpha/3\pi)Q$ is connected with torsion.

In that case the curl eqs. (49) yield two linearly independent solutions, representing a right- (E_+,H_+) and left-handed (E_-,H_-) wave; putting

$$E_\pm = E_1 + iE_2 = A_\pm e^{i(kx_3 - \omega_\pm t)} \tag{50}$$

from the modified Maxwell eqs.(49) we get the following dispersion relations for the two polarization modes

$$\omega_\pm^2 = (c^2/\varepsilon^2)k(k \mp \eta\varepsilon^2) \tag{51}$$

Notice that, in the absence of torsion ($\eta \longrightarrow 0$), we have $\omega_+ = \omega_- = ck/\varepsilon$ and then the frequencies are only red-shifted: $\omega \longrightarrow \omega/\varepsilon$ because of the action of the static gravitational potential, but the dispersion relation is the same for the two polarization modes, like vacuum.

It is well known, in fact, that a static gravitational field can influence only direction and magnitude of the velocity of light, unlike the gravitational field of a rotating body, which can also produce a rotation of the plane of polarization (see [33]) (see also [34,35,36]).

We find that the plane of polarization of a wave propagation in a torsionic background, rotates then of an angle $\Delta\vartheta$ which is directly proportional to the length Δl of the path covered in this "medium"

$$|\Delta\vartheta| = |\Omega| \, \Delta t = (1/2)\eta \, \Delta l \tag{52}$$

2) <u>dispersion</u>

The second important effect induced by torsion is dispersion: the velocity of an e.m. wave in a space with torsion, depends on its frequency and polarization mode. From (51) we can deduce, in fact, (considering a linear polarized wave sufficiently far from the central body i.e. $\varepsilon \Longrightarrow 1$) that the phase velocity, $v = \omega/k$, is given

$$v_{\pm} = c\left[1 \mp (\eta/k)\right]^{1/2} = v(k) \tag{53}$$

In general we can say that a space with curvature and torsion behaves like a dispersive medium described by an index of refraction $n(k,x)$ such that

$$n^2\omega^2 = c^2k^2 \tag{54}$$

from (54) we get

$$n_{\pm}^2 = \varepsilon^2\left[1 \pm (kc^2/\omega^2)\eta\right] = \varepsilon^2(1 \pm \lambda n_{\pm}^2\eta) \tag{55}$$

from which

$$n_{\pm}^2(k,x) = \varepsilon^2/(1 \mp \varepsilon^2\lambda\eta) \tag{56}$$

or, to the first order in G:

$$n_{\pm}^2 = 1 + 4GM/rc^2 \pm \lambda\eta \tag{57}$$

In the absence of torsion ($\eta \Longrightarrow 0$) then $n_+ = n_- = \varepsilon(r)$, i.e. the index of refraction is the same for the two polarization modes and it is only a function of the distance from the gravitating body. A light ray traveling at a distance R from its center is deflected by the well known angle δ

$$\delta = \varepsilon^2(R) - 1 = 4GM/Rc^2 \tag{58}$$

In the presence of torsion the gravitational index of refraction is altered by the factor $(1 \mp \lambda\eta\varepsilon^2)^{-1}$ as shown by (56). The angle of deflection δ', computed using the approximate expression (57) for the refractive index, turns out to be, to the first order in the torsionic contributions

$$\delta' = 4GM/Rc^2(1 \mp \lambda\eta) \qquad (59)$$

The joint action of a gravitational field and a torsionic background can induce, therefore, the spatial splitting Δx of the different wave-lengths contained in a wave-packet, on the plane normal to the propagation direction. The splitting is proportional to the distance l travelled through this medium

$$\Delta x = l\ \Delta\delta = \mp\ (4GM/Rc^2)\eta\ l\ \Delta\lambda \qquad (60)$$

The things that so far we have said are valid in the Einstein -Cartan theory, but we can also introduce a propagating torsion, imposing that the axial-vector part of the torsion tensor (that is (47)) is determoned by the pseudoscalar potential φ according to $Q_\beta = \partial\varphi$. As one of the consequence, we have that an electric charge interacting with a constant torsionic background, can produce a static magnetic field even in its rest system [34] (see also [35] and [36]).

4. Neutrino oscillations

We like to say also some words on the influence of torsion on neutrino oscillations [37].

Remember, first of all, that the possibility of oscillations from one neutrino type to another was first discussed by Pontecorvo [38] in 1967. It is well known that neutrino oscillations occur when the weak-interaction eigenstates are not the eigenstates of the total Hamiltonian [39]; therefore to have oscillations in vacuum (i.e.in the absence of interactions), it is required a nondiagonal (and obviously nonzero) mass matrix, that is the weak interaction eigenstates must be different from the mass eigenstates and the mass of the various types (e-, μ- and τ-neutrinos) must not be all degenerate. In the case of torsion, however, the oscillations are possible also if the neutrino masses are all degenerate and even if neutrinos are massless. The fact that oscillations are possible also if neutrinos are massless is very important because, in general, people says that if we find neutrino oscillations the neutrinos must have a mass different from zero. Briefly the Lagrangian density for a massless and uncharged spinor field ν is given by ($\hbar$ = c= 1)

$$\mathcal{L} = -(-g)^{1/2}[(1/2)(\bar{\nu}\gamma^{\mu}\nu_{;\mu} - \bar{\nu}_{;\mu}\gamma^{\mu}\nu) - i(3/4)Q^{\mu}\bar{\nu}\gamma_{\mu}\gamma_5\nu] \quad (61)$$

(see also ref.[32]) where g is the determinant of the metric tensor $g_{\mu\nu}$ and the ";" denotes the Riemann covariant derivative using only the Christoffel symbols, and Q^{μ} is the totally antisymmetric part of the torsion tensor

$$Q_{\mu} = (1/3!)\eta_{\mu\nu\alpha\beta}Q^{\nu\alpha\beta} \quad (62)$$

where $\eta_{\mu\nu\alpha\beta}$ is the completely antisymmetric symbol, defined by $\eta_{1234} = (-g^{1/2})$.

In ref.[37] we have shown that the various leptonic numbers are not separately conserved, thus allowing the possibility of 'flavour' mixing (that is the mixing of ν_{μ} with ν_e and ν_{τ}); the torsion eigenstates then are different from the weak eigenstates and then the torsion Hamiltonian density $\mathcal{H}_T = -\mathcal{L}_T$ is not diagonal in ν_a and ν_b and may be written [37]

$$\mathcal{H} = -iQ^{\mu}[k_a\bar{\nu}_a\gamma_{\mu}\gamma_5\nu_a + k_b\bar{\nu}_b\gamma_{\mu}\gamma_5\nu_b + k_{ab}(\bar{\nu}_a\gamma_{\mu}\gamma\nu_b + \bar{\nu}_b\gamma_{\mu}\gamma_5\nu_a)] \quad (63)$$

where the k coefficients are dimensionless real parameters and we have put $k_{ab} = k_{ba}$ to preserve CP invariance (see [39]).

We can diagonalize $\mathcal{H}_T$ by the transformation

$$|\nu_w\rangle = R(\theta_T)|\nu_T\rangle \quad (64)$$

$$|\nu_w\rangle = \begin{pmatrix} \nu_a \\ \nu_b \end{pmatrix}, \quad |\nu_T\rangle = \begin{pmatrix} \nu_{1T} \\ \nu_{2T} \end{pmatrix} \quad (65)$$

where $|\nu_w\rangle$ are weak eigenstates and $|\nu_T\rangle$

$$R(\theta_T) = \begin{pmatrix} \cos\theta_T & \sin\theta_T \\ -\sin\theta_T & \cos\theta_T \end{pmatrix} \quad (66)$$

and the angle θ is given by

$$tg2\theta_T = 2k_{ab}/(k_a - k_b) \qquad (67)$$

Eq.(84) becomes then

$$\mathcal{H} = - iQ^{\mu}(k_1\bar{\nu}_{1T}\gamma_{\mu}\gamma_5\nu_{1T} + k_2\bar{\nu}_{2T}\gamma_{\mu}\gamma_5\nu_{2T}) \qquad (68)$$

where

$$k_{1,2} = (1/2)(k_b + k_a \pm \sqrt{(k_b - k_a)^2 + 4k_{ab}^2}\ ) \qquad (69)$$

The torsion eigenstates are then ν_{1T} and ν_{2T} ; they can be obtained from eigenstates ν_a and ν_b through a rotation of an angle θ_T (see eq.(64)) and their torsion interaction energy is

$$E_T = \int d^3x(-g)^{1/2}\mathcal{H}_T \qquad (70)$$

In the case of nearly constant torsional background produced by a density of spins inside a macroscopical body one has

$$Q_4 = 0, \qquad Q_i = 4\pi G\Sigma_i \qquad (71)$$

($i = 1,2,3$) where Σ_i is the number of spins in a unitary volume aligned along the x_i direction.

From eq.(68),(70),(71) one has

$$E_{1T} = - 2k_1\mathbf{Q}\cdot\mathbf{S}, \qquad E_{2T} = - 2k_2\mathbf{Q}\cdot\mathbf{S} \qquad (72)$$

where

$$S_i = -(i/2)\int d^3x\sqrt{-g}\ \nu^+\gamma_i\gamma_4\gamma_5\nu \qquad (\hbar = 1) \qquad (73)$$

is the neutrino spin along x_i. Massless neutrinos have their spins aligned along the momentum axis $\hat{p} = p/|p|$ (in the opposite direction as the neutrino is left-handed) and then $\mathbf{S} = - \hat{p}/2$ and eq.(72) gives

$$E_{1T} = k_1\mathbf{Q}\cdot\hat{p}, \qquad E_{2T} = k_2\mathbf{Q}\cdot\hat{p} \qquad (74)$$

The energy is maximum when the torsion field and the neutrino momentum are parallel, it is minimum when they are antiparallel (see [37]. Then we have oscillations between the weak eigenstates

ν_a and ν_b.

The oscillation length, i.e.the distance required to produce a phase change of 2π is given by ($\Delta E_T = E_{1T} - E_{2T} = (k_1 - k_2)Q\cdot\hat{p}$)

$$l_T = (2\pi/\Delta E_T) = 1/\{2G\Sigma\cdot\hat{p}(k_1 - k_2)\} \qquad (75)$$

In ref.[37] is shown that the oscillation length, for example in the case of a ferromagnetic macroscopical body with a number $\rho \sim 10^{24}$ of elementary spins $\hbar/2$ aligned in a unitary volume, is $\geq 10^{43}$cm and then the effects are pratically zero. However if we consider superdense matter and also suppose that ordinary Newton gravity is replaced by strong gravity, we can have $l_T \geq 10^{-12}$cm. and therefore torsion can play a fundamental role in the presence of strong gravity, i.e.inside a collapsed matter and during the early stage of the Universe.

5. Neutron Interferometry

The phase shift on a neutron in a neutron interferometer. Gravitational phase shift was found in COW experiment [40]

$$|\Delta\Phi_G| = 4\pi\ (m/\hbar)(g/2v)A \qquad (76)$$

where m is gravitational mass, g the acceleration due to gravity, v is velocity of neutrons moving in field and A is area enclosed by interferometer beam paths. The explicit expression for $|\Delta\Phi|$ will be

$$|\Delta\Phi_G| = |\Delta\Phi_0|\{1 + (1/2)(gH/c^2)\} \qquad (77)$$

where

$$|\Delta\Phi_0| = 4\pi(m_0/\hbar)(g/2v)A \qquad (78)$$

(with m the rest mass of neutron) ($m = m_0(1 + 2gH/c^2)$; gH/c^2 is the gravitational potential energy and H is height difference of paths.

If torsion is considered we will have an additional term like $3G^2S^2/c^4R^4$, so in the experimental arrangement in which H = 0 we have a net result ($gH/c^2 = 0$)

$$|\Delta\Phi_T| = |\Delta\Phi_0|\left\{1 \pm 3GS^2/c^4R^4 \right\} \qquad (79)$$

But in the interaction of torsion with spins we have a principally new picture, namely, polarization plane rotation [41].

In a work with Pronin [42] to seee the effect of torsion on the interferometer with polarized neutrons in an external torsion field, we have separate the spin-torsion interaction effect and restrict ourselves by consideration to Minkowski-Cartan space, supposing that $Q_\mu = (0, \vec{Q})$, where $\vec{Q}$ is a constant vector, $g_{\mu\nu} = \eta_{\mu\nu} = \mathrm{diag}(1,-1,-1,-1)$. Then Pauli equation for spinors will be

$$i\hbar \frac{\partial \psi}{\partial t} = \left\{- \frac{\hbar^2}{2m} \bar{\nabla}^2 + \xi(\bar{\sigma}\cdot\bar{Q})\right\} \psi \qquad (80)$$

(where ξ is the spin-torsion interaction constant). We assume that the neutron beams (I and II) are polarized in the antiparallel direction to z-axis. Then the spinor normalized functions will be

$$\psi_I = |\downarrow_z\rangle = \begin{bmatrix}0\\1\end{bmatrix}, \qquad \psi_{II} = e^{i\vartheta}|\downarrow_z\rangle = e^{i\vartheta}\begin{bmatrix}0\\1\end{bmatrix} \qquad (81)$$

here ϑ is phase shift.

If the external torsion field equals zero, the degree of polarization of the beam after interferometry is

$$\bar{P} = \frac{\psi^+ \bar{\sigma} \psi}{\psi^+ \psi} = \left\{0, 0, -1\right\} \qquad (82)$$

where $\psi = \psi_I + \psi_{II}$.

But in the situation, when one (or both) neutron beam(s) interacts with torsion, the states ψ_I and ψ_{II} should be changed as

$$\psi'_I = \sqrt{\varepsilon}\ |\uparrow_z\rangle + \sqrt{1-\varepsilon}\ |\downarrow_z\rangle, \quad \psi'_{II} = e^{-i\vartheta}\left\{\sqrt{\varepsilon}\ |\uparrow_z\rangle + \sqrt{1-\varepsilon}\ |\downarrow_z\rangle\right\} \qquad (83)$$

and one finds

$$\bar{P}' = \frac{\psi^{+\prime} \bar{\sigma} \psi'}{\psi^{+\prime} \psi'} = \left\{2\ [\varepsilon(1-\varepsilon)]^{1/2}, 0, -1+2\varepsilon \right\} \qquad (84)$$

and here $\psi' = \psi'_I + \psi'_{II}$.

For example, when $\varepsilon = 1/2$, we have $\bar{P}' = \{1,0,0\}$. and we observe the effect of the polarized rotation plane due to quantum interferometry, which is caused by the previous interaction with torsion.

Let us give some estimation of the effect: we consider $\varepsilon \simeq \xi|\bar{Q}|$ where $|\bar{Q}|$ is proportional to polarized particles density n. In laboratory experiments it will be convenient to use the polarized particle beams in the accelerator. Then in optimal situation ($\varepsilon = 1/2$) we can find the upper limit on the spin-torsion interaction constant, because

$$\xi \leq 10^{17}/2n \qquad (85).$$

6. Torsion and quantum gravity

Now we return back to the fundamental property of the torsion, namely the non-closure property that is the fact that the integral

$$l^{\alpha} = \oint Q_{\beta\gamma}{}^{\alpha}\, dA^{\beta\gamma} \neq 0 \qquad (86)$$

over a closed infinitesimal contour is different from zero, can be treated as defects in space-time.

Gravity in fact, according to Einstein, is not a force but is the curvature of space-time and then we cannot quantize gravity like other forces but we must try to quantize the space-time itself.

We know also that is impossible, as demonstrated by Treder and Borzeszkowski [43], to find commutation rules corresponding to Rosenfeld's inequality relations

$$\begin{aligned} g_{ik}\, L_o^{\,2} &\geq \hbar G/c^3 \\ \Gamma^{i}{}_{kl} L_o^{\,3} &\geq \hbar G/c^3 \end{aligned} \qquad (87)$$

(L_o denotes the dimension of the spatial region over which the average value of g_{ik} and $\Gamma^{i}{}_{kl}$ is measured) because the quantities "field strength" and "length" appearing in (108) cannot be defined independently of each other and moreover the "field strength" Γ^{i}_{kl} are not tensorial quantities.

This problem may not arise if torsion is considered, since in this case the asymmetric part of the connection i.e. $\Gamma^{\alpha}_{[\beta\gamma]}$, i.e.the torsion tensor $Q_{\beta\gamma\cdot}^{\ \ \alpha}$ is a true tensorial quantity and moreover is independent from $g_{\mu\nu}$. But let consider this problem in some details.

Now from eq.(86) and remembering that torsion is linked to the intrinsic spin, and spin is quantized, we can say that the defects in space-time topology should occur in multiples of the Planck length $(\hbar G/c^3)^{1/2}$ i.e. [44]

$$l^{\alpha} = \oint Q_{\beta\gamma\cdot}^{\ \ \alpha} dx^{\beta}\wedge dx^{\gamma} = n\ (\hbar G/c^3)^{1/2} n^{\alpha} \qquad (88)$$

(n is an integer and n^{α} a unit vector).

Eq. (88) is analogous to the well known relation $\oint p dq = n\hbar$, i.e. the Bohr-Sommerfeld relation and would define a minimal fundamental length and a minimal time, i.e.the Planck length entering through the minimal unit of spin or action $\hbar$. So $\hbar$ has to deal with the intrinsic defect built into torsion structure of space time. We can write Eq.(88) also as Landau-Peierls type relations:

$$\Delta Q\cdot L_0^2 \ \geq\ (\hbar G/c^3)^{1/2} \qquad (89)$$

Following ref.[44] we are led to commutation relations such as:

$$[\ Q_{\beta\gamma\cdot}^{\ \ \alpha},\ dx^{\beta}\wedge dx^{\gamma}] = i(\hbar G/c^3)^{1/2} n^{\alpha} \qquad (90)$$

and to corresponding uncertainty relations:

$$\Delta Q^{\alpha}_{\ \cdot\beta\gamma}\ (\Delta x^{\beta} \wedge \Delta x^{\gamma}) \ \geq\ (\hbar G/c^3)^{1/2} n^{\alpha} \qquad (91)$$

Therefore, the Einstein-Cartan theory of gravitation, in contrast with Einstein's General Relativity Theory, should provide genuine quantum-gravity effects.

In this way we have quantized torsion field but not the background. We believe that there is the possibility to include also the metrical background (i.e. the curvature).

In order to show this possibility, we will see that R and Q can be conjugate variables.

We can say that, since curvature causes relative acceleration between neighbouring test particles, we have momentum uncertainty related to curvature as:

$$ma^{\mu}ds = \Delta p^{\mu} = m\, R^{\mu}{}_{\alpha\beta\gamma}\, \frac{dx^{\alpha}}{ds}\, dx^{\beta} n^{\gamma} = m\, R\, \frac{dS}{ds}\, k^{\mu} \qquad (92)$$

while we have position uncertainty connected with torsion.

So as position fluctuations are given by torsion, momentum fluctuation are due to curvature and we can interpret quantum effects (and then uncertainty principle) as consequences of space-time deformation. We can also say that torsion is responsible of the fluctuations of the position $\Delta x^{\mu} \cong Q^{\mu}dS$ while the defect angle is due to curvature $\Delta\vartheta \cong RdS$ (see [44,b]). We can write

$$\Delta p^{\mu}\cdot\Delta x \;\geq i\, \hbar \qquad (93)$$

where

$$\Delta x^{\mu} = Q\; dS\; k^{\mu} \qquad (94)$$

and

$$\Delta p^{\mu} = m\, R\, \frac{dS}{ds}\, k^{\mu} \;. \qquad (95)$$

We see that Q (torsion) and R (curvature) play the role of conjugate variables of the geometry (gravitational field), thus enabling us to write commutation relations between curvature and torsion (analogous to $[x,p] = i\hbar$) as:

$$\left[Q\; ,\; R\right] = i(\hbar G/c^{3})^{-3/2} \qquad (96)$$

or, more explicity, as:

$$\left[Q^{\alpha}{}_{\mu\nu}\; \Delta x^{\mu} \wedge \Delta x^{\nu}\; ,\; R_{\beta\mu\nu\rho}\; \Delta x^{\mu} \wedge \Delta x^{\nu} \wedge k^{\rho}\right] = i(\hbar G/c^{3})\delta^{\alpha}_{\beta} \qquad (97)$$

(where Δx^{μ}...refer to change in coordinates and k^{ρ} is vector joining neighbouring particles).

Curvature and torsion in this context appear to be conjugate variables for which we can write commutation relations like (96).

7. Quantum gravity in real space-time?

At this point we like to introduce the space-time algebra. In fact there is still a difficulty: it is well known that when we consider the early universe we have to deal both with elementary particle physics using quantum theory and with cosmology using general relativity; but general relativity is developed in real space-time while quantum theory needs a complex manifold. How can we conciliate general relativity with quantum theory? We think that it is important to quantize gravity in the real space-time. Then, first of all, the answer lies in a reformulation of Dirac's theory in terms of space-time geometric calculus without any complex number.

In other words the contradiction comes when, in order to take into account both mass and spin, it seems at first sight that we have to do with two different spaces: a real space-time where we describe the curvature, due to the mass, with tensors, and complex space-time where we describe torsion due to the spin, with spinors. But this is not completely satisfactory: one would like to describe these two fundamental physical properties, mass and spin, in a unique manifold [45], the real space-time that is, if we like to overcome this difficulty, we will try to describe spinors in the real space-time: this can be done through Hestenes algebra [46, 47]. For that reason we like to summarize briefly some aspect of multivector algebra.

Many works have been done at this respect by the Cambridge group (see [48,49]). We do not go through these works; only we say that the geometric algebra, with the multivector concept and the interpretation of imaginary units as generator of rotations, places tensors and spinors on the same foot: both are described in a real space-time [50,51,52].

Briefly, only two words, the basic idea of geometric algebra is to extend the algebra of scalars to an algebra of vectors and this is done by introducing an associative (Clifford) product over a graded linear space (where scalars are identified with the grade 0 elements of this space and vectors with the grade 1 elements). If a and b are two vectors, then one write the Clifford product as the juxtaposition ab (see for instance [48]).This product ab

(called geometric product) decomposes into a symmetric and an antisymmetric part, which define the inner and outer products between vectors that is $ab = a \cdot b + a \wedge b$ where

$$a \cdot b \equiv (1/2)(ab + ba) \qquad (98)$$

$$a \wedge b \equiv (1/2)(ab - ba) \qquad (99)$$

$a \cdot b$ is a scalar but $a \wedge b$ is neither a scalar nor a vector: it defines a new geometric element called a bivector (grade2) and may be regarded as a directed plane segment, which specifies the plane containing a and b. One can notice that for a and b parallel we have $ab = ba$ while $ab = -ba$ for a and b perpendicular. We can go on with this procedure defining higher grade elements and then a basis for the linear space [46]. In particular we are interested in the space-time algebra that is in the geometric algebra of the space-time. Here instead to consider the algebra generated from the Dirac γ-matrices, we consider the algebra generated by four orthogonal vectors $\{\gamma_\mu\}$, $\mu = 0\ldots3$ where a full basis for the space-time algebra is provided by the set

1	γ_μ	$\{\sigma_k, i\sigma_k\}$	$i\gamma_\mu$	
1-scalar	4-vectore	6-bivectors	4-pseudovectors	
grade 0	grade 1	grade 2	grade 3	
i				(100)
1-pseudoscalar				
grade 4				

where $\sigma_k \equiv \gamma_k\gamma_0$ k = 1...3 and the unit pseudoscalar of spacetime is

$$i \equiv \gamma_0\gamma_1\gamma_2\gamma_3 = \sigma_1\sigma_2\sigma_3 \qquad (101)$$

On this ground one proceeds to a reformulation of Dirac's theory in terms of space-time geometric calculus without any complex number. We can write Dirac equation as:

$$\hbar\, \nabla\psi\, \gamma_2\gamma_1 - (e/c)A\,\psi = mc\psi\gamma_0 \qquad (102)$$

where ∇ is the 4-dimensional generalization of gradient operator which, taking into account of the metric, is

$$\nabla \equiv \gamma^{\mu}\partial_{\mu} \quad , \quad \partial_{\mu} \equiv \partial/\partial x^{\mu} \quad , \quad A = A_{\mu}\gamma^{\mu} = A^{\mu}\gamma_{\mu} \tag{103}$$

and ψ is connected with Dirac column spinor Ψ by $\Psi = \psi u$ being u the unit column spinor.

Here we can observe that considering Dirac equation in in flat Minkowski space-time, we have

$$\eta_{\mu\nu} = (1/2)(\gamma_{\mu}\gamma_{\nu} + \gamma_{\nu}\gamma_{\mu}) = \gamma_{\mu}\cdot\gamma_{\nu} \tag{104}$$

but in the presence of a gravitational field should be generalized as

$$g_{\mu\nu} = (1/2)(\gamma_{\mu}\gamma_{\nu} + \gamma_{\nu}\gamma_{\mu}) = \gamma_{\mu}\cdot\gamma_{\nu} \tag{105}$$

In conclusion as an exemple we like to consider the geometric product of Dirac vectors i.e.

$$\gamma_{\mu}\gamma_{\nu} = \gamma_{\mu}\cdot\gamma_{\nu} + \gamma_{\mu}\wedge\gamma_{\nu} \tag{106}$$

which can be written as [51,52,53]

$$\gamma_{\mu}\gamma_{\nu} = g_{\mu\nu} + \sigma_{\mu\nu} \tag{107}$$

where $g_{\mu\nu}$ is the metric tensor while $\sigma_{\mu\nu}$ is connected with spin.

In fact, very briefly, without go in the detail, we know that the transformation law for spinor field $\psi(x)$ is written as

$$\psi'(x') = U(\Lambda)\psi(x) \tag{108}$$

where $U(\Lambda)$ is the usual 4 x 4 constant matrix representing the Lorentz transformation (the spinor indices are not written explicity). Λ is the Lorentz matrix involved in the vector Lorentz transformation in the flat tangent space $x'^{i} = \Lambda^{i}{}_{k}x^{k}$. Dirac equation $(j\gamma^{k}\partial_{k}\psi - m\psi = 0)$ is transformed as

$$\begin{aligned} j\gamma^{i}\,(\partial\psi'(x')/\partial x'^{i}) - m\,\psi'(x') &= \\ = j\gamma^{i}\Lambda_{i}{}^{k}\partial_{k}U\,\psi - m\,U\,\psi &= 0 \end{aligned} \tag{109}$$

It is well known that multiplying from the left by U^{-1} and imposing on the Dirac equation form invariance under a Lorentz transformation, we obtain the condition for the matrix U

$$U^{-1}\gamma^i U = \gamma^k \Lambda^i_{\ k} \tag{110}$$

Considering an infinitesimal transformation

$$\Lambda_{ik} \simeq \eta_{ik} + \omega_{ik} \tag{111}$$

(where $\omega_{ik} = \omega_{[ik]}$), we have

$$U = 1 + (1/2)\omega_{ik}S^{ik} \tag{112}$$

The ω_{ik} are six constant infinitesimal parameters and S^{ik} are the generators of the infinitesimal Lorentz transformation which in order that (13) be fulfilled must satisfy

$$S^{ik} = \gamma^{[i}\gamma^{k]} = (1/2)(\gamma^i\gamma^k - \gamma^k\gamma^i) \tag{113}$$

where γ are the Dirac matrices. Moreover the eq.(113) is connected with spin. In fact we know that [52] $\sigma_1\sigma_2 = \sigma_1 \wedge \sigma_2$ is the generator of rotation in 2-space and we know also that in 3-space $\sigma_1\sigma_2$, $\sigma_2\sigma_3$, $\sigma_3\sigma_1$ are connected with spin [51,52,53]. So we have that in space-time $\gamma_\mu \wedge \gamma_\nu$ are the generators of rotations (eq.113) and then are also connected with spin! So writing

$$\sigma_{\mu\nu} = (1/2)(\gamma_\mu\gamma_\nu - \gamma_\nu\gamma_\mu) , \tag{114}$$

as from Hestenes geometric product we have

$$\gamma_\mu\gamma_\nu = \gamma_\mu \cdot \gamma_\nu + \gamma_\mu \wedge \gamma_\nu \tag{115}$$

we can also write

$$\gamma_\mu\gamma_\nu = g_{\mu\nu} + \sigma_{\mu\nu} \tag{116}$$

that is eq.(107)

Eq.(107) seems to include automatically supersymmetry because we have commutator (fermionic field) and anticommutator (bosonic field): $g_{\mu\nu}$ is connected with bosons and $\sigma_{\mu\nu}$ is connected with fermions, and they are given simultaneously. In other words the Hestenes geometric algebra [46] with the concept of multivectors and with the precise geometrical interpretation of imaginary numbers, seems to be very important: the unit imaginary j appearing in the Dirac, Pauli and Schrödinger equations has a geometrical interpretation in terms of rotation in real space time [46] so one has to do with real space-time, spinors and tensors are treated in a unified way (and this allows us to write eq.(107)).

At last, using the multivector algebra, we like to rewrite the commutation relations (96) for curvature R and torsion Q

$$\left[Q\ ,\ R\right] = i(\hbar G/c^3)^{-3/2} \qquad (96)$$

in a real space-time in the case of totally antisymmetric torsion.

Now it is well known that the Dirac field theory gives a spin density tensor totally antisymmetric. Then considering the spin-torsion coupling we have to do with antisymmetric part of the torsion tensor. This fact suggests that, in the ambit of Einstein-Cartan theory, the torsion tensor $Q^{\alpha\beta\gamma}$ be completely antisymmetric. In this regards we can notice that also Alfred Yu, although for other reason [45], takes $Q^{\alpha\beta\gamma}$ totally antisymmetric: he in fact likes to satisfy from the beginning the principle of equivalence.

Now we are in position to introduce [54] the tri-vector $Q = Q^{\alpha\beta\gamma}\gamma_\alpha\wedge\gamma_\beta\wedge\gamma_\gamma$ as element of space-time algebra, where $\{\gamma_\alpha\}$ are the base vectors for which we have $\gamma_\alpha\gamma_\beta = g_{\alpha\beta} + \gamma_\alpha\wedge\gamma_\beta$.

Moreover, given the curvature bivector

$$\Omega^{\alpha\beta} = (1/2)R^{\alpha\beta\mu\nu}\gamma_\mu\wedge\gamma_\nu \qquad (117)$$

one can form the tri-vectors

$$R^\alpha = \Omega^{\alpha\beta}\wedge\gamma_\beta \qquad (118)$$

We have then the trivectors Q and R^α. Consider now the antisymmetric part of the geometric product

$$[Q,R^\alpha] = (1/2)(QR^\alpha - R^\alpha Q) \qquad (119)$$

This type of product between two tri-vectors gives a bivector (for example it is easy to verify $(\gamma_0\wedge\gamma\wedge_1\gamma_2)\wedge(\gamma_0\wedge\gamma_1\wedge\gamma_3) = \gamma_2\wedge\gamma_3$ remembering that $\gamma_0\gamma_1\gamma_2 = i\gamma_3$ where 'i' here indicates the pseudoscalar unit $\gamma_0\gamma_1\gamma_2\gamma_3$).

In the language of geometric algebra, the imaginary unit of

complex numbers is substituted by a bivector; then we can have commutation relations of canonic type:

$$Q\wedge R^{\alpha} = (1/2)Q^{\mu\nu\beta}R^{\alpha\tau}{}_{\nu\beta}\gamma_{\mu}\wedge\gamma_{\tau} \propto L_{Pl}^{-3} \qquad (120)$$

which contains the bivector $\gamma_{\mu}\wedge\gamma_{\tau}$ that replaces the unit imaginary in the spin space and is linked to the spin so that can be written in terms of Planck length.

$$Q \wedge R^{\alpha} = (1/2)\left[Q^{\mu\nu\beta}\gamma_{\mu}\wedge\gamma_{\nu}\wedge\gamma_{\beta}\ ,\ R^{\alpha\tau\nu\beta}\gamma_{\tau}\wedge\gamma_{\nu}\wedge\gamma_{\beta}\right] =$$

$$= (1/2)Q^{\mu\nu\beta}R^{\alpha\tau}{}_{\nu\beta}\gamma_{\mu}\wedge\gamma_{\tau} \propto L_{Pl}^{-3} \qquad (121)$$

As we must be in a spin plane, we have to consider the case of $\mu = 1$ and $\tau = 2$

$$\left[Q^{1\nu\beta}\gamma_{\mu}\wedge\gamma_{\nu}\wedge\gamma_{\beta}\delta^{\mu}_{1}\ ,\ R^{2\alpha\nu\beta}\gamma_{\tau}\wedge\gamma_{\nu}\wedge\gamma_{\beta}\delta^{\tau}_{2}\right] =$$

$$(1/2)Q^{1\nu\beta}R^{2\alpha}{}_{\nu\beta}\gamma_{\mu}\wedge\gamma_{\tau}\delta^{\mu}_{1}\delta^{\tau}_{2} \propto \gamma_{1}\wedge\gamma_{2} = L_{Pl}^{-3} \qquad (122)$$

This relation is very satisfactory because it is coherent with the Dirac theory in the space-time algebra. In fact the bivector $\gamma_1\wedge\gamma_2$ substitutes the imaginary unit of complex numbers and is strictly related with the spin analogously to the torsion (see, for instance,[51,52]).

In order that the commutation relation have this form, the following six conditions should be satisfied:

$$\begin{aligned} &Q^{1\nu\beta}R^{2\alpha}{}_{\nu\beta}\gamma_{[1}\gamma_{2]} = L_{Pl}^{-3} \quad \text{(one relation)} \\ &Q^{\mu\nu\beta}R^{\alpha\tau}{}_{\nu\beta}\gamma_{[\mu}\ \gamma_{\tau]} = 0 \quad \text{with } \mu \neq 1,\ \tau \neq 2 \quad \text{(five relations)} \end{aligned} \qquad (123)$$

In conclusion the six conditions represent a "choice of gauge" with respect to the local Lorentz rotations, according to the fact that the choice of the spin plane is arbitrary.

This matter is still in progress and is under consideration together with Datta and Ronchetti [54].

References

[1] E.Cartan - Compt.Rend.174, 437, 593 (1922)

[2] E.Cartan - Ann.Ec.Norm.40, 325 (1923)

[3] E.Cartan - Ann.Ec.Norm.41, 1 (1924)

[4] E.Cartan - "Sur les variables à connexion affine et la théorie de la relativité généralisée" ed.Gauthier Villard 1955

[5] A.Trautman - "Theory of Gravitation", preprint IFT/72/25, Warsaw Uni.: read at the Symposium "On the Development of the Physicist's Conception of Nature', Miramare, Trieste 1972

[6] F.W.Hehl, P.von der Heyde, G.D.Kerlick and J.M.Nester Rev.Mod.Phys.48, 393 (1976)

[7] V.de Sabbata and M.Gasperini - "Introduction to Gravitation" World Scientific Publishing Co Pte Ltd., Singapore (1985)

[8] V. de Sabbata and M.Gasperini - Lettere Nuovo Cimento 27, 133 (1980)

[9] P.A.M.Dirac - Proc.Roy.Soc.A165, 199 (1938)

[10] P.A.M.Dirac - Proc.Roy.Soc.A338, 439 (1974)

[11] C.Isham, A.Salam and J.Strathdee - Nature Phys.Sci. 244, 82(1973)

[12] V.de Sabbata nad M.Gasperini - in "Origin of Galaxies" ed.by Venzo de Sabbata, World Sci.Singapore 1981 pg.181

[13] V.de Sabbata and P.Rizzati - Lettere Nuovo Cimento 20, 525 (1977)

[14] V.de Sabbata - in "Gravitational Measurements, Fundamental Metrology and Constants" ed. by Venzo de Sabbata and V.N.Melnikov, Kluwer Acad.Publ. 1988 pag.115

[15] P.M.S.Blackett - Philos.Trans.Roy.Soc.London Ser. A245, 309 (1952)

[16] V.de Sabbata, V.N.Melnikov and P.I.Pronin - Progress of Theoretical Physics (invited paper) 88, 623 (1992)

[17] V.de Sabbata and M.Gasperini - Lett.Nuovo Cimento 25, 489 (1979)

[18] V.de Sabbata and M.Gasperini - Lett.Nuovo Cimento 27, 289 (1980)

[19] A.Trautman - Nature Phys.Sci.242, 7 (1973)

[20] F.W.Hehl - "Four Lectures on Poincaré Gauge Field Theory" in Proceedings of the 6th Course of the Int.School of Cosmology and Gravitation on 'Spin, Torsion, Rotation and Supergravity' ed. by Peter G.Bergmann and Venzo de Sabbata, Plenum Press, New York, 1980, 5 - 61

[21] M.Novello - Phys.Lett., 59A, 105 (1976)

[22] S.Hojman, M.Rosenbaum and M.P.Ryan - Phys.Rev., D17, 3141 (1978)

[23] F.W.Hehl, P.von der Heyde, G.D.Kerlick and J.M.Nester - Rev.Mod.Phys., 48, 393 (1976)

[24] G.D.Kerlick - Ann.Phys., 99, 127 (1976)

[25] F.W.Hehl, P.von der Heyde and G.D.Kerlick - Phys.Rev., D10, 1066 (1974)

[26] V.de Sabbata and Gasperini - Nukleonika 25, No.11-12,1980

[27] V.de Sabbata and M.Gasperini - Lett.Nuovo Cimento 28, 181 (1980)

[28] V.de Sabbata and M.Gasperini - Lett.Nuovo Cimento 28, 229 (1980)

[29] V.de Sabbata and M.Gasperini - Physics Letters 77A,300 (1980)

[30] V.de Sabbata and M.Gasperini - Phys.Rev.23D, 2116 (1981)

[31] V.de Sabbata and M.Gasperini - Physics Letters 83A,115 (1981)

[32] V.de Sabbata and M.Gasperini - Lett.Nuovo Cimento 30, 193 (1981)

[33] J.Plebanski - Phys.Rev.118, 1396 (1960)

[34] V.de Sabbata and M.Gasperini - Lett.Nuovo Cimento 28, 234 (1980)

[35] W.Adamowicz and A.Trautman - Bull.Acad.Pol.Sci. 23, 339 (1975)

[36] V.de Sabbata and M.Gasperini - Gen.Rel.Grav.10, 825 (1979)

[37] V.de Sabbata and M.Gasperini - Il Nuovo Cimento 66A, 479 (1981)

[38] B.Pontecorvo - Z.Eksp.Teor.Fiz.,53, 1717 (1967) (English translation:Sov.Phys.JEPT, 26, 984 (1968)

[39] S.Bilenky and B.Pontecorvo - Phys.Rep.C, 41, 225 (1978)

[40] R.Colella, A.W.Overhauser and S.A.Werner - Phys.Rev.Lett. 34, 1472 (1975)

[41] V.de Sabbata and M.Gasperini - Lett.Nuovo Cimento 33, 363 (1982)

[42] V.de Sabbata, P.I.Pronin and C.Sivaram - International Journal of Theoretical Physics 30, 1671 (1991)

[43] H.-H.v.Borzeszkowski and H.-J.Treder - Ann.der Physik 46, 315 (1989); see also "On Quantum Gravity" PRE-EL 88-05, Potsdam 1988, and H.-H.v.Borzeszkowski and H.-J.Treder, The Meaning of Quantum gravity, Dordrecht, Reidel Publ.Comp. 1988.

[44] V.de Sabbata - a) Il Nuovo Cimento 107A, 363 (1994); see also b) "Torsion, String Tension and Quantum Gravity" in "String Quantum Gravity and Physics at the Planck Energy Scale" ed.by N.Sanchez, World Sci.Publ.,Singapore 1993 p.543

[45] X.Yu - Astrophys.Space Sci.154, 321 (1989); see also "General relativity on spinor-tensor manifold" in "Quantum Gravity" ed.by Peter G.Bergmann, Venzo de Sabbata and Hans Jürgen Treder, World Sci.Singapore, p.382-411 (1986)

[46] D.Hestenes - "Clifford Algebra to Geometric calculus" D.Reidel Publ.Co. Dordrecht, Holland 1982

[47] D.Hestenes - "New Foundation for Classical Mechanics" D.Reidel Publ.Co. Dordrecht, Holland 1985

[48] S.Gull, A.Lasenby, and C.Doran - Found.Phys. 23, 1175, (1993)

[49] C.J.L.Doran, A.N.Lasenby and S.F.Gull - Found.Phys.23, 1239, (1993)

[50] D.Hestenes - J.Math.Phys.16, 556, (1975)

[51] B.K.Datta, V.de Sabbata, L.Ronchetti - "Quantization of Gravity in Real Space-Time" *preprint* (accepted for publ.in Nuovo Cimento B: will appear ~ 1998)

[52] B.K.Datta - "Physical Theories in Space-Time Algebra" in "Quantum Gravity" ed.by P.G.Bergmann, V.de Sabbata and H.-J.Treder, World Sci.Singapore, p.54 (1996)

[53] V.de Sabbata - "Twistors, Torsion and Tensor-Spinors Space-Time" ed.by P.G.Bergmann, V.de Sabbata and H.-J.Treder, World Sci.Singapore, p.80 (1996)

[54] B.K.Datta, V.de Sabbata, L.Ronchetti - "Commutation relations between Torsion and Curvatura in the case of totally Antisymmetric Torsion", *preprint*

MODERN PERSPECTIVES ON NEWTONIAN GRAVITY

GEORGE T. GILLIES
Department of Physics
University of Virginia
Charlottesville, Virginia 22901
E-mail: gtg@virginia.edu

ABSTRACT

The discrepancies among the most recently obtained experimental results for the absolute value of the Newtonian gravitational constant, G, emphasize the need for continued work aimed at obtaining a *bona fide* theoretical prediction of this fundamental constant. A brief overview of the experimental situation is presented, along with some commentary on the searches for variations in G. The status of attempts at arriving at a theoretical prediction of G are discussed, and the role of G in tests of hypothetical spin-couplings in gravity is considered. Some ideas for future work in this area are offered.

1. Introduction

One of the most difficult tasks encountered in experimental physics is the measurement of the absolute value of the Newtonian gravitational constant, G, which scales the magnitude of the gravitational force, F, acting between two masses, m and M, separated by a distance, r, according to the classical inverse-square-law expression

$$F = GmM/r^2 \qquad (1)$$

This elegant statement describes virtually all observable gravitational phenomenon except, for example, under the extremes of matter at very high density. The subtle corrections offered by general relativity (of which the "Newtonian limit" is a special case) are manifested typically as very weak effects that can only be resolved by careful, intricate measurements. The determination of G is made hard by the fact that the gravitational force is of long range and cannot be shielded and, therefore, it cannot be studied in complete isolation. Moreover, it is by far the weakest of the four known forces, and prone to masking by competing effects in any particular experimental arrangement. Also, since G has no known interdependence with any other fundamental constant, it cannot be evaluated in terms of other quantities and, as a result, it is difficult to incorporate G into the structure of the rest of modern physics.

2. The Absolute Value of G

The presently accepted value of G, established by CODATA during the 1986 adjustment of the values of the fundamental constants, is

$$G = (6.672\ 59 \pm 0.000\ 85) \times 10^{-11}\ m^3s^{-2}kg^{-1}$$

The experimental basis for this result, and the background relevant to the measurement of G in general, is discussed at length in a recent review[1]. In the decade since the CODATA value was established, several new experimental results have been forthcoming, but a significant discrepancy exists between them, and the experimental picture is still uncertain at the level between $1:10^3$ and $1:10^4$. All of the most recent results are cited and discussed in detail elsewhere[1]. (See Table 1 below for a listing of a few of the latest experimental findings.)

Table 1

Authors	Method	G (10^{-11} $m^3s^{-2}kg^{-1}$)
Fitzgerald 1995[2]	nulled torsion balance	6.6659 ± 0.0006
Michaelis et al. 1995/96[3]	torsion balance (tungsten)	6.71540 ± 0.00056
Michaelis et al. 1995/96[3]	torsion balance (Zerodur™)	6.7174 ± 0.0020
Bagley and Luther[4]	torsion pendulum	6.6740 ± 0.0007

In spite of the continuing level of disagreement between the findings from individual experiments, much ingenuity has gone into the designs of the individual laboratory efforts. In particular, for torsion pendulum-based measurements of G, the recognition of the difficulties introduced by anelasticity in the fiber suspending the balance beam has prompted the introduction of several new measurement strategies, some of which are just now being carried out by groups with experiments in progress. If agreement at the level of $1:10^4$ to $1:10^5$ can be forthcoming from the newest generation of high precision experiments, then it will be difficult to presume that the discordant results from the existing recently completed efforts are all free from the influences of any undiscovered systematic effects.

The often-discussed hope of carrying out a measurement of the absolute value of G in space has led to the proposal of many different methods for possibly attempting this task[5], but the great financial cost of actually doing such an experiment (and the very limited availability of launch resources) has so far prevented such an experiment from going forward. Moreover, the elimination of one set of experimental problems (eg., terrestrial gravitational noise), simply opens the door for others of a different nature (eg., management of the spacecraft's thermal environment). Careful study and planning will be critical to the success of any such mission.

3. Searches for Variations in G

There have been approximately 40 different studies carried out over the last 15 years for the purpose of revealing a time-dependency in G. The potential presence of such an effect, and resulting size of it, if observed and confirmed, would provide important information that would be of direct use in evaluating the existing multi-dimensional theories. While some phenomenological studies claim that the size of any $\dot{G}/G$ effect is negligibly small (see Burša[6] for instance, who contends that $\dot{G}/G < 5 \times 10^{-17}$ per year), the existing solar system (and other) observational evidence tends to cluster around measured values that are on the order of 10^{-12} per year (see Ref. 1 for a thorough listing of all the recent observational values). The situation is somewhat complicated by the fact that if the sizes of masses is varying continuously too, perhaps via some mechanism not unlike the "multiplicative or additive creation cosmologies" of Dirac, then the observations are really of the variation of the product of GM where M is the mass of the astronomical body or bodies under observation. Several direct laboratory measurements of both $\dot{G}/G$ and the temporal constancy of rest masses have been proposed, and some have actually been attempted, but none have yet been carried out at cosmologically interesting levels of precision.

A variety of other tests for variations in G have been carried out during the course of the 20th century[7,8]. These include experimental searches for gravito-nuclear couplings, searches for the dependence of G on the state of electrification or magnetization of matter, searches for directive action of the gravitational force via the weighing of crystals, and many other such possibilities. To date, all such endeavors have ended in null results, but interest remains high in this line of work. This has been spurred recently by predictions of a possible temperature dependence in G which is said to lie just beyond the existing experimental limits on the measured size of such an effect[9]. Interest has also remained high in the possibility of detecting gravitational shielding or screening, not unlike that claimed to have been observed in the beam-balance measurements of shielding made by Majorana[10]. The latest observational limits[11] on the size of the "extinction coefficient" that scales such an effect, however ($\leq 10^{-21}$ cm^2 g^{-1}), established by a re-analysis of lunar laser ranging data, would seem to rule out the existence of this phenomena, at least in the way that it was originally envisioned.

Much of the recent work done on a distance-dependence in the gravitational force was stimulated by the conjecture that there may be new weak forces that are composition dependent. This suggestion has led to an intense effort to explore the universality of free fall, within the context of a search for violations of the weak equivalence principle (WEP). Ever tighter limits on the exactness of the WEP have been established, and while these results now preclude the existence of new, weak, long-range forces of the type that were originally conceived, the overall product of this effort has been a significant strengthening of the foundations of the experimental basis of general relativity.

4. Attempts to Derive a Theoretical Estimate of G

The Newtonian constant is virtually always referred to as a "fundamental" constant of nature. There is a school of thought, though, that concedes the possibility that it might be a calculable quantity. If this were the case, then the philosophical fundamentality of G might be challenged, particularly if it ends up appearing as a factor in some overall coupling constant that arises within a grand unified-field theory (GUT) that incorporates gravity. Interestingly, in either case, the presence of a *bona fide*, accepted prediction for the absolute value of G, or its presence in an expression for a GUT coupling constant, would do much to motivate the quest for an accurately measured value.

The various existing attempts at a theoretical estimate of G usually have as their starting point general relativity. The classical Lagrangian/matter-action expressions are then used to generate a model that contains either G or a G-related factor (such as the Planck mass). Such efforts are not necessarily aimed at deriving a numerical value for G; rather, they seek to develop a theoretical statement that is consistent with the observed behavior of matter in gravitational fields, consistent with the structure, symmetry and sizes of the existing forces, and consistent with the cosmological conditions of the early universe. (There have been many attempts to "build" a numerical value for G from complicated ratios of the other fundamental constants, but most of these attempts are based on the *ad hoc* introduction of either new fields or some other new effects, and are thus not typically able to be incorporated into the standard model.) A review of the various theoretical predictions of G is given in Ref. 1. They begin with the conjecture by Sakharov that gravity can arise from zero-point fluctuations, and continue with the work by Adler, Zee and others on the derivation of an "induced" gravitational constant from changes in the matter stress-energy tensor or from scale-invariant gauge theories. Still others have addressed this issue by evaluating the evolution of the Planck mass relative to other coupling constants in the Lagrangian or by evaluating the action of a Weyl spinor interacting with a gravitational field. Most recently, 't Hooft, Damour and others have re-examined an early argument of Landau that G might be related to the fine structure constant through an exponential expression involving parameters that could be motivated by instanton physics. An actual numerical value was generated in this case, and it was in surprisingly good agreement with the accepted CODATA value (see Ref. 1 for details).

An interesting observation about many of these efforts is that the Planck mass and G are used almost interchangeably in them as the quantity of interest. Secondly, stochastic processes, in one form or another, almost always play an important role. These include zero-point fluctuations of the vacuum, chaotic inflation scenarios, fluctuations in quantum gravity, and/or the zitterbewegung of particle motions. Perhaps the most evident implication of these observations is that a theoretical description of G is intrinsically coupled with the quantization of the gravitational field and the non-deterministic behavior of it at the quantum level. As greater insight is gained into the role of gravity in a grand unified-field theory, the theoretical derivation of G may become a more tractable task.

5. The Role of G in Searches for a Spin Dependence in the Gravitational Force

Several papers in these Proceedings are given over to theoretical investigations of the role that spin might play in arriving at a deeper understanding of the force of gravity. Experimental papers by Ritter et al., also in these Proceedings, examine the motivations for experiments in this field and lay the groundwork for interpreting the results in terms of a parameter that scales the size of any anomalous spin-dependent effect that might be uncovered.

The experiments fall into three broad classes: (1) those that use torsion pendulums to measure the interactions of macroscopic spin-polarized test masses, (2) those that employ precision magnetometry of ferromagnetic samples, and (3) those that examine the interaction of nuclear spins within the field of the Earth. Examples of the first type of experiments are those of Ritter et al.[12,13], who probed for the existence of a spin-spin coupling in their first study and a spin-mass coupling in their second. Their experimental arrangements were, in essence, those of a standard torsion pendulum used to make a measurement of the relative value of G, with stability of the apparatus being an important concern. The goals were to see if a small anomalous effect might be resolved in time against the relatively large (but presumably sufficiently static) gravitational background signal. In one interpretation of the results, the upper limit on the strength of any such coupling was expressed in terms of a fractional value of G. While this was a useful approach in terms of developing an intuitive feel for the sensitivity of the measurement, such an interpretation by itself was not meant to provide any deep insights into the physical basis of the problem. Even so, this approach does allow those interested in these topics to conveniently compare the strength (in terms of the fractional value of G) of the upper limit on a spin-dependent anomaly against the size of other weak effects that can be similarly expressed. If a non-negligible spin-dependent (or some other type of) anomaly were ever found, though, an "effective" value of G could be defined that takes it into account and modifies the statement of the Newtonian potential appropriately, as was done with the addition of a Yukawa-type term into models of composition dependent gravity. (This type of expedient step would not preclude the need to see how such a result would impact, eg., the WEP or other aspects of general relativity.)

6. Conclusions

Measurements of the Newtonian gravitational constant, while very difficult, do provide us with unique information about several questions that are fundamental to the development of a clearer picture of the physics of the gravitational force. Determination of the absolute value of G has historically been a task of interest largely because of the metrological challenges involved, but developments in grand unified-field theories may yet provide an important fundamental impetus for carrying out high-precision, high-reliability experimental studies of this elusive constant of nature. Searches for variation of G in time, in space and with respect to other physical quantities have led to no confirmed observations of such phenomenon, but the theoretical underpinnings for the compactification of extra dimensions in multi-dimensional theories do motivate the possibility of a time-dependent G, and a very substantial effort has gone into the search for such an effect, but with only null results to date. The results of certain classes of experimental tests of hypothetical spin-dependent gravitational couplings can be interpreted relative to the strength of G, but other approaches lead to deeper insight into the physics of the spin-related couplings. An accepted theoretical estimate of G does not yet exist, but progress towards such a possibility is being made.

7. Acknowledgments

Partial financial support for the preparation of the review article (Ref. 1) upon which this paper is based was provided by the U.S. National Institute of Standards and Technology (NIST) under Contract No. 43NANB613947 to the University of Virginia. The author thanks Dr. Barry Taylor of the U.S. NIST for making this possible. He also thanks his colleagues Dr. T. J. Quinn, Director of the International Bureau of Weights and Measures, for much assistance with the original work, and Prof. Venzo de Sabbata of the University of Bologna for much encouragement. He thanks C. A. H. Nelson for stimulating discussions. Finally, he thanks the "Ettore Majorana Centre for Scientific Culture" for the opportunity to co-direct and participate in the XVth Course of the International School of Cosmology and Gravitation, and for the generous hospitality in Erice during the Course.

References

1. G. T. Gillies, *Rep. Prog. Phys.* **60** (1997) 151.
2. M. P. Fitzgerald, *Bull. Am. Phys. Soc.* **40** (1995) 975. (See also Ref. 1)
3. W. Michaelis, H. Haars, and R. Augustin, *Metrologia* **32** (1995/96) 267.
4. C. H. Bagley and G. G. Luther, *Phys. Rev. Lett.* **78** (1997) 3047.
5. A. J. Sanders and G. T. Gillies, *Riv. Nuovo Cimento* **19** (1996) 1.
6. M. Burša, *Studia Geoph. Geod.* **28** (1984) 360.
7. G. T. Gillies, *Metrologia* **24**(Supplement) (1987) 1.
8. G. T. Gillies, *Am. J. Phys.* **58** (1990) 525.
9. A. K. T. Assis and R. A. Clement, *Nuovo Cimento B* **108** (1993) 713.
10. Q. Majorana, *J. Phys. Radium* **9** (1930) 314.
11. D. H. Eckhardt, *Phys. Rev. D* **42** (1990) 2144.
12. R. C. Ritter, C. E. Goldblum, W.-T. Ni, G. T. Gillies, and C. C. Speake, *Phys. Rev. D* **42** (1990) 977.
13. R. C. Ritter, L. I. Winkler, and G. T. Gillies, *Phys. Rev. Lett.* **70** (1993) 701.

Quantum Tests of Space–Time Structures

Claus Lämmerzahl
Fakultät für Physik der Universität Konstanz,
Postfach 5560 M674, D - 78434 Konstanz, Germany
e–mail: claus@spock.physik.uni-konstanz.de

August 11, 1997

Abstract

In the first part a short review of the theoretical, experimental, and phenomenological work concerning tests of the foundations of Einstein's General Relativity, namely Einstein's Equivalence Principle (EEP), is given. The various predictions of the violation of EEP due to the attempt to unify gravity with other forces and quantum gravity is presented. Then a list of experiments testing the various aspects of EEP and basics of GR is given and discussed in connection to the quantum theory used in these experiments. At last the various phenomenological test theories are shortly outlined.

Due to the fact that most of the experiments are essentially resting upon quantum effects a formalism is developed which allows a systematic and complete analysis of experiments designed for testing the validity of EEP in the quantum domain. This test theory generalizes the test theory of Haugan by introducing all possible spin effects in a systematic way. One important point is that in the quantum domain developed here there are more possibilities to violate EEP than in the classical domain even if one uses spin–polarized matter. By means of this test theory experiments which may be used to test possible violations of EEP are discussed. These experiments are: atom beam interferometry, Hughes–Drever experiments, $g-2$ experiments, and torsion balance experiments with spin–polarized matter.

1 Introduction

1.1 Einstein's Equivalence principle

Einstein's Equivalence Principle (EEP) is at the very foundation of our present description of gravitation which is Einstein's General Relativity (GR). The mathematical structure of GR consists of a Riemannian manifold, that is, a differentiable manifold endowed with a space–time metric. This metric has its physical significance in (i) giving the measured distance between two space–time events and (ii) leading to the equation of motion, the geodesic equation, for freely falling point particles in the gravitational field.

EEP consists of three parts, namely the Weak Equivalence Principle (WEP), Local Lorentz Invariance (LLI), and Local Position Invariance (LPI). WEP states that all freely falling test bodies are falling along the same trajectory. This is the reason for the possibility to geometrize the gravitational interaction: The path of a particle is a property of space–time and is not connected with any property of the particle. LLI expresses the fact that all experimental results are independent of the relative velocity and relative orientation of two laboratories where the same experiment takes place. For example, the ratio of the frequencies of two atomic clocks does not depend on their (common) velocity or their (common) spatial orientation. The independence from the relative orientation amounts to a statement about the isotropy of space: If all physical phenomena are independent of the the relative orientation then this can be considered as a property of space, namely the isotropy of space. The validity of LLI excludes any non–scalar gravitational field. However, even in the case LLI is violated it is possible to have a Lorentz–covariant theory of gravitation (for example Einstein–Cartan theory). LPI states that experimental results are independent of the position in a gravitational field where the experiments are carried through. That means especially that the ratio of the frequencies of two atomic clocks does not depend on their (common) position in the gravitational field. WEP, LLI, and LPI, are exactly fulfilled in Einstein's GR as well as in the Brans–Dicke scalar–tensor theory, Brans and Dicke (1961).

No single experimental result gives evidence for a violation of EEP. All laboratory and solar systems experiments and astrophysical observations, like binary systems, are in perfect agreement with Einstein's GR. However, possible solutions of current problems in fundamental physics like the attempt to quantize gravity or to unify gravity with the other interactions, always predict a violation of EEP. Therefore any better and more accurate test of the EEP, whether it will confirm or violate the EEP, will be important for the development of the theoretical description and the understanding of gravity.

1.2 Violations of Einstein's Equivalence principle?

1.2.1 General considerations

While the equivalence principle is the most important building principle for Einstein's GR this principle is predicted to be violated in unifying theories like gauge theory of gravity, Klein–Kaluza theory, string and superstring theory. The main reason for that is that all these theories predict more gravitational fields than the single spin–2 metrical field $g_{\mu\nu}$ of GR. These additional fields violate LLI and LPI and thus EEP. In addition, considerations from high energy physics also lead to additional gravitational fields like the axion field. (See also the contribution of R. Ritter in this volume.) We list a few example of proposals leading to a violation of EEP:

1. Speculations about the violation of the discrete symmetries P, C, and T symmetry in gravitational interaction lead to the following possible interactions involving the spin of the participating bodies, see Leitner and Okubo (1964):

$$V(r) = U(r)\left[1 + A_1(\boldsymbol{\sigma}_1 \pm \boldsymbol{\sigma}_1)\cdot\widehat{\boldsymbol{r}} + A_2(\boldsymbol{\sigma}_1 \times \boldsymbol{\sigma}_2)\cdot\widehat{\boldsymbol{r}}\right] \tag{1}$$

   where $\boldsymbol{\sigma}_{1,2}$ are the spins and $\boldsymbol{r}$ is the distance between the two bodies, $\widehat{\boldsymbol{r}}$ is the corresponding unit vector, and $U(r)$ the Newtonian potential. In the case that one body (for example the earth) is unpolarized, then

$$V(r) = U(r)\left(1 + A\boldsymbol{\sigma}\cdot\widehat{\boldsymbol{r}}\right) \tag{2}$$

   From hyperfine splittings of the hydrogen ground state we have $A \leq 10^{-11}$ for protons and $A \leq 10^{-7}$ for electrons, Leitner and Okubo (1964).

   In Hari Dass (1976, 1977) the above ansatz is generalized to involve the velocity of the particles:

$$V(r) = U_0(r)\left[1 + A_1\boldsymbol{\sigma}\cdot\widehat{\boldsymbol{r}} + A_2\boldsymbol{\sigma}\cdot\frac{\boldsymbol{v}}{c} + A_3\widehat{\boldsymbol{r}}\cdot\left(\boldsymbol{\sigma}\times\frac{\boldsymbol{v}}{c}\right)\right] \tag{3}$$

   which is still CPT–invariant. Here $\boldsymbol{\sigma}\cdot\boldsymbol{r}$ violates P and T, $\boldsymbol{\sigma}\cdot\boldsymbol{v}$ violates P and C, and $\boldsymbol{r}\cdot(\boldsymbol{\sigma}\times\boldsymbol{v})$ violates C and T. See also the note of Peres (1978).

2. A general feature of all gauge approaches to gravitation, e.g. Poincaré gauge theory Hehl, Heyde, and Kerlick (1976), Hehl, Lemke, and Mielke (1991), and Audretsch, Hehl, and Lämmerzahl (1992), supergravity Nieuwenhuizen (1982), or a gauge theory of the full linear group $Gl(4,\mathbb{R})$ (see the contribution of Y. Ne'eman in this volume) leads to a space–time torsion (and, in the last case, also to a nonmetricity). These additional gravitational fields are usually coupled minimally in a matter Lagrangian and influence the field equation and thus the dynamics of the fields. These additional gravitational fields of course define preferred directions and thus violate LLI.

3. Starting from higher–dimensional Klein–Kaluza theories the dimensional reduction scheme also leads to the interaction, Bars and Visser (1986), Barr and Mohapatra (1986),

$$V(r) = -G\frac{m_1 m_2}{r}\left(1 + g e^{-m_\sigma r}\right) + \frac{g_V^2}{4\pi}(B_1 - \epsilon Z_1)(B_2 - \epsilon Z_2)\frac{e^{-m_V r}}{r} \tag{4}$$

($M_{1,2}$ are two masses, $B_{1,2}$ and $Z_{1,2}$ their Baryon and atomic number, m_V and m_σ two ranges corresponding to a vector and a scalar coupling, g_V and g_σ corresponding coupling constants, and M_0 a normalization mass). These interactions first violate Newton's inverse square law and second are dependent on the composition of the bodies.

4. In the model of Nielsen and Picek (1983) the scalar Higgs field is assumed to propagate according to a metric which is different to the metric which governs the propagation of the gauge fields. This model leads to a modification of the metric for the propagation of a massless fermion: The metric for the left–handed fermion is different from the metric for the right–handed one. This leads in a non–relativistic limit to a spin–momentum coupling. See also the discussion in Fischbach, Haugan, Tadić, and Cheng (1985) who showed that this model leads to a violation of WEP.

5. Supersymmetry breaking at low energies gives additional forces created by moduli (gravitationally coupled scalar fields) gives Yukawa–like couplings, Dimopoulos and Giudice (1996), of the same form as the first term in eqn (4).

6. A solution of the the strong *PC* puzzle (for a review see e.g. Kim (1987)) can be given by an axion interaction which lead to new macroscopic forces, Moody and Wilczek (1984). Axions are also a candidate for dark matter in the universe, see e.g. Kolb and Turner (1990). This axion interaction leads to additional potentials between scalar and spinorial matter: mass–mass coupling

$$V(r) = -\kappa_0 \frac{e^{-r/\lambda}}{r}, \tag{5}$$

spin–mass coupling

$$V(r) = \hbar^3 D \boldsymbol{\sigma} \cdot \hat{\boldsymbol{r}} \left(\frac{1}{\lambda r} + \frac{1}{r^2}\right) e^{-r/\lambda}, \tag{6}$$

and spin–spin coupling

$$V(r) = \hbar^3 T \left((\boldsymbol{\sigma}_1 \cdot \boldsymbol{\sigma}_2) \left(\frac{1}{\lambda^2 r} + \frac{1}{r^3} + \frac{4\pi}{3}\delta^3(r)\right) - (\boldsymbol{\sigma}_1 \cdot \hat{\boldsymbol{r}})(\boldsymbol{\sigma}_2 \cdot \hat{\boldsymbol{r}}) \left(\frac{1}{\lambda^2 r} + \frac{3}{\lambda r^2} + \frac{3}{r^3}\right)\right) e^{-r/\lambda} \tag{7}$$

where λ is the range of this potential and σ^i the usual Pauli spin-matrices. D and T are in units mass^{-1}.

7. Spin–spin interactions of the structure (7) can arise (in the non–relativistic limit) effectively from arion couplings (see e.g. Vorobyov and Gitarts (1988), and references cited therein) which are not of electromagnetic origin, or from theories with propagating torsion where torsion is created by the elementary particle spin, Hayashi and Shirafuji (1979) or Hammond (1995). This means that even in the case one shields all electromagnetic fields, there will be an influence of one spin on the other.

8. The breaking of LLI also comes in by considering quantum field theory for non–trivial boundary conditions. For example, inside a capacitor the velocity of light is modified, see Scharnhorst (1990) and Barton and Scharnhorst (1993), or the neighborhood of a metallic plate induces a tensorial (anomalous, see below) inertial mass tensor, Golestanian and Kardar (1997).

9. From quantum gravity: The low–energy limit of the $E_8 \otimes E_8$ superstring model Lazarides, Panagiotakopoulos, and Shafi (1986) also gives additional gravitational interactions.

The general features of these gravitational interactions derived from the above–mentioned unifying theories are:

1. Violation of the inverse square law by
   (a) Yukawa–like interaction potentials $V = -G\frac{m_1 m_2}{r/\lambda}\alpha\lambda e^{-r/\lambda}$ with a certain range λ and strength α which are characteristic for the underlying theory.
   (b) additional r^{-n} potentials for $n \neq 1$.

   These violations can also be interpreted as a r–dependent gravitational constant G (see the contribution of G. Gillies in this volume).

2. Additional spin–interactions which are not of electromagnetic origin. These can be devided into the following cases:
   (a) spin–spin interactions.
   (b) spin–direction interaction.
   (c) spin–velocity interaction.

1.2.2 On space–time torsion

According to the topic of this conference we ask the question in which way torsion, which is a gravitational field in addition to the space–time metric and thus a field which violates EEP, couples to matter fields. Only by means of the influence of torsion to matter fields one is able to detect the existence of torsion.

A hypothetical space–time torsion which usually appears in gauge theories of gravitation, for example in the Poincaré–gauge theory, Hehl et al. (1976), Hehl et al. (1991), and Audretsch et al. (1992), or in supergravity Nieuwenhuizen (1982) couples to the spin of elementary particles (there is no coupling to angular momentum as has been shown by Yasskin and Stoeger (1980)). If one performs a minimal coupling procedure for the Dirac Lagrangian in Riemann–Cartan space–time, then the non–relativistic limit of the corresponding Dirac equation gives, beside the usual Pauli equation, two additional couplings to torsion Lämmerzahl (1997a): First, a coupling between the space part $\boldsymbol{K}$ of the axial torsion vector K_μ to the spin: $\boldsymbol{K} \cdot \boldsymbol{\sigma}$, and second, a coupling between spin and momentum with the time–component of the axial torsion as coefficient: $K_{(0)}\boldsymbol{\sigma} \cdot \boldsymbol{p}$:

$$i\hbar\frac{\partial}{\partial t}\psi = -\frac{\boldsymbol{p}^2}{2m}\psi - e\phi\psi + \frac{e\hbar}{2mc}\boldsymbol{H}\cdot\boldsymbol{\sigma}\psi + \frac{\hbar}{m}K_{(0)}\boldsymbol{\sigma}\cdot\boldsymbol{p}\psi - mU\psi - \hbar c\boldsymbol{K}\cdot\boldsymbol{\sigma}\psi \qquad (8)$$

Here, $\boldsymbol{H}$ and ϕ is an additional magnetic field and electrostatic potential. The same equation, but without the $K_{(0)}$–term has been derived also by e.g. Hayashi and Shirafuji (1979), Bagrov, Buchbinder, and Shapiro (1992) and Hammond (1995).

It is clear from this Hamiltonian that the spatial part of the axial torsion leads to an additional spin precession which may be detected e.g. in Hughes–Drever–type experiments or $g-2$ experiments. For the first one an estimate from already performed experiments can be given Lämmerzahl (1997a): $K \lesssim 1.5 \times 10^{-15}\,\mathrm{m}^{-1}$. For $g-2$ experiments an even more tighten constraint can be given (see below). The component $K_{(0)}$ leads to a torsion–induced spin–momentum coupling which can be tested using atom beam interferometry and leads to a constraint of $K_{(0)} \lesssim 10^{-2}\,\mathrm{m}^{-1}$ (Lämmerzahl (1997a) and see below).

The coupling of torsion to the electromagnetic field is more intricate: First, we accept that the application of the "comma–goes–to–semicolon–rule" makes sense only on the level of the physically measurable quantities, that is, on the level of the field strengths (for another concept, see e.g. Andrade, Oguri, Lopes, and Hammond (1992)). Indeed, applying this rule to the definition of the field strength in terms of the vector potential leads to a gauge–dependent theory and also to a non–uniqueness of the Aharonov–Bohm effect. That means also that the homogeneous Maxwell equations loose their validity. However, even in the case that one minimally couples the special relativistic Maxwell equations to Riemann–Cartan geometry leads to field equations which do not lead to current conservation. Only if the space–time torsion is restricted to the axial part only, and if this axial part is the gradient of a scalar potential, then current conservation is fulfilled, Puntigam, Lämmerzahl, and Hehl (1997) and Lämmerzahl, Puntigam, and Hehl (1997). Indeed, this kind of coupling is also the result of a QED analysis of DeSabbata and Gasperini (1981) where the renormalized photon propagator gets parts of the axial torsion coupled to the virtual Dirac particles. In addition, this kind of coupling may also be responsible for a recently observed anisotropy of the electromagnetic radiation in the universe, Nodland and Ralston (1997), where, however, other explanations are possible, for example Obukhov, Korotky, and Hehl (1997). For simplicity, we do not assume in the following any anomalous coupling of the electromagnetic field to non–metrical fields (compare also footnote 5).

1.3 Experimental tests of Einstein's Equivalence principle

Here we review some tests of the various parts of EEP. A survey of the literature shows that except the torsion balance and free fall tests of the WEP as well as some experiments on LLI, all experiments designed for testing the various parts of EEP use at least parts which can be described only by quantum Theory. For example, Hughes–Drever type experiments testing the isotropy of space observe the (hypothetical) orientation dependence of Zeeman levels. For tests of LPI, which amounts in tests of the gravitational red shift, usually atomic clocks are used which are characterized by certain transitions between atomic energy levels. Therefore, most of the experiments testing the space–time structure are *quantum tests of EEP*. In addition, since all experiments made on earth are in fact done in a noninertial frame moving in a gravitational field, each high precision experiment can be used as a test of the underlying space–time structure.

In Table 1 to Table 3 we give an (incomplete) list of experiments testing the various aspects of WEP, LLI, and LPI. Thereby we distinguish between pure quantum tests like spectroscopy, semi–quantum tests where one uses elements of quantum mechanics but also purely classical notions, for example the classical force on spin–polarized matter, and purely classical tests, like the torsion balance tests of the WEP.

One should also bear in mind that (in the same way as the WEP has to be tested for all kinds of matter) all basic structures of the space–time geometry have to be tested for all kinds of matter. For example, it is not enough to test the isotropy of light propagation and the independence of the velocity of light from the velocity of the frame of reference in order to state that locally space–time possesses a Minkowskian structure. It is still possible that other propagation phenomena, like those of particles with internal structures, may show a violation of this space–time structure (see below the discussion of the generalized Dirac equation (14)). Only if *all* phenomena show the *same* features then one can speak about a specific geometrical structure. Otherwise each phenomenon possesses its own geometry, for example its own metric which governs the dynamical propagation. This would lead to a bi–metric or multi–metric theory, which clearly violates LLI (see e.g. the model of Nielsen and Picek (1983) or the $TH\epsilon\mu$–formalism). Consequently, it is important to tests each geometrical notion with all kinds of matter. In this connection it is of help if, for example, by means of the coupling of charged matter with the electromagnetic field, one can test properties of the electromagnetic field by testing properties of bound systems. This is done, for example, in the $TH\epsilon\mu$ formalism.

tested notion	experiment	method	experimental realization	estimate
isotropy of space	Hughes, Robinson, and Beltran-Lopez (1960)	anisotropy influences energy levels: splitting of Zeeman line	NMR with ^{7}Li	$\delta m/m \leq 10^{-20}$
	Drever (1961)	dito	NMR with ^{7}Li	$\delta m/m \leq 2 \times 10^{-23}$
	Prestage et al. (1985)	dito	NMR with ^{9}Be$^+$	$\delta m/m \leq$
	Lamoreaux et al. (1986) (also Lamoreaux et al. (1989, 1990))	dito	NMR with ^{201}Hg	$\delta m/m \leq 2 \times 10^{-28}$
	Chupp et al. (1989)	dito	NMR with ^{21}Ne	$\delta m/m \leq 5 \times 10^{-30}$
spin-direction (mass) (see eqn (6))	Velyukhov (1968)	gravitational dipole influences proton Larmor frequency	NMR experiments	$\approx 31\,$Hz (appears to be wrong)
	Young (1969)	dito	free–precession proton magnetometer	$\leq 0.3\,$Hz
	Wineland and Ramsey (1972)	spin–position interaction influences energy levels	ground–state hyperfine separation in deuteron	$\delta\nu < 10^{-4}$Hz $D \leq 6.7 \times 10^{-8}$kg^{-1}
	Wineland et al. (1991)	spin–dependent potential influences energy levels	NMR	$D(^9\text{Be}) \leq 9.0 \times 10^{-9}$kg^{-1} $D(e) \leq 4.5 \times 10^{-5}$kg^{-1} $D(n) \leq 2.7 \times 10^{-8}$kg^{-1}
	Venema et al. (1992)	dito	nuclear spin precession	$D(n) \leq 1.7 \times 10^{-9}$kg^{-1}
	Youdin et al. (1996)	nearby mass induces splitting of spin states	comparison of frequencies of two different (Hg and Cs) magnetometers	$\frac{g_s g_p}{\hbar c}(e) \leq 2.3 \times 10^{-29}$, $\frac{g_s g_p}{\hbar c}(n) \leq 3.6 \times 10^{-29}$
LLI: spin–spin interaction	Ramsey (1979)	nearby spins influence energy levels	Molecular spectra	$\alpha \leq 5 \times 10^{-5}$
	Aleksandrov et al. (1983)	dito	NMR	$\alpha \leq 10^{-10}$
	Ansel'm and Neronov (1985)	dito	NMR	$\alpha \leq 10^{-10}$
LLI: spin–momentum $\kappa\boldsymbol{\sigma} \cdot \widehat{\boldsymbol{v}}$	Berglund et al. (1995)	splitting of energy levels due to velocity direction	relative frequency of Hg and Cs magnetometers	$\kappa(e) \leq 2 \times 10^{-4}\,$Hz, $\kappa(n) \leq 1.1 \times 10^{-7}\,$Hz

Table 1: Pure quantum tests of the isotropy of space, and spin–direction and spin–momentum interaction.

tested notion	experiment	method	experimental realization	estimate
WEP $\eta = \Delta a/a$	Koester (1976)	wavelength of neutron wave depends on gravitational acceleration	neutron gravity refractometer	$\eta < 5 \times 10^{-4}$
	Witteborn and Fairbank (1967)	path of electron	measurment of time of flight of electron	$\eta \leq 0.09$
LLI: spin–velocity interaction $\boldsymbol{\sigma} \cdot \boldsymbol{v}$	Phillips and Woolum (1969)	velocity of earth induces a time–varying torque	torsion pendulum	$\delta E \leq 4.5 \times 10^{-11}$ eV
	Phillips (1987)	dito	torsion pendulum	$\delta E \leq 8.5 \times 10^{-18}$ eV
LLI: spin–spin interaction $V = \alpha_s \frac{\mu_e \sigma \cdot \mu_e \sigma}{r}$	Graham and Newman (1985)		torsion pendulum	$\alpha_s \leq 3 \times 10^{-10}$
	Vorobyov and Gitarts (1988)	direct spin–spin interaction between electrons	induced ferromagnetism	$\alpha_s \leq 5 \times 10^{-14}$
	Bobrakov et al. (1991)	dito	induced ferromagnetism	$\alpha_s \leq 8.5 \times \times 10^{-15}$
	Hawkins 1989			
	Ritter et al. (1990)	dito	torsion balance	$\alpha_s \leq 9 \times 10^{-12}$
	Pan, Ni, and Chen (1992)	dito	torsion balance	$\alpha_s \leq 1.5 \times 10^{-12}$
	Chui and Ni (1993)	direct spin–spin interaction between electrons	induced paramagnetism	$\alpha_s \leq 5 \times 10^{-14}$
	Ni et al. (1994)	dito	dito	$\alpha_s \leq 1.2 \times 10^{-14}$
LLI: spin–mass interaction $\boldsymbol{\sigma} \cdot \boldsymbol{r}$	Hsieh et al. (1989)	spin–direction influences force	weighing polarized bulk matter	$\eta \leq 10^{-8}$
	Ritter, Winkler, and Gillies (1993)	electron	torsion balance	$\frac{g_s g_p}{4\pi\hbar c} \leq 3 \times 10^{-27}$
LLI: constancy of c: $\Delta c = \epsilon \left(\frac{v}{c}\right)^2$	Hils and Hall (1990)	comparison of clock with length scale	time given by I_2 absorption line and length by Fabry–Perot	$\epsilon \leq 7 \times 10^{-5}$
LPI: $\Delta\nu = (1-\alpha)\frac{U}{c^2}$	Vessot et al. (1979, 1980)	redshift experiment	two identical atomic clocks in different potential	$\alpha \leq 7 \times 10^{-5}$
	Turneaure et al. (1983)	comparison experiment	two non–identical atomic clocks in same potential	$\alpha \leq 10^{-2}$
	Godone, Novero, and Tavella (1995)	redshift experiment	non–identical atomic clocks	$\alpha \leq 7 \times 10^{-5}$

Table 2: List of semi quantum tests. These tests use quantum aspects of matter but also classical notions; for example: a net spin which is of quantum origin, exerts a force on a pendulum which is measured by a purely classical procedure.)

tested notion	experiment	method	experimental realization	estimate
WEP: $\eta = \frac{\Delta a}{a}$	Newton	gravitational acceleration of bulk matter depends on composition	pendulum with different substances in the field of the earth	$\eta \leq 10^{-3}$
	Bessel (1832)	dito	dito	$\eta \leq 2 \times 10^{-5}$
	Eötvös, Pekár, and Fekete (1922)	dito	torsion balance with different substances in field of the sun	$\eta \leq 5 \times 10^{-8}$
	Roll, Krotkov, and Dicke (1964)	dito	torsion balance with Al and Au in the field of the sun	$\eta \leq 10^{-11}$
	Braginsky and Panov (1971)	dito	torsion balance with Al and Pt in the field of the sun	$\eta \leq 2 \times 10^{-12}$
	Niebauer, McHugh, and Faller (1987)	dito	free fall interferometer with Cu-U in the field of the earth	$\eta \leq 5 \times 10^{-10}$ $\alpha\lambda \leq 8\,\mathrm{m}$
	Kuroda and Mio (1989, 1990)	dito	free fall interferom. with Al-Be, Al-Cu, Al-C in field of earth	$\eta \leq 2 \times 10^{-10}$ $\alpha\lambda \leq 9\,\mathrm{m}$
	Romaides et al. (1994)	dito	free fall in field of the earth	$\alpha\lambda \leq 10^{-1}$
	Su et al. (1994)	dito	torsion balance with Be-Cu and Be-Al	$\eta \leq 10^{-12}$
	Gundlach et al. (1997)	dito	U acts as source on Cu and Pb	$\Delta a \leq 6 \times 10^{-13} \frac{\mathrm{m}}{\mathrm{sec}^2}$
LLI: isotropy of space	Michelson and Morley (1887)	velocity of light depends on orientation	Michelson interferometer	
	Joos (1930)	dito	dito	
	Jaseja et al. (1964)	dito	two optical masers in orthogonal orientation and interferom.	
	Brillet and Hall (1979)	comparison of solid state length with optical length		$\Delta c/c \leq 5 \times 10^{-15}$
LLI: constancy of c: $\Delta c = \epsilon \left(\frac{v}{c}\right)^2$	Kennedy and Thorndyke (1932)	velocity of light in moving laboratory		$\epsilon \leq 2 \times 10^{-2}$
	Ives and Stillwell (1938, 1941)	frequency depends on velocity of source	high speed hydrogen canal rays	

Table 3: Classical tests using bulk matter (with no net spin and mass multipoles) and propagation of electromagnetic waves.

1.4 Theoretical frameworks for describing tests of Einstein's Equivalence Principle

According to what we remarked in Section 1.2 there are many theoretical predictions for a violation of EEP or a violation of Einstein's GR. However, this are considerations within certain theoretical models. In order to be able to make a proper interpretation of experimental tests one needs a model–independent theoretical frame for a description of these tests. In other words, one needs a model–independent theory which allows all possible violations of principles underlying GR. Such kind of a theory is called a *test theory* and is necessarily phenomenological.

A test theory is a theoretical framework which is broad enough to allow all possible violations of principles which one wants to test experimentally. The various parts violating these principles are characterized by parameters which may be constrained by experiment provided the experiment gives a null–result. The quality of the constraints is given by the accuracy of these experiments.

1.4.1 The PPN formalism

One such test theory which is still within the metric description of gravitation, is the PPN formalism (generalization of Eddington–Robertson formalism, see Misner, Thorne, and Wheeler (1973), and Will (1993)). This test theory describes all possible relations between the various components of the space–time metric and possible sources which are characterized by rest mass, velocity, pressure and so on. The experimental bounds on the PPN parameters from experiments and astrophysical observations are given in Will (1993). Possible limits from atom beam interferometry are presented in Lämmerzahl (1994, 1996a).

1.4.2 Haugan's approach

This general kind of a test theory which is model independent has its beginning with a consideration of Cocconi and Salpeter (1958) concerning a possible realization of Mach's principle. Mach's principle states that the inertial properties of matter are influenced by the nearby matter which is usually present in a certain direction. Consequently, the inertial mass of a body should have a tensorial character leading to a modified kinetic energy of a point particle $E = \frac{1}{2}M_{ij}v^i v^j$. Since the largest mass in our vicinity is the Milky Way the largest difference between effects involving the components of M_{ij} should be for experiments with a given intrinsic direction parallel and perpendicular to the direction towards the center of our galaxy. This is the case for the Zeeman splitting of energy levels with a given magnetic field defining the quantization axis. A first estimate on this difference $\Delta M/M \leq 10^{-14}$ has been given in Cocconi and Salpeter (1960).

This consideration has been generalized by Haugan (1979) who made the ansatz for the energy of a composite body

$$H = \frac{m}{2}\left(\delta^{ij} + \frac{\delta m_{\text{i}ij}}{m}\right) v^i v^j + (m\delta_{ij} + \delta m_{\text{g}ij})U^{ij}(x) \tag{9}$$

where $\delta m_{\text{i}ij}$ and $\delta m_{\text{g}ij}$ are anomalous inertial and gravitational mass tensors. v^i is the velocity of the body, and U^{ij} is a gravitational potential tensor which trace gives the Newton potential: $\delta_{ij}U^{ij} = U$. The anomalous mass tensors may depend on the used matter as well as on its quantum state. It is easy to see, Haugan (1979), Will (1993), that the anomalous inertial mass tensors gives rise to violations of LLI, while the anomalous gravitational mass tensors breaks LPI[1]. The corresponding acceleration is easily calculated

[1]The corresponding tests looking for a ΔM are claimed to be tests of the spatial isotropy of space, see Phillips (1987), Prestage, Bollinger, Itano, and Wineland (1985), Lamoreaux, Jacobs, Heckel, Raab, and Fortson (1986), or of LLI, see Chupp et al. (1989). In a strict sense this is not true. One cannot test the isotropy of space using one kind of particles only. Since each particle in general has its own mass and also possibly its own anomalous mass tensor, one only shows that for one kind of particle space appears to be isotropic. In order to claim that 'space' is isotropic this experiment has to be performed for

to be

$$a^i = \delta^{ij} g_j - \left(\frac{\delta m_{\rm i}^{ij}}{m} g_j - \delta^{ij} \frac{\delta m_{\rm g}{}_{kl}}{m} \frac{\partial U^{kl}}{\partial x^j} \right) \tag{10}$$

It is clear that LLI violating anomalous inertial mass tensors as well as LPI violating anomalous gravitational mass tensors lead to a violation of WEP.

1.4.3 The $TH\epsilon\mu$ formalism

Another kind of test theory is the $TH\epsilon\mu$ formalism (see Lightman and Lee (1973), Will (1993)). In this formalism it is allowed that the maximum velocity of massive particles is different from the velocity of light. This is described by a phenomenological bi–metric theory, one metric for the propagation of light, and the other metric for the propagation of particles. It is instantaneously clear that there is a special frame of reference so that both light cones are isotropic. This obviously breaks LLI. A possible space–dependence of these phenomena breaks LPI so that in general the weak EP and thus EEP is violated. By calculating the Hamiltonian for a bound system it turns out that there appear anomalous inertial and gravitational mass tensors due to this anomalous interaction between electrons and protons, Will (1974), Gabriel and Haugan (1990).

This test theory for the EEP has been widely used for the interpretation of many experiments. For example, within this theory the atomic energy levels have been calculated Will (1974), Turneaure, Will, Farrel, Mattison, and Vessot (1983), and Gabriel and Haugan (1990). These results have been applied to the interpretation of Hughes–Drever experiments and also to the Vessot–Levine red–shift experiment, because the latter tests the red shift of atomic clocks which are given by transitions between energy levels. In addition, this formalism has been used to calculate effects due to second quantization of the electromagnetic field, like the Lamb shift, the anomalous magnetic moment of the electron etc in order to look for violation of EEP due to 2nd quantization, Alvarez and Mann (1995a, 1995b, 1996a, 1996b, 1997).

An enlarged version of this $TH\epsilon\mu$–formalism has been given by Horvath, Logiudice, Riveros, and Vucetich (1988) for the standard model. In the same way as in the usual $TH\epsilon\mu$–formalism the characteristic maximal velocities of the Maxwell field and of massive point particle is different, now these velocities are different for the fermions, the Higgs–fields, the gauge fields, and the ghost fields. It is clear that this breaks LLI and also the free fall of a composite macroscopic body depends on its composition and inner structure.

---

all kinds particles and fields. (Another line of reasoning is: For one individual field it is always possible to find a coordinate system so that the principal (kinetic) part of the field equation is proportional to the Laplacian which due to its $O(3)$ symmetry is clearly isotropic. This can be done for each particle. However, these coordinate systems then in general do not coincide.)

Another point is the connection of a possibly anisotropic mass with properties of space. This is well justified if one takes over the constructive point of view that space is something which is *explored* by the dynamics of matter. If the dynamics of the matter is universal with respect to some type of effects then this property of the dynamics can be assigned to be a property of space (or space–time), which in the case considered is the isotropy of space. This is a basic principle for establishing Lorentz transformations as a symmetry of space–time. A corresponding postulate, namely that no particle has an anomalous inertial mass, requires some universal property of the particles and has the status of an equivalence principle. (This is true even in the case that the isotropy of space–time is broken due to a non–inertial motion of the observer.)

Consequently, all Hughes–Drever type experiments do not test a property of space, but instead only *search* for a violation of LLI or for a violation of the isotropy of space–time. (The same is of course true for tests of the weak equivalence principle.) If however, one single experiment will produce some positive effect, then our notion of space–time geometry will break down. In addition, having these points in mind it is well justified to interpret Hughes–Drever experiments as experiments which test (in the sense described above) parts of the EEP. Furthermore, having all these restrictions in the applicability of the used notions in mind, one can say that Hughes–Drever experiments are tests for the isotropy of space in the same sense as Eötvös type experiments are tests of the WEP.

1.4.4 Ni's formalism

In Ni (1977) a restricted test theory for a violation of EEP with validity of WEP is given. The difference to usual Einstein's GR is an additional pseudoscaler field φ which appears in an additional term of the Maxwell Lagrangian $\varphi F_{\mu\nu}\overset{*}{F}{}^{\mu\nu}$ where $\overset{*}{F}{}^{\mu\nu}$ is the dual of $F_{\mu\nu}$. This term does not violate WEP but acts on the polarization of test bodies[2].

1.4.5 A quantum test theory

While the test theories described above are using mainly classical notions (and of course the corresponding quantized theory), we are deriving in this work a test theory which is quantum from the very beginning. One main reason for the necessity of this approach is that there are more possibilities to violate EEP in our test theory than on any other test theory which starts on the classical level. There may be effects in the quantum domain which have no counterpart in the classical domain. Since, as a quick inspection of all the tests listed above shows, most of the tests are using at least parts operating quantum mechanically, it seems natural to construct a more powerful *quantum test theory*.

Consequently, in order to be able to draw a proper interpretation of all the quantum tests mentioned above, one has to have a test theory at hand which is purely quantum mechanical. This is the purpose of this contribution. We show how such a quantum test theory can be derived by using quantum notions only. Thereby we are *model independent* in the spirit of Haugan's test theory. After setting up this test theory we apply it to three classes of experiments, namely to (i) atom beam interferometry, (ii) Hughes–Drever type experiments, and (iii) to $g-2$ experiments. We restrict to *laboratory tests*. In some cases astrophysical observations may lead to better results. On the other hand, this kind of tests is indirect since in many cases one has to assume properties of the participating bodies.

In the course of derivation of our quantum test theory no geometrical notion except that of a differentiable manifold, should be used. Instead, the geometrical content of our gravitational fields should be derived from experiment. Only if all the parameters which describe the violations of EEP can be set to zero, then gravity can be described by a Riemannian geometry. If one of the parameters which we derive below appears to be non zero, then we may arrive at a Riemann–Cartan space–time as the geometrical description for the gravitational interaction. In this case we have besides the metric another gravitational field, namely torsion, which in general violates EEP.

The first step in our derivation of our quantum test theory we derive a general dynamical equation from basic physical principles. These principles which are assumed to apply for quantum fields are (i) the well–posedness of the Cauchy problem , (ii) the superposition principle, (iii) finite propagation speed, and (iv) a conservation law. The resulting general dynamical theory is given by a generalized Dirac equation, that is, a hyperbolic first order system of partial differential equations. It seems to be a very severe violation of widely accepted present day basic principles of physical theories if one abandons one of the above requirements. The next step is to make contact with Newtonian physics, that is, to make a non–relativistic approximation as well as a specification of the gravitational fields so that one gets as dynamical equation for the quantum matter the usual Schrödinger equation coupled to the Newtonian potential plus additional terms which describe couplings violating the EEP. That means, in the case the EEP holds, the dynamical equation is the usual Schrödinger equation. This dynamical equation in the non–relativistic limit which turns out to be a generalized Pauli equation is our new test theory for describing tests of the EEP. It is important that our test theory includes the elementary particle spin and its interaction with the metric and further hypothetical gravitational fields from the very beginning.

[2]The field φ acts as axion coupled to the electromagnetic field which can be interpreted as potential for an axial torsion, see Puntigam et al. (1997) and Section 1.2.2.

$$\left.\begin{array}{l}\text{well posed Cauchy–Problem}\\ \text{superposition principle}\\ \text{finite propagation speed}\\ \text{conservation law}\end{array}\right\} \Rightarrow \left\{\begin{array}{c}\text{generalized}\\ \text{Dirac–equation}\end{array}\right\} \Rightarrow \left\{\begin{array}{c}\text{generalized}\\ \text{Pauli equation}\end{array}\right.$$

Due to our most general approach our test theory includes in a systematic and complete manner all possibilities for a violation of the EEP on the non–relativistic level including the spin of particles. There seems to be no other way to violate the EEP than by the interactions described in our quantum test theory. Consequently, by a strong experimental restriction of all the EEP violating parameters in our test theory we can conclude that gravity can be described by a Riemannian space–time geometry.

It is the natural inclusion of spin in our approach and the fact that within our quantum test theory there are more possibilities to violate the EEP than in the corresponding classical theory which in our opinion makes it necessary to construct a genuine quantum test theory. It is not enough to have a classical test theory and carry through a quantization procedure. In this case the whole richness of quantum interactions will not be exhausted.

In the following we first indicate how to derive the generalized Pauli equation as quantum test theory for the description of quantum tests. This quantum test theory is the applied to the description of atom beam interferometry in gravitational field, of Hughes–Drever like experiments and to $g-2$ experiments.

This should be completed by a corresponding test theory for the equations governing the dynamics of the *electromagnetic fields*. This is important because the electromagnetic interactions is responsible for the existence of bound systems like atoms which are used in the experiments to test the EEP. While the usual Maxwell equations which are given by the equations for current conservation flux conservation as well as by a constitutive law, show up a very simple mathematical form and do not couple to any post–Riemannian structures, Puntigam et al. (1997), an axiomatic approach starting from the dynamics of the field strength may lead to equations, Lämmerzahl et al. (1997), leading to a possible birefringence behavior in gravitational fields as well as to velocities of light different from the velocity corresponding to the null cone of the generalized Dirac equation.

The advantages of this new quantum test theory compared with other test theories are:

1. There is no need for a preferred frame in which physics is isotropic, like in the Robertson–Mansouri–Sexl test theory for special relativity, see Robertson (1949), Mansouri and Sexl (1977a, 1977b, 1977c), or in the $TH\epsilon\mu$–formalism.

2. Spin is included from the very beginning in a consistent way.

3. It provides a systematic and (in the non–relativistic domain) complete way for the study of various anomalous effects.

2 A new quantum test theory

The derivation of our quantum test theory starts with quantum fields which we assume to be able to measure, at least in principle, by a set of measuring apparatus. This quantum field should be representable by a complex vector which depends on the position: $\varphi(x) \in \mathbb{C}^r$ for some $r \in \mathbb{N}$. These quantum field are now required to obey some basic principles which are testable by experiment. For doing so we assume that all physical processes take place within a 4–dimensional differential paracompact manifold. Our requirements are the following:

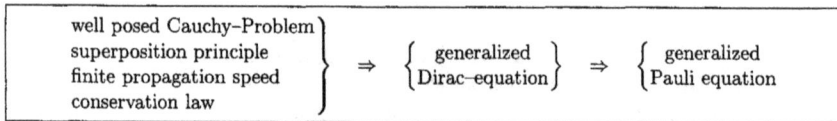

1. Well–posed Cauchy problem: We first require that there is a 3 + 1 splitting of our four–dimensional manifold. The manifold is therefore split into a set of hypersurfaces labeled with a parameter t which has nothing to do with any physical time. With respect to this splitting the dynamics of our quantum field under consideration should possess a deterministic evolution structure. That means, if φ is known on a hypersurface given by t_0 then we require that this will uniquely determine its value φ on any hypersurface with $t \geq t_0$. That means that there is a mapping
$$\varphi_{t_0} \to \varphi_t = U(t, t_0, \varphi_{t_0}) \tag{11}$$
2. Superposition principle: We require the superposition principle to be valid in the sense that is φ_{t_0} propagates to φ_t and ψ_{t_0} propagates to ψ_t then also any sum $\alpha\varphi_{t_0} + \beta\psi_{t_0}$ propagates to $\alpha\varphi_t + \beta\psi_t$. As a consequence, U is a linear operator: $U(t, t_0, \varphi_{t_0}) = U(t, t_0)\varphi_{t_0}$. Differentiation leads to an evolution equation, or the abstract Cauchy problem:
$$\frac{d}{dt}\varphi = G_t\varphi \tag{12}$$
where G_t is some linear operator (which still may be non–local, for example).
3. Finite propagation speed: According to all physical experiences all physical actions propagation with a finite velocity. Within our scheme this means: If one prepares on a hypersurface with t_0 a state φ_{t_0} with compact support then also for $t \geq t_0$ the propagated state should possess a compact support. A first consequence is that the operator G_t is local, which means that it is a differential operator in the coordinates within the hypersurfaces (see Lämmerzahl (1997b) for a more detailed treatment). Furthermore, it can be shown that this operator must be of first order. Consequently, we get from (12)
$$\frac{d}{dt}\varphi = G^{\hat{\mu}}\partial_{\hat{\mu}}\varphi + G\varphi \tag{13}$$
where we arranged the coordinates in such a way that $x^{\hat{\mu}}$, $\hat{\mu} = 1, 2, 3$ lie within the hypersufaces.
4. Conservation of probability: At last we require a conservation of probability $\frac{d}{dt}\int \varphi^+\varphi d^3x = 0$. From this we get the anti–hermiticity of the coefficients $G^{\hat{\mu}}$.

The final result of these requirements is a *first order hyperbolic system of partial differential equations*, which can be brought into the form of a *generalized Dirac equation* (GDE)
$$0 = i\tilde{\gamma}^\mu\partial_\mu\varphi - M\varphi \tag{14}$$
In our opinion it is very hard to avoid one of the above requirements. First, each of the above requirements is very well founded by physical experience, and second, by not using one of these requirements one obtains a very complicated theoretical frame (e.g. non–local equations, integro–differential equations) which is very hard to discuss.

Until now, the number of components φ is not specified. In order to be able to make contact with the usual Dirac equation describing the dynamics of spin–$\frac{1}{2}$ fields, we restrict ourselves to four components: $\varphi \in \mathbb{C}^4$.

We describe some implications of this GDE and and make some comments:

1. The characteristic equation for this first order system
$$0 = \det(\tilde{\gamma}^\mu k_\mu) \tag{15}$$

with $k_\mu = \partial_\mu \Phi$ where Φ is the characteristic surface Courant and Hilbert (1962), is a forth order polynomial in k_μ. Therefore, there are in general four solutions k_0 for given $k_{\hat{\mu}}$ which results in a splitting of the usual null cone (see Fig.1).

2. Due to the hyperbolicity of this first order system it is possible to define a characteristic velocity c solely in terms of the matrices $\tilde{\gamma}^\mu$.

3. The matrix M has to transform inhomogeneously under transformations $\varphi \to \varphi' = S\varphi$. However, the addition of a matrix M_0 which transforms homogeneously, does not change this: $M + M_0$ still transforms inhomogeneously. Therefore it is possible to split M into a part still transforming inhomogeneously which we call Γ and a part transforming homogeneously:
$$M = \Gamma + M_0 \tag{16}$$

4. In order to exhibit the dynamics of the generalized Dirac equation we perform a 3 + 1 splitting and solve this equation for the time derivative. We get ($\hat{\mu}, \hat{\nu} = 1,2,3$)
$$i\partial_0\varphi = i\tilde{\alpha}^{\hat{\mu}} c \partial_{\hat{\mu}}\varphi + \widetilde{\Gamma}\varphi + \widetilde{M}\varphi \tag{17}$$
with $\tilde{\alpha}^{\hat{\mu}} = (\tilde{\gamma}^0)^{-1}\tilde{\gamma}^{\hat{\mu}}$, $\widetilde{\Gamma} = (\tilde{\gamma}^0)^{-1}\Gamma$ and $\widetilde{M} = (\tilde{\gamma}^0)^{-1}M_0$. From the conservation of the probability the matrices $\tilde{\alpha}^{\hat{\mu}}$ are hermitian: $(\tilde{\alpha}^{\hat{\mu}})^+ = \tilde{\alpha}^{\hat{\mu}}$.

5. Since $\widetilde{M}$ has the dimension of time^{-1}, or, in conventional units, the dimension of mass velocity^2 we first extract a scalar parameter m of the dimension of a mass[3]: $\widetilde{M} = m\tilde{\beta}$ where $\tilde{\beta}$ is a 4×4-matrix.

6. We also introduce some c's in order to make $\tilde{\alpha}^{\hat{\mu}}$ and $\tilde{\beta}$ dimensionless and to give $\widetilde{\Gamma}$ the dimension time^{-1} ($\tilde{\beta}$ is now dimensionless because we introduced the mass parameter m).

7. We do not require that the matrices $\tilde{\alpha}^{\hat{\mu}}$ and $\tilde{\beta}$ fulfill a Clifford algebra. Indeed, the non–closure of the Clifford algebra[4] is the reason for violation of LLI. In terms of the generalized γ matrices in (14) we have the general structure
$$\tilde{\gamma}^\mu\tilde{\gamma}^\nu + \tilde{\gamma}^\nu\tilde{\gamma}^\mu = 2g^{\mu\nu} + X^{\mu\nu} \tag{18}$$
where $g^{\mu\nu} = \frac{1}{4}\mathrm{tr}(\tilde{\gamma}^\mu\tilde{\gamma}^\nu)$ and $X^{\mu\nu}$ is some traceless 4×4 matrix which has no physical relevance. Only if the $\tilde{\gamma}^\mu$ fulfill a Clifford algebra, then $X^{\mu\nu} = 0$.

8. The last point is to introduce the coupling to the electromagnetic field. Here we assume that this will be done via the minimal coupling procedure $\partial_\mu \to \nabla_\mu = \partial_\mu + eA_\mu$ ($A_\mu \leftrightarrow (\phi, A_i)$) where we assume that the electromagnetic field fulfills the usual Maxwell equations[5].

[3] Please note that there is no means to introduce $\hbar$. This is consistent with the fact that in the usual Schrödinger equation coupled to the Newtonian potential division by $\hbar$ gives that the only effective constant in this equation is the ratio $m/\hbar$. Indeed, this ratio can be measured only, compare Weiss, Young, and Chu (1994).

[4] It should be noted that the fact that the matrices $\tilde{\gamma}^\mu$ do not fulfill the usual Clifford algebra it is still possible to show that they fulfill a generalized Clifford algebra (see Lämmerzahl (1990, 1997b)). That means that the Clifford algebra gives notice of the geometry of the corresponding partial differential equation, in our case (14).

[5] This is a point which has to be improved. First it should be shown that the electromagnetic field couples in a minimal way to quantum matter and second, it is a unjustified assumption that the Maxwell equations should be valid. In fact, application of the scheme used here to derive the generalized Dirac equation, to the electromagnetic field one arrives at some generalized Maxwell equation, see Lämmerzahl et al. (1997). In general the solution of these generalized Maxwell equations show (in the same way as the generalized Dirac equation, see below) violation of LLI and LPI and thus EEP. It is this generalized Maxwell equation which should be used to calculate the center–of–mass Hamiltonian and Hamiltonian for the energy levels for bound system. These Hamiltonians are used to describe atom interferometers and atomic clocks which are used to test the space–time structure.

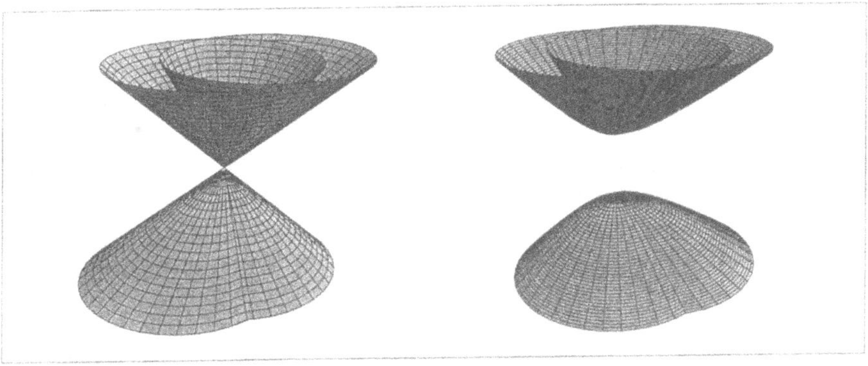

Figure 1: Structure of null cone and mass shell for fields obeying a GDE for which the usual Clifford algebra is not fulfilled. Each component of the null cone and of the mass shell corresponds to a specific polarization state. Usually, that is, without anomalous terms, there is only one null cone and one mass shell.

The final result of the generalized Dirac equation in 3 + 1 form is

$$i\partial_0\varphi = \widetilde{\alpha}^{\hat{\mu}} c\nabla_{\hat{\mu}}\varphi + c\widetilde{\Gamma}\varphi + mc^2\widetilde{\beta}\varphi + e\phi\varphi \tag{19}$$

An important point is that in general the null cones and mass shells of this GDE are not unique: In general there are birefringence effects in vacuum, see Fig.1. It is not difficult to show that if and only if there is no splitting of the null cone and of the mass shell, then $X^{\mu\nu} = 0$.

Since the general form of the dynamical equation is too complicated for the discussion of specific experiments, we are doing a non–relativistic approximation. In order to do so we have to specify the gravitational fields[6] in the generalized Dirac equation. All gravitational fields are contained in the matrices $\widetilde{\alpha}^{\hat{\mu}}$, $\widetilde{\Gamma}\varphi$, and $\widetilde{\beta}$. It is clear that the dominant part of all gravitational fields is the Newtonian potential. Therefore we assume that the x–dependence of the above matrices is due to the Newtonian potential only, and all other gravitational fields are taken to be constant over the dimension of the experiment. Analogously to the ansatz of Haugan (1979), the Newtonian part of the gravitational field is the U as well as the U^{ij}. In the following we restrict to stationary situations. Therefore:

$$\begin{aligned}
\widetilde{\beta}(x) &= \overset{0}{\widetilde{\beta}} + \overset{1}{\widetilde{\beta}}\frac{1}{c^2}\left(U(x) + \frac{\delta m'_{gij}}{m}U^{ij}(x)\right) \\
\widetilde{\alpha}^i(x) &= \overset{0}{\widetilde{\alpha}}{}^i + \overset{1}{\widetilde{\alpha}}{}^i\frac{1}{c^2}\left(U(x) + \frac{\delta m'_{gij}}{m}U^{ij}(x)\right) \\
\Gamma(x) &= \overset{0}{\Gamma} + \overset{1}{\Gamma}\frac{1}{c^2}\left(U(x) + \frac{\delta m'_{gij}}{m}U^{ij}(x)\right) + \overset{1}{\Gamma}{}^i\frac{1}{c^2}\partial_i U(x)
\end{aligned}$$

[6]Here we use the notion *gravitational field* for all interactions which are not given by the electroweak and strong interactions. Therefore all hypothetical new interactions ("fifth force") and interactions modifying the usual gravitational laws are called gravitational. Whether these new interactions fulfill any equivalence principle so that can be geometrized in some way is not considered at this stage of reasoning.

The matrices $\overset{0}{\widetilde{\beta}}$, $\overset{1}{\widetilde{\beta}}$, $\overset{0}{\widetilde{\alpha}^i}$, $\overset{1}{\widetilde{\alpha}^i}$, $\overset{0}{\Gamma}$, $\overset{1}{\Gamma}$, and $\overset{1}{\Gamma^i}$ are constant. It is possible to choose a coordinate system so that $\overset{0}{g}{}^{\mu\nu} = \eta^{\mu\nu}$ where $\overset{0}{g}{}^{\mu\nu}$ is the 2nd rank tensor introduced in (18) with the matrices $\overset{0}{\widetilde{\gamma}}{}^{\mu}$ instead of $\widetilde{\gamma}^{\mu}$. $\delta m'_{\mathrm{g}ij}/m$ is a dimensionless parameter which describes violations of LPI. Hypothetical violations of LLI are contained in the matrices $\overset{0}{\widetilde{\beta}}$, $\overset{1}{\widetilde{\beta}}$, and $\overset{0}{\Gamma}$.

Using this ansatz and performing a non–relativistic approximation á la Foldy–Wouthuysen, we get a *generalized Pauli–equation* (GPE), see Lämmerzahl (1996b, 1997b) and Lämmerzahl and Bleyer (1997):

$$
\begin{aligned}
i\frac{\partial}{\partial t}\varphi = & -\frac{1}{2m}\Bigg(\delta^{ij} \underbrace{-\frac{\delta m_{\mathrm{i}}^{ij}}{m} - \frac{\delta \bar{m}_{\mathrm{ik}}^{ij}\sigma^k}{m}}_{\substack{\text{spin–dependent tensor}\\ \text{of anom. inertial mass}}}\Bigg)\nabla_i\nabla_j\varphi + \underbrace{\left(cA^i_j + \frac{1}{m}a^i_j\right)\sigma^j i\nabla_i\varphi}_{\text{spin–momentum–coupling}} \\
& + \Bigg[e\,\phi(\boldsymbol{x}) + \frac{e}{2mc}\Big(\sigma^i + \underbrace{F^i_k\sigma^k + K^i}_{\text{anomalous magnetic dipole}}\Big)H_i(\boldsymbol{x}) \\
& + m\,U(\boldsymbol{x}) + \underbrace{\boldsymbol{C}\cdot\boldsymbol{\sigma}\, m\, U(\boldsymbol{x}) + \delta m_{\mathrm{g}ij}U^{ij}(\boldsymbol{x})}_{\substack{\text{spin–dependent tensor}\\ \text{of anom. gravitational mass}}} + \underbrace{c\boldsymbol{T}\cdot\boldsymbol{\sigma}}_{\substack{\text{torsion–like}\\ \text{coupling}}} + \underbrace{mc^2\boldsymbol{B}\cdot\boldsymbol{\sigma}}_{\substack{\text{spin–dep.}\\ \text{`rest–mass'}}}\Bigg]\varphi
\end{aligned}
\tag{20}
$$

Here σ^i are the usual Pauli–matrices and the coefficients $\delta m_{\mathrm{i}}^{ij}$, $\delta \bar{m}_{\mathrm{ik}}^{ij}$, A^i_j, a^i_j, $\boldsymbol{C}$, and $\boldsymbol{B}$ come in by projecting the various combinations of products of the matrices $\widetilde{\alpha}^i$ and $\widetilde{\beta}$ onto the 'large' components, Lämmerzahl and Bleyer (1997). ϕ and $\boldsymbol{H}$ are the electrostatic potential and the magnetic field. In the GPE we already transformed away terms of the form $C\varphi$ and $C^i\partial_i\varphi$ where C and C^i are constant.

It is clear that this GPE generalizes the ansatz of Haugan (1979) by the various spin–couplings to the gravitational potential. Terms giving a coupling between the spin and the Newtonian potential have first been introduced by Hari Dass (1976, 1977) (compare eqn (3)) and Peres (1978). Our coupling $\boldsymbol{C}\cdot\boldsymbol{\sigma}\, m\, U(\boldsymbol{x})$ agrees with the non–relativistic term in the three couplings Hari Dass considered. In our scheme we also treat couplings to non–Newtonian gravitational fields.

The anomalous magnetic dipole has a similar structure as the corresponding quantity derived by Alvarez and Mann (1995a) using QED within the $TH\epsilon\mu$–formalism. However, since our coupling with the magnetic field is not proportional to the spin, we do not introduce an anomalous gyromagnetic ratio tensor, but instead an anomalous magnetic dipole.

Here each term is uniquely characterized by its dependence on the mass m and the velocity c. We call the next–to–last term "torsion–like" because in the case of the non–relativistic limit of the Dirac equation in Riemann–Cartan space–time this term is due to the space–part of the axial torsion (compare eqn (8)). The time–part of the axial torsion gives rise to a spin–momentum coupling given by A^i_j.

An important point is that the classical acceleration which can be calculated from (20) by means of a classical approximation ($\boldsymbol{S}$ is the spin of the particle under consideration)

$$
a^i = -\left(\delta^{ij} + \frac{\delta m_{\mathrm{i}}^{ij}}{m} + \left(\frac{\delta \bar{m}_{\mathrm{ik}}^{ij}}{m} + \delta^{ij}C_k\right)S^k\right)\partial_j U - \delta^{ij}\frac{\delta m_{\mathrm{g}kl}}{m}\partial_j U^{kl} \tag{21}
$$

does not depend on all the EEP–violating parameters which appear in the generalized Pauli equation: A^i_j, a^i_j, $\boldsymbol{B}$, and $\boldsymbol{T}$ do not appear in (21). Therefore, *on the quantum level there exist more possibilities to violate EEP than on classical level.* This is the case even if one uses macroscopic matter with a net spin–polarization. That means that by doing classical tests of the equivalence principle one cannot rule out that there may occur EEP–violation on the quantum level.

Beside the path we also can get the information about the dynamical behavior of the spin expectation value in the classical approximation:

$$\frac{d}{dt}\boldsymbol{S} = \boldsymbol{\Omega} \times \boldsymbol{S} \tag{22}$$

with

$$\Omega_i := \frac{1}{2m}\frac{\delta\bar{m}^{kl}_{\mathrm{ii}}}{m}p_k p_l + \left(\frac{1}{m}a^j_i + cA^j_i\right)p_i + mc^2 B_i + cT_i + C_i mU(x) \tag{23}$$

That means that besides $\delta m^{ij}_{\mathrm{i}}$ and δm_{gkl} all the anomalous parameters influence the spin precession. In other words: Only if one takes the path (21) *and* the dynamics of the spin (23) into account one can make statements about the complete set of parameters characterizing the violation of EEP[7]. While the precession of the net polarization of a macroscopic body is very difficult to observe the corresponding quantum tests (Hughes–Drever, $g-2$) are much more sensitive.

Since we are considering the non–relativistic approximation only, we cannot conclude that if all the parameters characterizing the anomalous coupling vanish then EEP is valid. However, what we can say is the following:

- Any deviation from the usual Schrödinger equation implies a LLI–, LPI–, or WEP–violation

$$\left.\begin{array}{rll} \delta m^{ij}_{\mathrm{i}}, \delta\bar{m}^{ij}_{\mathrm{i}k}, A^i_j, a^i_j, \boldsymbol{B}, \boldsymbol{T} & \neq 0 \ \Rightarrow & \text{violation of LLI} \\ \delta m_{gij}, C_i & \neq 0 \ \Rightarrow & \text{violation of LPI} \end{array}\right\} \Rightarrow \quad \text{violation of EEP}$$

  It is clear that the parameters which indicate a breaking of LLI are connected with that part of the generalized $\tilde{\gamma}^\mu$–matrices which lead to a violation of the Clifford algebra.

- The non-vanishing of one parameter implies that gravity not describable by a space-time metric.

3 Comparison with experiment

There are three classes of pure quantum experiments which we want to discuss within the frame of our quantum test theory. These quantum experiments are (i) atomic beam interferometry, (ii) Hughes–Drever experiments, and (iii) $g-2$ experiments. While for the last one the generalized Pauli equation (20) can be used, for the first two one has to calculate the Hamiltonian for a bound system, for example, for the Hydrogen atom. The total two– or many–particle Hamiltonian can be split within some approximation into a center–of–mass part and an internal part. The center–of–mass motion can be tested by atomic interferometry and the internal part by spectroscopy, for example, by Hughes–Drever experiments.

[7] This observation supports the idea that one has to treat mass and spin, as it is suggested by the group theoretical analysis of the Poincaré–group, on equal footing, Trautman (1972), Hehl et al. (1976).

one–particle Hamiltonian	$\longrightarrow$	$g-2$ experiments
	center of mass motion $\longrightarrow$	atom interferometry
	$\nearrow$	
two–particle Hamiltonian		
	$\searrow$	
	energy eigenvalues $\longrightarrow$	Hughes–Drever experiments

Except for the description of the $g-2$ experiment we have to calculate the center–of–mass Hamiltonian and the Hamiltonian for the energy levels from a two– or many–particle Hamiltonian whereby all particles are charged an may interact not only with one another but also with an external electromagnetic field. This is necessary in order to be able to perform atom beam interferometry or to describe the Zeeman effect. We do not derive these Hamiltonians but merely quote the result (see Lämmerzahl (1997b) and Lämmerzahl and Bleyer (1997)) where we specialize to the case that the first particle has no spin

1. The center–of–mass Hamiltonian
$$H_{\text{c-m}} = -\frac{\hbar^2}{2}(M_{\rm i}^{-1})^{ij}\frac{\partial}{\partial X^i}\frac{\partial}{\partial X^j} + \frac{m_2}{m}N^i_{2j}\sigma^j i\frac{\partial}{\partial X^i} + M_{{\rm g}ij}U^{ij}(\boldsymbol{X}) + P_{2i}\sigma^i \tag{24}$$
with $M_{\rm i} = (m_1+m_2)\delta_{ij} + \delta m_{{\rm i}1ij} + \delta m_{{\rm i}2ij}$, $M_{\rm g} = m_1 + m_2 + \delta m_{{\rm g}1ij} + \delta m_{{\rm g}2ij}$, $N_2^i = (m_2 c A^i_{2j} + a^i_{2j})/(m_1+m_2)$, and $P_{2i} = mc^2 B_{2i} + cT_{2i} + m_2 C_{2i} U(\boldsymbol{X})$. Note that $m_{{\rm i}2ij}$ and $m_{{\rm g}2ij}$, and therefore the center–of–mass definition depends on the spin of the second particle.

2. The Hamiltonian describing the energy levels of an electromagnetically bound system:
$$H_{\rm E} = H_0 + H_{\text{E, em}} + H_{\text{E, Einst.}} + H_{\text{E, non-Einst.}} \tag{25}$$
with
$$H_0 = -\frac{1}{2m_{\rm red}}\delta^{ij}\frac{\partial}{\partial x^i}\frac{\partial}{\partial x^j} + \Phi'(x) \tag{26}$$
$$H_{\text{E, em}} = -\frac{e_1}{2m}\left(\frac{m_1}{m_2} - \frac{m_2}{m_1}\right)H^l\delta^{ij}\epsilon_{ilk}x^k i\hbar\frac{\partial}{\partial x^j} + \frac{e_1}{m_{\rm red}}H_k\sigma^k - e_1 x^i\nabla_i\phi(\boldsymbol{x}_c) \tag{27}$$
$$H_{\text{E, Einst.}} = \frac{1}{2}m_{\rm red}x^k x^l\nabla_k\nabla_l U(\boldsymbol{x}_c) \tag{28}$$
$$\begin{aligned} H_{\text{E, non-Einst.}} = & -\frac{\hbar^2}{2}\left(\frac{m^{ij}_{\rm i1}}{m_1^2} + \frac{m^{ij}_{\rm i2}}{m_2^2}\right)\frac{\partial}{\partial x^i}\frac{\partial}{\partial x^j} + N_2^i i\frac{\partial}{\partial x^i} \\ & + m_{\rm red}x^k\left(\left(\frac{\delta m_{{\rm g}2ij}}{m_2} - \frac{\delta m_{{\rm g}1ij}}{m_1}\right)\nabla_k U^{ij}(\boldsymbol{x}_c) + \delta_{kp}\left(\frac{\delta m^{pl}_{\rm i2}}{m_2} - \frac{\delta m^{pl}_{\rm i1}}{m_1}\right)\nabla_l U(\boldsymbol{x}_c)\right) \\ & + (m_{{\rm g}1ij} + m_{{\rm g}2ij})U^{ij}(\boldsymbol{x}_c) + \left(c^2 m_2 B_{2k} - m_2 C_i U(\boldsymbol{x}_c) + cT_{2k}\right)\sigma^k \end{aligned} \tag{29}$$
where Φ' is some effective binding potential, $m_{\rm red} = m_1 m_2/(m_1+m_2)$, and $\boldsymbol{x}_c$ is the position of the atom.

By performing *classical* tests like the torsion balance tests of the WEP (e.g. Su et al. (1994)) or of anomalous spin–couplings (e.g. Ritter et al. (1993)) we can, employing eqn (21), get estimates on the parameters $\boldsymbol{C}$, $\delta \bar{m}^{ij}_{\rm ik}$, and $\delta m_{{\rm g}kl}$.

Both Hamiltonians have the same structure as the one–particle Hamiltonian (20). In order to keep the notation simple, we calculate the following experiments within the notation of a one–particle Hamiltonian.

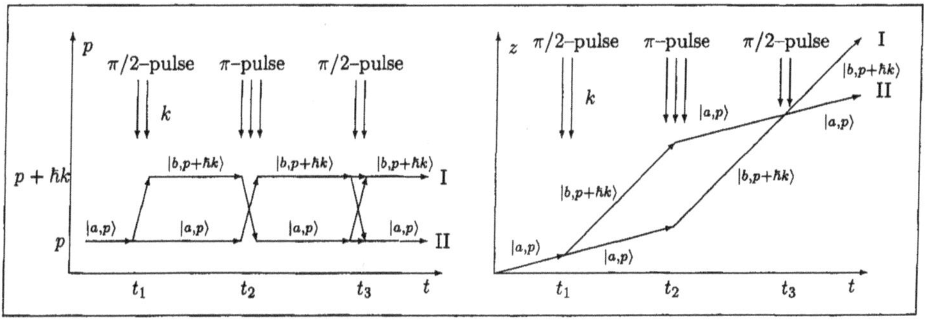

Figure 2: Geometry of the Kasevich–Chu interferometer in momentum space (left) and configuration space (right). $|a, \boldsymbol{p}\rangle$ and $|b, \boldsymbol{p} + \hbar\boldsymbol{k}\rangle$ denote the ground and excited atomic state with momentum $\boldsymbol{p}$ and $\boldsymbol{p} + \hbar\boldsymbol{k}$, respectively. The measured quantity is the number of atoms in the excited state (port I) or in the ground state (port II). For this measurement no resolution in configuration space is required, and the true geometry of the interference experiment is given by a closed loop in momentum space. In configuration space the atoms are characterized by the momentum of wave packets (no plane wave solution). Only in the case of wave packets one can speak of a closed loop in configuration space. However, in a real experiment the width of the wave packets is much larger than the distance between the maxima of the two split wave packets, so that in praxi no resolution of the two wave packets is possible.

3.1 Atom beam interferometry

In the following we always calculate possible effects for the atom beam interferometer of Kasevich and Chu (1991), Chu and Kasevich (1992). This interferometer works as follows: A laser pulse with wavelength $\boldsymbol{k}$ interacts with a beam of atoms with momentum $\boldsymbol{p}$ so that after this laser pulse half of the atoms have the original momentum $\boldsymbol{p}$ and the other half the momentum $\boldsymbol{p} + \hbar\boldsymbol{k}$. This is called a $\pi/2$–pulse. After a time T a second laser pulse inverses the population: all atoms with momentum $\boldsymbol{p}$ have after this pulse the momentum $\boldsymbol{p} + \hbar\boldsymbol{k}$ and vice versa. A third laser pulse after another time T again acts as $\pi/2$–pulse, see Fig.2.

In the following we always assume that the experiments will give a null result and take into account the accuracy of the Kasevich–Chu atom beam interferometer.

3.1.1 Spin–flip–experiment

In this experiment we prepare an atomic state in a specific spin state. We assume a beam splitting in spin space so that half of the atoms stay in the original spin state and the other half have a state with the opposite spin eigenvalue. After a time Δt the state with the 'flipped' spin is flipped again to the original state. In the meantime both state accumulated a different energy due to the interaction of the spin with external fields. This results in the phase shift

$$\delta\phi^{\text{atom}} = \frac{1}{\hbar}\left[H(p, S) - H(p, -S)\right]\Delta t \tag{30}$$

If we use the GPE then all spin–dependent terms contribute to the phase shift. If, in addition, we assume that a possible experiment using the atom beam interferometer of Kasevich and Chu (1991), Chu and Kasevich (1992) gives a null–result then the accuracy of this experiment gives us estimates on the various parameters. The parameter describing the spin–momentum coupling is limited by

$$|A^i_j| \leq 10^{-17} \tag{31}$$

which generalizes the result obtained in Audretsch, Bleyer, and Lämmerzahl (1993). For the other parameters we get better results with Hughes–Drever and $g-2$ experiments.

3.1.2 Measurement of acceleration

The theory of the atom beam interferometer of Kasevich and Chu (1991), Chu and Kasevich (1992) gives that an acceleration $\boldsymbol{a}$ leads to the phase shift $\delta\phi^{\text{atom}} = -\boldsymbol{k}\cdot\boldsymbol{a}T^2$. If we take the acceleration (21) in a spherically symmetric gravitational field (see e.g. Will (1993)), then we get as phase shift

$$\delta\phi^{\text{atom}} = -(1+\alpha)\,kT^2 g\cos(\vartheta + b) \tag{32}$$

with

$$\alpha := \frac{1}{5}\left(\frac{\delta m_{gzz}}{m} + 4\frac{\delta m_{gxx}}{m}\right) - \frac{\delta m_{izz}(S)}{m}, \qquad b := \frac{6}{5}\frac{\delta m_{gzx}}{m} + \frac{\delta m_{izx}(S)}{m} \tag{33}$$

where $\delta m^{ij}_{\text{i}}(S) := \delta m_{\text{i}ij} + \delta\bar{m}^{ij}_{\text{i}k}S^k + \delta^{ij}C_kS^k$. with two different kinds of atoms and null result:

$$\alpha, b \leq 10^{-7}\,. \tag{34}$$

With the very precise results from Hughes–Drever experiments (see below) on $\delta\bar{m}^{ij}_{\text{i}k}$ and C_k we get estimates for the anomalous gravitational mass tensor only. If the corresponding experiment will be carried through, this result will be three orders of magnitude better than current results from redshift experiments, Godone et al. (1995).

Despite the fact that these estimates are quite encouraging and in favor to do these experiments, even better results can be expected if one uses Bose condensates or atomic lasers Bordé (1995), Spreeuw, Pfau, Janicke, and Wilkens (1995).

3.2 Hughes–Drever type experiments

In this type of experiments one measures possible anomalous shifts of transitions frequencies between atomic energy levels. Since these energy levels are calculated from (25) they are in general dependent of the anomalous couplings.

The typical Hughes–Drever experiments deal with nuclear transitions where the atomic nucleus consists in a $J=0$ core and a valence proton with angular momentum $L=1$ and spin $S=\frac{1}{2}$. That means, the nucleus possesses the states $|J, M_J\rangle$ with $J=3/2$. Using these states and the usual approximation scheme we get for the transition frequencies

$$\begin{aligned} E(\tfrac{3}{2},\tfrac{3}{2}) - E(\tfrac{3}{2},\tfrac{1}{2}) &= \delta + \bar{\delta}_1 + \bar{\delta}_2 \\ E(\tfrac{3}{2},\tfrac{1}{2}) - E(\tfrac{3}{2},-\tfrac{1}{2}) &= \bar{\delta}_2 \\ E(\tfrac{3}{2},-\tfrac{1}{2}) - E(\tfrac{3}{2},-\tfrac{3}{2}) &= -\delta + \bar{\delta}_1 + \bar{\delta}_2 \end{aligned}$$

with

$$\begin{aligned} \delta &:= -\frac{\hbar^2}{3a^2}\left(\frac{\delta m^{xx}_{\text{i}}}{m^2} + \frac{\delta m^{yy}_{\text{i}}}{m^2} - 2\frac{\delta m^{zz}_{\text{i}}}{m^2}\right) \\ \bar{\delta}_1 &:= -\frac{\hbar^2}{a^2}\left(\frac{\delta\bar{m}^{xx}_{\text{i}z}}{m^2} + \frac{\delta\bar{m}^{yy}_{\text{i}z}}{m^2} - 2\frac{\delta\bar{m}^{zz}_{\text{i}z}}{m^2}\right) \\ \bar{\delta}_2 &:= -\frac{5\hbar^2}{3a^2}\frac{\delta\bar{m}^{zz}_{\text{i}z}}{m^2} + \frac{2}{3}(mc^2B_z + mC_zU) + \frac{\hbar}{c}T_z \end{aligned}$$

V_{nucl} = harmonic oscillator potential, a = radius of nucleus.

If we use the fact that current experiments showed no effect of splitting of the energy levels nor any shifts of it, then we get from the accuracy of the experiments, Chupp et al. (1989) (see also Prestage et al. (1985) and Lamoreaux et al. (1986)) which is given by $\delta E < 0.45 \cdot 10^{-6}$ Hz, the following estimates:

$$\left|\frac{\delta \bar{m}^{xx}_{\mathrm{iz}}}{m}\right| \leq 5 \cdot 10^{-30}, \quad \left|\frac{\delta m^{xx}_{\mathrm{i}}}{m}\right| \leq 5 \cdot 10^{-30}, \quad |B_z| \leq 3 \cdot 10^{-31}, \\ |C_z| \leq 3 \cdot 10^{-24}, \quad |T_z| \leq 1.5 \cdot 10^{-15}\ \mathrm{m}^{-1} \tag{35}$$

Here we used for radius of the atomic nucleus $a = 1.5$ fm and the gravitational potential $U/c^2 \approx 10^{-7}$ of our galaxy. The estimate on $|T_z|$ can be interpreted as representing the up to now most stringent estimate on a hypothetical axial torsion vector, Lämmerzahl (1997a).

3.3 $g-2$ experiments

In $g-2$ experiments one measures the projection of the polarization of the electrons in direction of their motion, that is, $\boldsymbol{S} \cdot \boldsymbol{n}$ where $\boldsymbol{n}$ unit vector in direction of the velocity, see e.g. Kinoshita (1990). These experiments are high precision experiments which are designed to test the anomalous g factor which is calculated from QED. This experiment can be described using the Pauli equation with relativistic corrections. However, we neglect these correction for simplicity.

The dynamics of the polarization $\boldsymbol{S}$ and the velocity $\boldsymbol{v}$ is easily calculated from the Heisenberg equations of motion with the Hamiltonian given by the GPE. The result is in the case of vanishing gravitational and electrostatic potential, $U = 0$ and $\phi = 0$,

$$a^i = \left(\delta^{ij} + \frac{\delta m^{ij}_{\mathrm{i}}}{m} + \frac{\delta \bar{m}^{ij}_{\mathrm{ik}}}{m} S^k\right) \frac{e}{m} v^l \epsilon_{ljm} H^m + \left(\frac{\delta \bar{m}^{ij}_{\mathrm{ik}}}{m} v_j - c\left(\frac{1}{m}\lambda^i_k + c\Lambda^i_k\right)\right) \frac{e}{2mc\hbar} H_l \epsilon^{lkp} S^p \tag{36}$$

The general structure of these dynamical equations is

$$\partial_t v^i = \overset{av}{\Omega}{}^i_j v^j + \overset{as}{\Omega}{}^i_j S^j + \overset{avs}{\Omega}{}^i_{jk} v^j S^k + A^i \tag{37}$$

$$\partial_t S^i = \overset{ss}{\Omega}{}^i_j S^j + \overset{svs}{\Omega}{}^i_{jk} v^j S^k + \overset{svvs}{\Omega}{}^i_{jkl} v^j v^k S^l \tag{38}$$

with

$$\overset{av}{\Omega}{}^i_j = \left(\delta^{ik} + \frac{\delta m^{ik}_{\mathrm{i}}}{m}\right) \frac{e}{m} \epsilon_{jkm} H^m \tag{39}$$

$$\overset{as}{\Omega}{}^i_j = -\frac{\delta \bar{m}^{ik}_{\mathrm{1j}}}{m}\left(\frac{e}{m}\partial_k \phi^M + \partial_k U\right) - \delta^{ik} C_j \partial_k U - c\left(\frac{1}{m}\lambda^i_k + c\Lambda^i_k\right) \frac{e}{2m} H_k(x) \epsilon^{kmj} \tag{40}$$

$$\overset{avs}{\Omega}{}^i_{jk} = \frac{\delta \bar{m}^{il}_{\mathrm{ik}}}{m} \frac{e}{m} \epsilon_{jlm} H^m \tag{41}$$

$$\overset{ss}{\Omega}{}^i_j = \epsilon^{ijk} \frac{e}{mc} H_k(x) + 2\epsilon^{ijk}\left(\frac{e}{2mc} H_l K^l_k + mc^2 B_k + cT_k + C_k m U(x)\right) \tag{42}$$

$$\overset{svs}{\Omega}{}^i_{jk} = \epsilon^{ikl}\left(a^j_l + cmA^j_l\right) \tag{43}$$

$$\overset{svvs}{\Omega}{}^i_{jkl} = \epsilon^{ilm} \frac{1}{2} \delta \bar{m}^{jk}_{\mathrm{im}} \tag{44}$$

where $\overset{ss}{\Omega}$, $\overset{svs}{\Omega}{}^i_{jk}$, and $\overset{svvs}{\Omega}{}^i_{jkl}$ are antisymmetric in the first and last indices. Only $\overset{av}{\Omega}{}^i_j$ and $\overset{ss}{\Omega}{}^i_j$ consist in non-anomalous terms. In this case they are proportional. If one takes QED into account one has to replace

$$\overset{ss}{\Omega}{}^i_j = \epsilon^{ijk} g \frac{e}{mc} H_k(x) + 2\epsilon^{ijk} \left(\frac{e}{2mc} H_l K^l_k + mc^2 B_k + cT_k + C_k m U(x) \right) \tag{45}$$

where g is the anomalous g–factor.

Because the observed quantity is $\boldsymbol{S} \cdot \boldsymbol{n}$ it is useful to split $\boldsymbol{S} = (\boldsymbol{S} \cdot \boldsymbol{n})\boldsymbol{n} + \boldsymbol{S}_\perp$ and to obtain from (38) the dynamics of $(\boldsymbol{S} \cdot \boldsymbol{n})$ and $\boldsymbol{S}_\perp$ separately. If no anomalous interactions are present, one can eliminate the dynamics of $\boldsymbol{S}_\perp$ and gets as remaining equation of motion $\frac{d^2}{dt^2}(\boldsymbol{S} \cdot \boldsymbol{n}) = -\omega^2(\boldsymbol{S} \cdot \boldsymbol{n})$ with $\omega^2 = \omega_i \omega_i$ and $\omega_i = (g-2)\frac{e}{2mc} H_i$. By observation of $(\boldsymbol{S} \cdot \boldsymbol{n})$ one gets information of the anomalous g–term.

Also in the general case with anomalous couplings one gets from (37) and (38) the equations of motion for the observed quantity $(\boldsymbol{S} \cdot \boldsymbol{n})$. However, the precession frequency is different from that without anomalous couplings. The difference consists just in the anomalous terms so that from the accuracy of this type of experiment one gets and from the fact that no such anomalous terms have been observed, one gets the following (rough) estimates:

$$|K^{kl}_i| \leq 10^{-11}, \quad \left| \frac{\delta m^{ij}_{\rm I}}{m} \right| \leq 10^{-11}, \quad \frac{\delta \bar{m}^{kl}_{\rm Ii}}{m} \leq 2 \cdot 10^{-32} \frac{1}{(v/c)^2},$$
$$a^j_i \leq 3 \cdot 10^{-20} \frac{1}{v/c} {\rm m}^{-1}, \quad A^j_i \leq 10^{-32} \frac{1}{v/c}, \quad B^i \leq 10^{-32}, \quad T^i \leq 3 \cdot 10^{-20} {\rm m}^{-1}$$

which are in some cases better than those obtained from the Hughes–Drever experiments.

3.4 Quasiclassical tests of EEP

Here we want to indicate a description of tests which can be carried through with classical bulk matter with possible net spin. In this case we can use the quasi–classical limit of the GPE or eqn (21).

One main device for searching for anomalous couplings is the torsion pendulum. This device can be described with the help of the Hamilton function which one gets as classical limit from the GPE. One has to add to this Hamilton function the oscillator potential of the pendulum which comes from the fiber torsion constant. Elimination of the momentum in favor of the velocity and introducing the angular velocity gives a Hamilton function or a Lagrange function in terms of the angular momentum and the angle θ around the z–axis which is the quantity which is observed in this kind of experiment. From this it is standard to derive the equation of motion for the angle θ. The driving force now comes from the fact that in the laboratory system (that is, the rotating earth) the z–axis is fixed while the external fields (which are constant in another frame) have a sidereal time periodicity. This periodicity acts as driving force on this pendulum so that it can be observed from the dynamics of the angle θ. In the case that θ remains at the same initial position, one can infer from the accuracy of the apparatus estimates on all hypothetical interactions[8].

3.4.1 Tests of the WEP

The usual classical tests of the WEP use unpolarized bulk matter and detect a possible different force on two different masses by means of a torsion balance. The tested quantity in this case is $\delta m_{gij} U^{ij}(\boldsymbol{x})$ where $U^{ij}(\boldsymbol{x})$ is due to the gravitational field of either the sun or the earth (see Table 3).

Quasiclassical tests which employs quantum matter in a classical limit are the folowing tests which observe the free fall of quantum particles in the gravitational field of the earth: the velocity of freely falling atoms, Estermann, Simpson, and Stern (1947), the wavelength of freely falling neutrons, Koester (1967, 1976), and the time of flight of electrons, Witteborn and Fairbank (1967). These experiments measure (21) but with no good accuracy.

[8] A more explicit treatment will be given elsewhere.

3.4.2 Tests of anomalous spin–couplings

In these tests one uses polarized bulk matter as test masses. Due to the temporal variation of the fields coupling to the spin there will be a time-dependent torque acting on the pendulum. Again, if one do not observe any sidereal dependence of the angle θ then the accuracy of the experiments gives upper limits on the corresponding fields[7]. There are two cases which have to be distinguished: First, one looks whether there is any anomalous spin coupling at all, and second, one can use some model (like torsion theory, where spin creates torsion which in turn again acts on spin) and look for spin–spin couplings. The latter version is important if the range of the corresponding interaction is assumed to be small. This type of experiments have been listed in Table 2.

4 Concluding remarks

In this article we presented a new test theory for the description of experiments designed to test structures of the underlying space–time. This test theory uses only quantum notions and is not derived from the quantization of a classical theory. This is important because, as one can see from the classical limit, in classical physics there may be be less possibilities to violate basic principles than in quantum mechanics.

Another important point is that one has to perform test of the space–time structure for all possible kinds of matter. For example, even in the case that the electromagnetic propagation phenomena prove to be isotropic it may be that some spinorial matter possesses anisotropies.

In the case of bound systems it may happen that anomalous terms may be due to the matter field or due to the anomalous behaviour of the Maxwell field (what we neglected in the present tretament). On the one hand, in our present treatment the appearance of anomalous inertial mass tensors is solely due to the fact that the generalized Dirac matrices do not fulfill a Clifford algebra and thus lead to a splitting of the corresponding null cone. On the other hand, calculations within the $TH\epsilon\mu$ formalism where there is no splitting of the null cone for matter fields, also lead to anomalous inertial mass tensors, Will (1993). Therefore it will be a theoretical task to suggest experiments which can distinguish between the various sources for effective anomalous couplings.

Acknowledgments

I thank J. Audretsch, Ch. Bordé, R. Ritter and St. Schiller for discussions and especially R. Ritter for a valuable guide to the literature. Also travel support of the "Optik-Zentrum" of the University of Konstanz is acknowledged.

List of abbreviations

EEP = Einstein's equivalence principle
GDE = generalized Dirac equation (eqn (14))
GPE = generalized Pauli equation (eqn (20))
GR = General Relativity
WEP = weak equivalence principle
LLI = local Lorentz invariance
LPI = local position invariance

References

Aleksandrov, E., Ansel'm, A., Pavlov, Y., & Umarkhodzhaev, R. (1983). A restriction on the existence of a new type of fundamental interaction (the "arion" long-range interaction) in an experiment on spin precession of mercury nuclei. *Sov. Phys. JETP*, *58*, 1103.

Alvarez, C., & Mann, R. (1995a). The equivalence principle and anomalous magnetic moment experiments. *Phys. Rev., D 54*, 7097.

Alvarez, C., & Mann, R. (1995b). Testing the equivalence principle by lamb shift energies. *Phys. Rev., D 54*, 5954.

Alvarez, C., & Mann, R. (1996a). Testing the equivalence principle in the quantum regime. *preprint*, Honorable mention in the Gravity Research Foundation Essay Contest.

Alvarez, C., & Mann, R. (1996b). Testing the equivalence principle using atomic vacuum energy shifts. *Mod. Phys. Lett., A 11*, 1757.

Alvarez, C., & Mann, R. (1997). The equivalence principle in the non–baryonic regime. *Phys. Rev., D 55*, 1732.

Andrade, L. Garcia de, Oguri, V., Lopes, M., & Hammond, R. (1992). On the energy splitting of spacetral lines induced by the torsion field. *Nuovo Cim., B 107*, 1167.

Ansel'm, A., & Neronov, Y. (1985). Restriction on the existence of spin–spin coupling of nonelectromagnetic origin in experiments on the measurement of the gyromagnetic ratios of the proton and deuteron. *Sov. Phys. JETP, 61*, 1154.

Audretsch, J., Bleyer, U., & Lämmerzahl, C. (1993). Testing Lorentz invariance with atomic beam interferometry. *Phys. Rev., A 47*, 4632.

Audretsch, J., Hehl, F., & Lämmerzahl, C. (1992). Matter wave interferometry and why quantum objects are fundamental for establishing a gravitational theory. In J. Ehlers & G. Schäfer (Eds.), *Relativistic gravity research. with emphasis on experiments and observations* (Vol. Lecture Notes in Physics **410**, p. 368). Berlin: Springer–Verlag.

Bagrov, V., Buchbinder, I., & Shapiro, I. (1992). On the possible experimental manifestations of the torsion field at low energies. *Sov. J. Phys., 35*, 5.

Barr, S., & Mohapatra, R. (1986). Range of feeble forces from higher dimensions. *Phys. Rev. Lett., 57*, 3129.

Bars, I., & Visser, M. (1986). Feeble intermediate–range forces from higher dimensions. *Phys. Rev. Lett., 57*, 25.

Barton, G., & Scharnhorst, K. (1993). QED between parallel mirrors – light signals faster than light, or amplification by vacuum. *J. Phys. A: Math. Gen., 26*, 2037.

Berglund, C., Hunter, L., Krause, D., Prigge, E., & Ronfeldt, M. (1995). New limits on local lorentz invariance from hg and cs magnetometers. *Phys. Rev. Lett., 75*, 1879.

Bessel, F. (1832). Versuche über die kraft, mit welcher die erde körper von verschiedener beschaffenheit anzieht. *Ann. Physik und Chemie, 25*, 401.

Bobrakov, V., Borisov, Y., Lasakov, M., Serebov, A., Tal'daev, R., & Trofimova, A. (1991). An experimental limit on the existence of the electron quasimagnetic (arion) interaction. *JETP Lett., 53*, 294.

Braginsky, V., & Panov, V. (1971). Verification of the equivalence of inertial and gravitational mass. *Sov. Phys. JETP, 34*, 464.

Brans, C., & Dicke, R. (1961). Mach's principle and a relativistic theory of gravitation. *Phys. Rev., 124*, 925.

Brillet, A., & Hall, J. (1979). Improved laser test of the isotropy of space. *Phys. Rev. Lett., 42*, 549.

Ch.J., B. (1995). Amplification of atomic fields by stimulated emission of atoms. *Phys. Lett., A 205*, 217.

Chu, S., & Kasevich, M. (1992). Measurement of the gravitational acceleration of an atom with a light-pulse atom interferometer. *Appl. Phys., B 54*, 321.

Chui, T., & Ni, W.-T. (1993). Experimental search for an anomalous spin–spin interaction between electrons. *Phys. Rev. Lett., 71*, 3247.

Chupp, T., Hoara, R., Loveman, R., Oteiza, E., Richardson, J., & Wagshul, M. (1989). Results of a new test of local Lorentz invariance: A search for mass anisotropy in ^{21}Ne. *Phys. Rev. Lett., 63*, 1541.

Cocconi, G., & Salpeter, E. (1958). A search for anisotropy of inertia. *Nuovo Cim., X*, 646.

Cocconi, G., & Salpeter, E. (1960). Upper limit for the anisotropy of inertia from the Mössbauer effect. *Phys. Rev. Lett., 4*, 176.

Courant, R., & Hilbert, D. (1962). *Methods of mathematical physics*. New York: Interscience Publishers.

DeSabbata, V., & Gasperini, M. (1981). Gauge invariance, semiminimal coupling, and propagatimg torsion. *Phys. Rev., D 23*, 2116.

Dimopoulos, S., & Giudice, G. (1996). Macroscopic forces from supersymmetry. *Phys. Lett., B 379*, 105.

Drever, R. (1961). A search for the anisotropy of inertial mass using a free precession technique. *Phil. Mag., 6*, 683.

Eötvös, R., Pekár, D., & Fekete, E. (1922). Beiträge zum Gesetz der Proportionalität von Trägheit und Gravität. *Ann. Physik (Leipzig), 68*, 11.

Estermann, I., Simpson, O., & Stern, O. (1947). The free fall of atoms and the measurement of the velocity distribution in a molecular beam of Caesium atoms. *Phys. Rev.*, *71*, 238.

Fischbach, E., Haugan, M., Tadić, D., & Cheng, H.-Y. (1985). Lorentz noninvariance and the Eötvös experiment. *Phys. Rev.*, *D 32*, 154.

Gabriel, M., & Haugan, M. (1990). Testing the Einstein Equivalence Principle: Atomic clocks and local Lorentz invariance. *Phys. Rev.*, *D 41*, 2943.

Godone, A., Novero, C., & Tavella, P. (1995). Null gravitational redshift experiments with nonidentical atomic clocks. *Phys. Rev.*, *D 51*, 319.

Golestanian, R., & Kardar, M. (1997). Mechanical response of vacuum. *Phys. Rev. Lett.*, *78*, 3421.

Graham, D., & Newman, R. (1985). . In M. MacCallum (Ed.), *Proceedings of the 11th international conference on general relativity and gravitation, stockholm* (p. 614). New York: Cambridge University Press.

Gundlach, J., Smith, G., Adelberger, E., Heckel, B., & Swanson, H. (1997). Short-range test of the equivalence principle. *Phys. Rev. Lett.*, *78*, 2523.

Hammond, R. (1995). Upper limit on the torsion coupling constant. *Phys. Rev.*, *52*, 6918.

Hari Dass, N. (1976). Test for C, P, and T nonconservation in gravitaton. *Phys. Rev. Lett.*, *36*, 393.

Hari Dass, N. (1977). Experimental tests for some quantum effects in gravitation. *Ann. Physics (N.Y.)*, *107*, 337.

Haugan, M. (1979). Energy conservation and the principle of equivalence. *Ann. Phys.*, *118*, 156.

Hayashi, K., & Shirafuji, T. (1979). New general relativity. *Phys. Rev.*, *19*, 3524.

Hehl, F., Heyde, P. von der, & Kerlick, G. (1976). General relativity with spin and torsion: Foundations and prospects. *Rev. Mod. Phys.*, *48*, 393.

Hehl, F., Lemke, J., & Mielke, E. (1991). Two lectures on fermions and gravity. In J. Debrus & A. Hirshfeld (Eds.), *Geometry and theoretical phyiscs* (p. XXX). Berlin: Springer–Verlag.

Hils, D., & Hall, J. (1990). Improved Kennedy–Thorndyke experiment to test special relativity. *Phys. Rev. Lett.*, *64*, 1697.

Horvath, J., Logiudice, E., Riveros, C., & Vucetich, H. (1988). Einstein equivalence principle and theories of gravitation: A gravitationally modified standard model. *Phys. Rev.*, *D 38*, 1754.

Hsieh, C.-H., Jen, P.-Y., Ko, K.-L., Li, K.-Y., Ni, W.-T., Pan, S.-S., Shih, Y.-H., & Tyan, R.-J. (1989). The equivalence principle experiment for spin–poilarized bodies. *Mod. Phys. Lett.*, *4*, 1597.

Hughes, V., Robinson, H., & Beltran-Lopez, V. (1960). Upper limit for the anisotropy of inertial mass from nuclear resonance experiments. *Phys. Rev. Lett.*, *4*, 342.

Ives, H., & Stillwell, G. (1938). An experimental study of the rate of a moving clock, I. *J. Opt. Soc. Am.*, *28*, 215.

Ives, H., & Stillwell, G. (1941). An experimental study of the rate of a moving clock, II. *J. Opt. Soc. Am.*, *31*, 369.

Jaseja, T., Javan, A., Murray, J., & Townes, C. (1964). Test of special relativity or of the isotropy of space by use of infrared masers. *Phys. Rev.*, *133*, A 1221.

Joos, G. (1930). Die Jenaer Wiederholung des Michelsonversuchs. *Ann. Physik (Leipzig)*, *7*, 385.

Kasevich, M., & Chu, S. (1991). Atomic interferometry using stimulated Raman transitions. *Phys. Rev. Lett.*, *67*, 181.

Kennedy, R., & Thorndyke, E. (1932). Experimental establishment of the relativity of time. *Phys. Rev.*, *42*, 400.

Kim, J. (1987). Light pseudoscalar, particle physics and cosmology. *Phys. Rep.*, *150*, 1.

Kinoshita, T. (1990). *Quantum electrodynamics*. Singapore: World Scientific.

Koester, L. (1967). Absolutmessung der kohärenten Streulängen von Wasserstoff, Kohlenstoff und Chlor sowie Bestimmung der Schwerebeschleunigung für freie Neutronen mit dem Schwerkraft–Refraktometer am FRM. *Z. Physik*, *198*, 187.

Koester, L. (1976). Verification of the equivalence principle of gravitational and inertial mass for the neutron. *Phys. Rev.*, *D 14*, 907.

Kolb, E., & Turner, M. (1990). *The early universe*. Redwood City: Addison-Wesley.

Kuroda, K., & Mio, N. (1989). Test of composition–dependent force by a free–fall interferometer. *Phys. Rev. Lett.*, *62*, 1941.

Kuroda, K., & Mio, N. (1990). Limits on a possible composition–dependent force by a Galilean experiment. *Phys. Rev.*, *42*, 3903.

Lämmerzahl, C. (1990). The geometry of matter fields. In deSabbata V. & A. J. (Eds.), *Quantum mechanics in curved space–time, NATO ASI series, series B: Physics* (Vol. 230, p. 23). New York: Plenum Press.

Lämmerzahl, C. (1994). Relativistic treatment of the Raman light pulse atom beam interferometer with applications in gravity theory. *J. Physique II (France)*, *4*, 2089.

Lämmerzahl, C. (1996a). Atom beam interferometry in gravitational fields. In A. Macias, T. Matos, O. Obregon, & H. Quevedo (Eds.), *Recent developments in gravitation and mathematical physics, Proceedings of the 1st Mexican school* (p. 223). Singapore: World Scientific.

Lämmerzahl, C. (1996b). Quantum tests of foundations of general relativity. *preprint*, Univ. of Konstanz.

Lämmerzahl, C. (1997a). Constraints on space–time torsion from Hughes–Drever experiments. *Phys. Lett. A*, *A 228*, 223.

Lämmerzahl, C. (1997b). A new quantum test theory for gravitational fields. In A. Garcias, C. Lämmerzahl, A. Macias, M. T., & D. Nuñes (Eds.), *Recent developments in gravitation and mathematical physics, Proceedings of the 2nd Mexican school* (p. http://kaluza.physik.uni-konstanz.de/SNP/Books.html). Konstanz: Science Network Publishing.

Lämmerzahl, C., & Bleyer, U. (1997). A quantum test theory for basic principles of special and general relativity with applications to atom beam interferometry and Hughes–Drever type experiments. *preprint*, Univ. of Konstanz.

Lämmerzahl, C., Puntigam, R., & Hehl, F. (1997). Reasons for the electromagnetic field to obey Maxwell's equations. *preprint*, *University of Cologne.*

Lamoreaux, S., Jacobs, J., Heckel, B., Raab, F., & Fortson, E. (1986). New limits on spatial anisotropy from optically pumped ^{201}Hg and ^{199}Hg. *Phys. Rev. Lett.*, *57*, 3125.

Lamoreaux, S., Jacobs, J., Heckel, B., Raab, F., & Fortson, E. (1989). Optical pumping technique for measuring small nuclear quadrupole shifts in $^1S_0$ atoms and testing spatial isotropy. *Phys. Rev.*, *A 39*, 1082.

Lamoreaux, S., Jacobs, J., Heckel, B., Raab, F., & Fortson, E. (1990). Erratum: Optical pumping technique for measuring small nuclear quadrupole shifts in $^1S_0$ atoms and testing spatial isotropy. *Phys. Rev.*, *A 42*, 3763.

Lazarides, G., Panagiotakopoulos, C., & Shafi, Q. (1986). Phenomenology and cosmology with superstrings. *Phys. Rev. Lett.*, *56*, 432.

Leitner, J., & Okubo, S. (1964). Parity, charge conjugation, and time reversal in the gravitational interaction. *Phys. Rev.*, *136*, B 1542.

Lightman, A., & Lee, D. (1973). Restricted proof that the weak equivalence principle implies the Einstein equivalence principle. *Phys. Rev.*, *D 8*, 364.

Mansouri, R., & Sexl, R. (1977a). A test theory of special relativity: I. simultaneity and clock synchronisation. *Gen. Rel. Grav.*, *8*, 497.

Mansouri, R., & Sexl, R. (1977b). A test theory of special relativity: Ii. first order tests. *Gen. Rel. Grav.*, *8*, 515.

Mansouri, R., & Sexl, R. (1977c). A test theory of special relativity: Iii. second order tests. *Gen. Rel. Grav.*, *8*, 809.

Michelson, A., & Morley, E. (1887). . *Am. J. Sci.*, *34*, 333.

Misner, C., Thorne, K., & Wheeler, J. (1973). *Gravitation*. San Francisco: Freeman.

Moody, J., & Wilczek, F. (1984). New macroscopic forces? *Phys. Rev.*, *D 30*, 130.

Ni, W.-T. (1977). Equivalence principles and electromagnetism. *Phys. Rev. Lett.*, *38*, 301.

Ni, W.-T., Chui, T., Pan, S.-S., & Cheng, B.-Y. (1994). Search for anomalous spin–spin interaction between electrons using a DC SQUID. *Physica*, *B 194*, 153.

Niebauer, T., McHugh, M., & Faller, J. (1987). Galileian test for the fifth force. *Phys. Rev. Lett.*, *59*, 609.

Nielsen, H., & Picek, I. (1983). Lorentz non–invariance. *Nucl. Phys.*, *B 211*, 269.

Nieuwenhuizen, P. van. (1982). Supergravity. *Phys. Rep.*, *68*, 189.

Nodland, B., & Ralston, J. (1997). Xxx. *Phys. Rev. Lett.*, *78*, 3043.

Obukhov, Y., Korotky, V., & Hehl, F. (1997). On the rotation of the universe. *preprint*, *University of Cologne.*

Pan, S.-S., Ni, W.-T., & Chen, S.-C. (1992). Experimental search for anomalous spin–spin interactions. *Mod. Phys. Lett.*, *7*, 1287.

Peres, A. (1978). Test of the equivalence principle with spin. *Phys. Rev.*, *D 18*, 2739.

Phillips, P. (1987). Test of spatial isotropy using a cyrogenic torsion pendulum. *Phys. Rev. Lett.*, *59*, 1784.

Phillips, P. R., & Woolum, D. (1969). A test of Lorentz invariance using a torsion pendulum. *Nuov. Cim.*, *64B*, 28.

Prestage, J., Bollinger, J., Itano, W., & Wineland, D. (1985). Limits for spatial anisotropy by use of nuclear–spin–polarized ^{9}be$^+$ ions. *Phys. Rev. Lett.*, *54*, 2387.

Puntigam, R., Lämmerzahl, C., & Hehl, F. (1997). Maxwell's theory on a post-riemannian spacetime and the equivalence principle. *Class. Qaunt. Grav.t*, *14*, 1347.

Ramsey, N. (1979). The tensor force between two protons at long range. *Physica*, *A 96*, 285.

Ritter, R., Goldblum, C., Ni, W.-T., Gillies, G., & Speake, C. (1990). Experimental test of equivalence principle with polarized masses. *Phys. Rev.*, *D 42*, 977.

Ritter, R., Winkler, L., & Gillies, G. (1993). Search for anomalous spin-dependent forces with a polarized-mass torsion pendulum. *Phys. Rev. Lett.*, *70*, 701.

Robertson, H. (1949). Postulates versus observations in the special theory of relativity. *Rev. Mod. Phys.*, *21*, 378.

Roll, P., Krotkov, R., & Dicke, R. (1964). The equivalence of inertial and passive gravitational mass. *Ann. Phys. (N.Y.)*, *26*, 442.

Romaides, A., Sands, R., Eckhardt, D., Fischbach, E., Talmadge, C., & Kloor, H. (1994). Second tower experiment: Further evidence for Newtonian gravity. *Phys. Rev.*, *D 50*, 3608.

Scharnhorst, K. (1990). On the propagation of light in the vacuum between plates. *Phys. Lett.*, *B 236*, 354.

Spreeuw, R., Pfau, T., Janicke, U., & Wilkens, M. (1995). Laser-like scheme for atomic-matter waves. *Europhys. Lett.*, *32*, 469.

Su, Y., Heckel, B., Adelberger, E., Gundlach, J., Harris, M., Smith, G., & Swanson, H. (1994). New test of the universality of free fall. *Phys. Rev.*, *D 50*, 3614.

Trautman, A. (1972). On the Einstein-Cartan equations I. *Bull. Acad. Pol. Sci. Ser. Sci. Math. Astr. Phys.*, *20*, 185.

Turneaure, J., Will, C., Farrel, B., Mattison, E., & Vessot, R. (1983). Test of the principle of equivalence by a null gravitational red-shift experiment. *Phys. Rev.*, *27*, 1705.

Velyukhov, G. (1968). Search for the gravitational moment of the proton. *Sov. Phys. JETP Lett.*, *8*, 229.

Venema, B., Majumder, P., Lamoreaux, S., Heckel, B., & Fortson, E. (1992). Search for a coupling of the earth's gravitational field to nuclear spin in atomic mercury. *Phys. Rev. Lett.*, *68*, 135.

Vessot, R., & Levine, M. (1979). A test of the equivalence principle using a space-borne clock. *Gen. Rel. Grav.*, *10*, 181.

Vessot, R., Levine, M., Mattison, E., Blomberg, E., Hoffmann, T., Nystrom, G., Farrel, B., Decher, R., Eby, P., Baughter, C., Watts, J., Teuber, D., & Wills, F. (1980). Test of relativistic gravitation with a space-borne hydrogen maser. *Phys. Rev. Lett.*, *45*, 2081.

Vorobyov, P., & Gitarts, Y. (1988). A new limit on the arion interaction constant. *Phys. Lett.*, *B 208*, 146.

Weiss, D., Young, B., & Chu, S. (1994). Precision measurement of $\hbar/m_{Cs}$ based on photon recoil using laser-cooled atoms and atomic interferometry. *Appl. Phys.*, *B 59*, 217.

Will, C. (1974). Gravitational red-shift measurements as tests of nonmetric theories of gravity. *Phys. Rev.*, *D 10*, 2330.

Will, C. (1993). *Theory and experiment in gravitational physics (revised edition)*. Cambridge: Cambridge University Press.

Wineland, D., Bollinger, J., Heinzen, D., Itano, W., & Raizen, M. (1991). Search for anomalous spin-deoendent forces using stored-ion spectroscopy. *Phys. Rev. Lett.*, *67*, 1735.

Wineland, D., & Ramsey, N. (1972). Atomic deuterium maser. *Phys. Rev.*, *A 5*, 821.

Witteborn, F., & Fairbank, W. (1967). Experimental comparison of the graviational force on freely falling electrons and metallic electrons. *Phys. Rev. Lett.*, *19*, 1049.

Yasskin, P., & Stoeger, W. (1980). Propagation equations for test bodies with spin and rotation in theories of gravity with torsion. *Phys. Rev.*, *D 21*, 2081.

Youdin, A., Krause, D., Jagannathan, K., & Hunter, L. (1996). Limits on spin-mass couplings within the axion window. *Phys. Rev. Lett.*, *77*, 2170.

Young, B. (1969). Search for a gravity shift in the proton Larmor frequency. *Phys. Rev. Lett.*, *22*, 1445.

INTEGRABILITY OF D-DIMENSIONAL COSMOLOGICAL MODELS

V. R. Gavrilov, V.N. Melnikov
Center for Gravitation and Fundamental Metrology, VNIIMS,
3-1 Ulyanova St., Moscow, 117313, Russia
e-mail: melnikov@fund.phys.msu.su

Abstract

The D-dimensional cosmological model on the manifold $M = R \times M_1 \times M_2 \times \ldots \times M_n$ describing the evolution of Einsteinian factor spaces M_i in the presence of multicomponent perfect fluid source is considered. The barotropic equation of state for mass-energy densities and the pressures of the components is assumed in each space. Integrable cases for pseudo-Euclidean Toda-like systems are reviewed. When the number of the non Ricci-flat factor spaces and the number of the perfect fluid components are both equal to 2, the Einstein equations for the model are reduced to the generalized Emden-Fowler (second-order ordinary differential) equation, which has been recently investigated by Zaitsev and Polyanin within discrete-group analysis. Using the integrable classes of this equation one generates new integrable cosmological models. The corresponding metrics are presented. The method is demonstrated for the special model with Ricci-flat spaces M_1, M_2 and the 2-component perfect fluid source.

PACS numbers: 04.20.J, 04.60.+n, 03.65.Ge

1 Introduction

Following the purpose to study the early universe we develop the multidimensional generalization [8]-[13], [16]-[21],[25],[26],[28] of the standard Friedman-Robertson-Walker world model. If the extra dimensions of the space-time manifold really exist, the unique conceivable site, where they might become dynamically important, seems to be possible. This is some early stage of the evolution. Usually within multidimensional cosmology (see, for instance, [1]-[2], [5]-[13],[15]-[26],[28],[31]-[36] and references therein) it is assumed the occurrence of the topological partition for the multidimensional space-time on the external 3-dimensional space and additional so called internal space (or spaces) due to the quantum processes at the beginning of this stage. In correspondence with such partition the space-time acquires the topology $M = R \times M_1 \times \ldots \times M_n$, where R is the time axis, one part of the manifolds $M_1, \ldots, M_n$ is interpreted as a 3-dimensional external space and the other part stands for internal spaces. Usually the internal spaces are compact, however the models with noncompact internal spaces are also discussed [14],[20], [29] [30]. The subsequent evolution of the multidimensional Universe is considered as classical admitting the description by means of the multidimensional Einstein equations. Achieving the integrability of these equations is the main goal of our investigation here. As the present world seems to be 4-dimensional, there is the assumption that the internal space(s) had contracted to extremely small sizes, which are inaccessible for experiment. This contraction accompanied by the expansion of the external space is described by some models (the first model of such type has been found in [6]) within multidimensional cosmology and is called the dynamical compactification.

We consider a mixture of several perfect fluid components as a source for the multidimensional Einstein equations. Such multicomponent systems are usually employed in 4-dimensional cosmology and in many cases are adequate types of matter for description some early epochs in the history of the universe [4].

The paper is organized as follows. In section 2 we describe the multidimensional cosmological model and obtain the Einstein equations in the form of the Lagrange-Euler equations following from some Lagrangian. Here we develop the n-dimensional vector formalism for the integrating equations of motion. Concluding Section 2 we present the review of the all known integrable models. In Section 3 we suggest the method for obtaining the new class of the integrable models on the manifold $M = R \times M_1 \times M_2$. The method is based on reducing the Einstein equations to the generalized Emden-Fowler (second-order ordinary differential) equation. The method is useful for any 2-component model on the manifold $M = R \times M_1 \times M_2$ except for the cases admitting the integration by more simple ways. The total number of the model components is equal to the sum of the number of non Ricci-flat spaces with the number of perfect fluid components. Integrable classes recently derived by Zaitsev and Polyanin for the generalized Emden-Fowler equation allow to generate new integrable cosmological models. Their metrics are presented. In Section 3 the method is applied for models with Ricci-flat spaces M_1, M_2 and 2-component perfect fluid.

2 The model and the equations of motion

Within n-factor spaces cosmological model D-dimensional space-time manifold M is considered here as a product of the time axis R and n manifolds $M_1, \ldots, M_n$, i.e.

$$M = R \times M_1 \times \ldots \times M_n, \tag{2.1}$$

The product of one part of the manifolds gives the external 3-dimensional space and the remaining part stands for so called internal spaces. The internal spaces are supposed to be compact. Further, for the sake of generality, we admit that dimensions $N_i = \dim M_i$ for $i = 1, \ldots, n$ are arbitrary.

The manifold M is equipped with the metric

$$g = -e^{2\gamma(t)} dt \otimes dt + \sum_{i=1}^{n} \exp[2x^i(t)] g^{(i)}, \tag{2.2}$$

where $\gamma(t)$ is an arbitrary function determining the time t and $g^{(i)}$ is the metric on the manifold M_i. We suppose that the manifolds $M_1, \ldots, M_n$ are the Einstein spaces, i.e.

$$R_{k_i l_i}[g^{(i)}] = \lambda_i g^{(i)}_{k_i l_i}, \quad k_i, l_i = 1, \ldots, N_i, \quad i = 1, \ldots, n, \tag{2.3}$$

where λ_i is constant. In the special case, when M_i is a space of constant Riemann curvature K_i the constant λ_i reads: $\lambda_i = K_i(N_i - 1)$ (here $N_i > 1$).

Using the assumptions (2.3) we obtain the following non-zero components of the Ricci tensor for the metric (2.2) [19]

$$R^0_0 = e^{-2\gamma} \left(\sum_{i=1}^{n} N_i (\dot{x}^i)^2 + \ddot{\gamma}_0 - \dot{\gamma}\dot{\gamma}_0 \right) \tag{2.4}$$

$$R^{m_i}_{n_i} = \left\{ \lambda_i \exp[-2x^i] + \left[\ddot{x}^i + \dot{x}^i(\dot{\gamma}_0 - \dot{\gamma}) \right] e^{-2\gamma} \right\} \delta^{m_i}_{n_i} \tag{2.5}$$

where we denoted

$$\gamma_0 = \sum_{i=1}^{n} N_i x^i. \tag{2.6}$$

Indices m_i and n_i in (2.4),(2.5) for $i = 1, \ldots, n$ run over from $(D - \sum_{j=i}^{n} N_j)$ to $(D - \sum_{j=i}^{n} N_j + N_i)$ $(D = 1 + \sum_{i=1}^{n} N_i = \dim M)$.

We consider a source of a gravitational field in the form of a multicomponent perfect fluid. The energy-momentum tensor of such source under the comoving observer condition reads

$$T_N^M = \sum_{\mu=1}^{\bar{m}} T_N^{M(\mu)}, \tag{2.7}$$

$$\left(T_N^{M(\mu)}\right) = \operatorname{diag}\left(-\rho^{(\mu)}(t), p_1^{(\mu)}(t)\delta_{l_1}^{k_1}, \ldots, p_n^{(\mu)}(t)\delta_{l_n}^{k_n}\right), \tag{2.8}$$

Furthermore we suppose that for any μ-th component of a perfect fluid the barotropic equation of state holds

$$\left(1 - h_i^{(\mu)}\right)\rho^{(\mu)}(t), \quad \mu = 1, \ldots, \bar{m}, \tag{2.9}$$

where $h_i^{(\mu)} = \text{const}$. It should be noted that each μ-th component admits different barotropic equations of state in different spaces $M_1, \ldots, M_n$. From the physical viewpoint it follows from the separation of internal spaces with respect to the external one and with respect to each other.

One easily shows that the equation of motion $\nabla_M T_0^{M(\mu)} = 0$ for the μ-th component of the perfect fluid described by the tensor (2.8) reads

$$\dot{\rho}^{(\mu)} + \sum_{i=1}^{n} N_i \dot{x}^i (\rho^{(\mu)} + p_i^{(\mu)}) = 0. \tag{2.10}$$

Using the equations of state (2.9), we obtain from (2.10) the following integrals of motion

$$A^{(\mu)} = \rho^{(\mu)} \exp\left[2\gamma_0 - \sum_{i=1}^{n} N_i h_i^{(\mu)} x^i\right] = \text{const}. \tag{2.11}$$

The Einstein equations $R_N^M - R\delta_N^M/2 = \kappa^2 T_N^M$ (κ^2 is the gravitational constant), can be written as $R_N^M = \kappa^2[T_N^M - T\delta_N^M/(D-2)]$. Further we employ the equations $R_0^0 - R/2 = \kappa^2 T_0^0$ and $R_{n_i}^{m_i} = \kappa^2[T_{n_i}^{m_i} - T\delta_{n_i}^{m_i}/(D-2)]$. Using (2.4)-(2.9), we obtain for them

$$\frac{1}{2}\sum_{i,j=1}^{n} G_{ij}\dot{x}^i\dot{x}^j + V = 0, \tag{2.12}$$

$$\begin{aligned} \lambda^i e^{-2x^i} + [\ddot{x}^i + \dot{x}^i(\dot{\gamma}_0 - \dot{\gamma})]e^{-2\gamma} &= -\kappa^2 \sum_{\mu=1}^{\bar{m}} A^{(\mu)} \left(h_i^{(\mu)} - \frac{\sum_{k=1}^{n} N_k h_k^{(\mu)}}{D-2}\right) \\ &\times \exp\left[\sum_{i=1}^{n} N_i h_i^{(\mu)} x^i - 2\gamma_0\right]. \end{aligned} \tag{2.13}$$

Here

$$G_{ij} = N_i\delta_{ij} - N_iN_j \tag{2.14}$$

are the components of the minisuperspace metric,

$$V = e^{2\gamma}\left(-\frac{1}{2}\sum_{i=1}^{n}\lambda^i N_i e^{-2x^i} + \kappa^2\sum_{\mu=1}^{\bar{m}} A^{(\mu)}\exp\left[\sum_{i=1}^{n} N_i h_i^{(\mu)} x^i - 2\gamma_0\right]\right). \tag{2.15}$$

The dependence on densities $\rho^{(\mu)}$ in (2.12),(2.13) has been canceled according to relations (2.11).

It is not difficult to verify that after the gauge fixing $\gamma = F(x^1, \ldots, x^n)$ equations of motion (2.13) may be considered as the Lagrange-Euler equations obtained from the Lagrangian

$$L = e^{\gamma_0 - \gamma}\left(\frac{1}{2}\sum_{i,j=1}^{n} G_{ij}\dot{x}^i\dot{x}^j - V\right) \tag{2.16}$$

under the zero-energy constraint (2.12).

Now we introduce the n-dimensional real vector space R^n. By $e_1, \ldots e_n$ we denote the canonical basis in R^n ($e_1 = (1, 0, \ldots, 0)$ etc.). Hereafter we use the following vectors: the vector we need to obtain

$$x = x^1(t)e_1 + \ldots + x^n(t)e_n, \tag{2.17}$$

the vector induced by the curvature of the space M_k

$$v_k = -\frac{2}{N_k}e_k = \sum_{i=1}^{n}\frac{-2}{N_k}\delta^i_k e_i \tag{2.18}$$

and the vector induced by the μ-th component of the perfect fluid

$$u_\mu = \sum_{i=1}^{n}\left(h_i^{(\mu)} - \frac{\sum_{k=1}^{n} N_k h_k^{(\mu)}}{D-2}\right)e_i. \tag{2.19}$$

Let $< .,. >$ be a symmetrical bilinear form defined on R^n such that

$$< e_i, e_j >= G_{ij}. \tag{2.20}$$

The form is nongenerated and the inverse matrix to (G_{ij}) has the components

$$G^{ij} = \frac{\delta^{ij}}{N_i} + \frac{1}{2-D}. \tag{2.21}$$

The form $< .,. >$ endows the space R^n with the metric, which signature is $(-,+,...,+)$ [17],[18]. By the usual way we may introduce the covariant components of vectors. For the vectors v_k and u_μ we have

$$v^i_{(k)} = -2\frac{\delta^i_k}{N_k}, \quad v^{(k)}_i = \sum_{i=1}^{n} G_{ij}v^j_{(k)} = 2(N_i - \delta^k_i), \tag{2.22}$$

$$u^i_{(\mu)} = h_i^{(\mu)} - \frac{\sum_{k=1}^{n} N_k h_k^{(\mu)}}{D-2}, \quad u^{(\mu)}_i = \sum_{i=1}^{n} G_{ij}u^j_{(\mu)} = N_i h_i^{(\mu)}. \tag{2.23}$$

The values of $< v_k, v_i >$, $< v_k, u_\mu >$ and $< u_\mu, u_\nu >$ are presented in Table 1.

$<.,.>$	v_j	u_ν
v_i	$4(\frac{\delta_{ij}}{N_i} - 1)$	$-2h_i^{(\nu)}$
u_μ	$-2h_j^{(\mu)}$	$\sum_{i=1}^n h_i^{(\mu)} h_i^{(\nu)} N_i + \frac{1}{2-D}[\sum_{i=1}^n h_i^{(\mu)} N_i][\sum_{j=1}^n h_j^{(\nu)} N_j]$

TABLE I. Values of the bilinear form $< .,. >$ for the vectors v_i and u_μ, induced by curvature of the space M_i and μ-th component of the perfect fluid correspondingly.

A vector $y \in R^n$ is called time-like, space-like or isotropic, if $< y, y >$ has negative, positive or null values correspondingly. Vectors y and z are called orthogonal if $< y, z >= 0$. It should be noted that the curvature induced vector v_i is always time-like, while the perfect fluid induced vector u_μ admits any value of $< u_\mu, u_\mu >$ (see Table 1).

Using the notation $< .,. >$ and the vectors (2.17)-(2.19), we may write the zero-energy constraint (2.12) and the Lagrangian (2.16) in the form

$$E = \frac{1}{2} < \dot{x}, \dot{x} > + V = 0, \tag{2.24}$$

$$L = e^{\gamma_0 - \gamma} \left(\frac{1}{2} < \dot{x}, \dot{x} > - V \right), \tag{2.25}$$

where

$$V = e^{2(\gamma - \gamma_0)} \left[-\frac{1}{2} \sum_{i=1}^{n} \lambda^i N_i e^{<v_i, x>} + \kappa^2 \sum_{\mu=1}^{\tilde{m}} A^{(\mu)} e^{<u_\mu, x>} \right]. \tag{2.26}$$

It is obvious from (2.26) that the term induced in the potential by the non-Ricci flat space M_i is similar to the term induced by μ-component of the perfect fluid. Due to this fact the non-zero curvature of the manifold M_i may be also called a component and now we use the notion of the component in such new sense. Further we employ the so called harmonic time gauge, which implies

$$\gamma(t) = \gamma_0 = \sum_{i=1}^{n} N_i x^i. \tag{2.27}$$

From the mathematical viewpoint the problem consists in integrability of the system with $n \geq 2$ degrees of freedom, described by the Lagrangian of the form

$$L = \frac{1}{2} < \dot{x}, \dot{x} > - \sum_{\mu=1}^{m} a^{(\mu)} e^{<b_\mu, x>}, \tag{2.28}$$

where $x, b_\mu \in R^n$. In (2.28) m denotes the total number of the components including the curvatures and the perfect fluid components. It should be noted that the kinetic term $< \dot{x}, \dot{x} >$ is not the positively definite bilinear form as it usually takes place in classical mechanics. Due to the pseudo-Euclidean signature $(-, +, ..., +)$ of the form $< .,. >$ such systems may be called pseudo-Euclidean Toda-like systems as the potential like that given in (2.28) defines Toda lattices citeToda well known in the classical mechanics. In [8],[19],[20] the following classes of the integrable pseudo-Euclidean Toda-like systems have been found

1. $m = 0$. This case corresponds to the vacuum multidimensional cosmological model on the manifold $M = R \times M_1 \times \ldots \times M_n$ with all Ricci-flat spaces M_i. The corresponding metric is a multidimensional generalization of the well-known Kasner solution [19].

2. m=1, the vector b_1 is arbitrary. The metrics for this 1-component case were obtained in [20]. This integrable class may be enlarged by the addition of new components inducing vectors collinear to the vector b_1.

3. $m \geq 2$, $n = 2$, $b_\mu = b + C_\mu b_0$, where b is an arbitrary vector and b_0 is an arbitrary isotropic vector, C_μ=const. This class was integrated in [20] only under the zero energy constraint.

4. $m \geq 2$, the vectors $b_1, \ldots, b_m$ are linear independent and satisfy the conditions $< b_\mu, b_\nu >= 0$ for $\mu \neq \nu$. This integrable class may be enlarged by the addition of the new components inducing vectors collinear to one from the orthogonal set $b_1, \ldots, b_m$. The corresponding cosmological models are studied in [8].

5. $m \geq 2$, the vectors $b_1, \ldots, b_m$ are space-like and may be interpreted as a set of admissible roots [3] of a simple complex Lie's algebra G. In this case the pseudo-Euclidean Toda-like system is trivially reducible to the Toda lattice associated with the Lie algebra G [34]. For $G = A_2 \equiv sl(3, C)$ the metric of the corresponding cosmological model was explicitly written in [8].

In the present paper we consider only 2-component ($m = 2$) pseudo-Euclidean Toda-like systems with 2 degrees of freedom ($n = 2$) under the zero energy constraint. The corresponding multidimensional cosmological models are 2-factor spaces, i.e.

$$M = R \times M_1 \times M_2 \tag{2.29}$$

and admit the following combinations of the components: curvature of M_1 and curvature of M_2 (vacuum models); curvature of M_1 or M_2 and 1-component perfect fluid; 2-component perfect fluid in Ricci-flat spaces M_1 and M_2. In our recent paper [10] we have integrated the vacuum model of the type (2.29) with 2 curvatures for the dimensions (dimM_1,dimM_2)=(6,3),(8,2),(5,5). Now we develop more general procedure useful for any combination of the 2 components.

3 Reducing to the generalized Emden-Fowler equation

Let us consider the equations of motion following from the Lagrangian (2.28) with $n = m = 2$ under the zero energy constraint. If the vectors b_1 and b_2 satisfy one of the following conditions

1. b_1 and b_2 are linearly dependent,

2. $< b_1, b_2 >= 0$, i.e. b_1 and b_2 are orthogonal ,

3. $< b_1 - b_2, b_1 - b_2 >= 0$, i.e vector $b_1 - b_2$ is isotropic,

the equations of motion are easily integrable and the corresponding exact solutions have been obtained in the papers [8],[20]. Now we aim to develop the integration procedure just for all remaining cases. Then, we suppose further that b_1 and b_2 do not satisfy any condition from 1-3.

Let us introduce in R^2 an orthogonal basis formed by the following two vectors

$$f_1 = (u_{22} - u_{12})b_1 + (u_{11} - u_{12})b_2, \quad f_2 = b_2 - b_1, \tag{3.1}$$

where we denoted

$$u_{\mu\nu} = < b_\mu, b_\nu >, \quad \mu, \nu = 1, 2. \tag{3.2}$$

According to our admission f_2 is not the isotropic vector, i.e.

$$< f_2, f_2 > = u_{11} + u_{22} - 2u_{12} \neq 0. \tag{3.3}$$

One may easily check that $u_{12}^2 - u_{11}u_{22} \geq 0$ for any vectors $b_1, b_2 \in R^2$ and $u_{12}^2 - u_{11}u_{22} = 0$ if and only if b_1 and b_2 are linearly dependent. Then in the case under consideration

$$< f_1, f_1 > = -(u_{12}^2 - u_{11}u_{22})(u_{11} + u_{22} - 2u_{12}) \neq 0, \tag{3.4}$$

$$< f_1, f_1 > / < f_2, f_2 > = -(u_{12}^2 - u_{11}u_{22}) < 0, \tag{3.5}$$

i.e. one from the orthogonal vectors f_1 and f_2 is space-like and the other is time-like.

The vector $x(t)$ we have to find decomposes as follows

$$x = \frac{< x, f_1 >}{< f_1, f_1 >} f_1 + \frac{< x, f_2 >}{< f_2, f_2 >} f_2. \tag{3.6}$$

For the new configuration variables

$$z(t) = \frac{< x, f_2 >}{2} + \ln \sqrt{\left|\frac{a^{(2)}}{a^{(1)}}\right|}, \tag{3.7}$$

$$y(t) = \frac{1}{2}\sqrt{-\frac{< f_2, f_2 >}{< f_1, f_1 >}} < x, f_1 > \tag{3.8}$$

the Lagrangian (2.28) and the corresponding zero-energy constraint look as follows

$$L = 2\beta\left(\dot{z}^2 - \dot{y}^2\right) - V(z, y), \tag{3.9}$$

$$E = 2\beta\left(\dot{z}^2 - \dot{y}^2\right) + V(z, y) = 0, \tag{3.10}$$

where the potential $V(z, y)$ has the form

$$V(z, y) = V_0 e^{2\alpha\beta y}\left(\mathrm{sgn}\left[a^{(1)}\right] e^{2\beta_1 z} + \mathrm{sgn}\left[a^{(2)}\right] e^{2\beta_2 z}\right). \tag{3.11}$$

In (3.9)-(3.11) the following constants are used

$$\alpha = \sqrt{u_{12}^2 - u_{11}u_{22}}, \quad \beta = (u_{11} + u_{22} - 2u_{12})^{-1}, \tag{3.12}$$

$$\beta_1 = -(u_{11} - u_{12})\beta, \quad \beta_2 = \beta_1 + 1 = (u_{22} - u_{12})\beta, \tag{3.13}$$

$$V_0 = |a^{(1)}|^{\beta_2} |a^{(2)}|^{-\beta_1}. \tag{3.14}$$

It should be mentioned that use of a basis in the form (3.1) provides the factorization of the potential (3.11) with respect to coordinates of the vector $x(t)$ (the additional linear transformation (3.7),(3.8) does not matter in this situation). Such factorization of the potential is essential for developing the following procedure proposed in [1]. Using the equation of motion following from the Lagrangian (3.9)

$$\begin{aligned} \ddot{z} &= -\frac{1}{2\beta} V_0 e^{2\alpha\beta y}\left(\beta_1 \text{sgn}\left[a^{(1)}\right] e^{2\beta_1 z} + \beta_2 \text{sgn}\left[a^{(2)}\right] e^{2\beta_2 z}\right), && (3.15)\\ \ddot{y} &= \frac{\alpha}{4} V(z,y), && (3.16) \end{aligned}$$

the zero-energy condition (3.10) written in the form

$$\dot{z}^2 = \frac{1}{2\beta}\frac{V(z,y)}{(\dot{y}/\dot{z})^2 - 1} = \frac{1}{2\beta}\frac{V(z,y)}{(\mathrm{d}y/\mathrm{d}z)^2 - 1} \tag{3.17}$$

and the relation

$$\frac{\mathrm{d}^2 y}{\mathrm{d}z^2} = \frac{\ddot{y} - \ddot{z}\frac{\mathrm{d}y}{\mathrm{d}z}}{\dot{z}^2} \tag{3.18}$$

we obtain the second-order ordinary differential equation

$$\frac{\mathrm{d}^2 y}{\mathrm{d}z^2} = \left[\left(\frac{\mathrm{d}y}{\mathrm{d}z}\right)^2 - 1\right]\left\{\frac{1}{2}\left(\beta_1 + \beta_2 + \frac{e^{2z} - \varepsilon}{e^{2z} + \varepsilon}\right)\frac{\mathrm{d}y}{\mathrm{d}z} + \alpha\beta\right\}, \tag{3.19}$$

where

$$\varepsilon = \text{sgn}\left[a^{(1)} a^{(2)}\right]. \tag{3.20}$$

We notice that due to the factorization of the potential the right side of the equation (3.19) does not contain y, so, in fact, the equation is the first-order one with respect to $\mathrm{d}y/\mathrm{d}z$.

This procedure is valid for the solutions such that $\dot{z} \not\equiv 0$. Under the zero energy constraint the solutions of (3.15),(3.16) with $\dot{z} \equiv 0$ give the following vector $x(t)$

$$x(t) = p \ln|t - t_0| + q, \tag{3.21}$$

where the constant vectors $p, q \in R^2$ are such that

$$\begin{aligned} &p = \frac{2}{\alpha^2} f_1, && (3.22)\\ &e^{<q,b_1>} = \frac{\beta_2}{a^{(1)}\alpha^2\beta} > 0, \quad e^{<q,b_2>} = -\frac{\beta_1}{a^{(2)}\alpha^2\beta} > 0. && (3.23) \end{aligned}$$

We note that the exceptional solution (3.21) exists only if the inequalities in (3.23) are satisfied. It should be mentioned, that the set of the equations (3.10),(3.15),(3.16) does not admit static solutions $\dot{z} = \dot{y} \equiv 0$ due to the condition (3.3). The solutions with $\dot{z} = \pm\dot{y}$ are also impossible, so using the relation (3.17) we do not lose any solutions of the set (3.10),(3.15),(3.16) except, possibly, the solution (3.21).

Let us suppose that one is able to obtain the general solution of the equation (3.19) in the parametrical form $z = z(\tau)$, $y = y(\tau)$, where τ is a parameter. Then using (3.6)-(3.8) we obtain the vector x as the function of the parameter τ

$$x(\tau) = \frac{2y(\tau)}{\alpha}(-\beta_2 b_1 + \beta_1 b_2) + 2\beta\left[z(\tau) - \ln\sqrt{\left|\frac{a^{(2)}}{a^{(1)}}\right|}\right](b_2 - b_1). \tag{3.24}$$

We recall that coordinates of the vector $x(\tau)$ in the canonical basis are the logarithms of the scale factors for the spaces M_1, M_2. The relation between the harmonic time t and the parameter τ may be always derived by integration of the zero-energy constraint written in the form of the separable equation

$$\mathrm{d}t^2 = 2\beta\frac{\left(\frac{\mathrm{d}y}{\mathrm{d}\tau}\right)^2 - \left(\frac{\mathrm{d}z}{\mathrm{d}\tau}\right)^2}{V(z(\tau), y(\tau))}\mathrm{d}\tau^2. \tag{3.25}$$

Thus the problem of the integrability by quadrature of the pseudo-Euclidean Toda-like systems with 2 degrees of freedom under the zero-energy constraint is reduced to the integrability of the equation (3.19).

For $\mathrm{d}y/\mathrm{d}z$ the equation (3.19) represents the first-order nonlinear ordinary differential equation. Its right side is the third-order polynom (with the coefficients depending on z) with respect to the $\mathrm{d}y/\mathrm{d}z$. An equation of such type is called Abel's equation (see, for instance [27],[37]). There are no methods to integrate arbitrary Abel's equation, however the equation (3.19) may be integrated for some values of the parameters $\alpha\beta$ and $\beta_1 + \beta_2$. First of all let us notice that the equation (3.19) has the partial integrals $y \pm z$ =const, which make the relation (3.17) singular and as was already mentioned are not partial integrals of the set (3.10),(3.15),(3.16). Existence of this partial solution of the Abel equation (3.19) allows to find the following nontrivial transformation

$$\mathrm{e}^{2z} = -\varepsilon\frac{X}{Y}\frac{\mathrm{d}Y}{\mathrm{d}X}, \tag{3.26}$$

$$y = \delta\left[z + \ln\left|\frac{Y}{X}\right| + \ln C\right], \quad \delta = \pm 1, \quad C > 0, \tag{3.27}$$

which reduces the Abel equation (3.19) to the so called generalized Emden-Fowler equation

$$\frac{\mathrm{d}^2 Y}{\mathrm{d}X^2} = X^n Y^m \left(\frac{\mathrm{d}Y}{\mathrm{d}X}\right)^l, \tag{3.28}$$

where the constant parameters n, m and l read

$$n = \frac{1}{2}(\beta_1 + \beta_2 - 2\delta\alpha\beta - 3) = \frac{-2u_{11} - u_{22} + 3u_{12} - \delta\sqrt{u_{12}^2 - u_{11}u_{22}}}{u_{11} + u_{22} - 2u_{12}}, \tag{3.29}$$

$$m = -\frac{1}{2}(\beta_1 + \beta_2 - 2\delta\alpha\beta + 3) = \frac{-u_{11} - 2u_{22} + 3u_{12} + \delta\sqrt{u_{12}^2 - u_{11}u_{22}}}{u_{11} + u_{22} - 2u_{12}}, \tag{3.30}$$

$$l = -\frac{1}{2}(\beta_1 + \beta_2 + 2\delta\alpha\beta - 3) = \frac{2u_{11} + u_{22} - 3u_{12} - \delta\sqrt{u_{12}^2 - u_{11}u_{22}}}{u_{11} + u_{22} - 2u_{12}}. \tag{3.31}$$

For our models the parameters in the generalized Emden-Fowler equation are not independent. It follows from (3.29),(3.30) that

$$n + m = -3. \tag{3.32}$$

In the special case $l = 0$ the equation (3.28) is known as the Emden-Fowler equation.

If the parameters l and m given by (3.31),(3.30) are such that $l = 0,\ m \neq -1$ there exists one more transformation

$$1 + \varepsilon e^{2z} = -\frac{2}{m-1}\frac{X}{Y}\frac{\mathrm{d}Y}{\mathrm{d}X}, \tag{3.33}$$

$$y = \delta\left[z - \frac{1}{m-1}\ln Y^2 + C\right], \quad \delta = \pm 1,\ C \in R, \tag{3.34}$$

which reduces the Abel equation (3.19) to the following integrable Emden-Fowler equation

$$\frac{\mathrm{d}^2 Y}{\mathrm{d}X^2} = Y^{\frac{m+3}{m-1}}. \tag{3.35}$$

There are no methods for integrating of the generalized Emden-Fowler equation with arbitrary independent parameters n, m and l also. However, the discrete-group methods recently developed in [37] allow to integrate by quadrature 3 two-parametrical classes, 11 one-parametrical classes and about 90 separated points in the parametrical space (n, m, l) of the generalized Emden-Fowler equation. For instance, the two-parametrical integrable classes arise when m and l are arbitrary and $n = 0$ or when n and l are arbitrary and $m = 0$. The one-parametrical class with $l = 0$ and $n + m = -3$ is also integrable by quadrature.

Let us suppose that the two components of the 2-factor spaces cosmological model under consideration induce such vectors b_1 and b_2 that (corresponding to the model) the generalized Emden-Fowler equation (3.28) with parameters defined by (3.2),(3.29)-(3.31) is integrable in the parametrical form $X = X(\tau)$, $Y = Y(\tau)$, where τ is a parameter. Then, using the parameter τ as the new time coordinate we obtain by the formulas (3.26),(3.27),(3.24),(3.25) the following final result for the metric (2.2)

$$g = -f^2(\tau)[a_1(\tau)]^{2N_1}[a_2(\tau)]^{2N_2} d\tau \otimes d\tau + [a_1(\tau)]^2 g^{(1)} + [a_2(\tau)]^2 g^{(2)}, \tag{3.36}$$

where we denoted

$$f(\tau) = \sqrt{\frac{2|\beta|}{V_0} C^{n+l}} \left[\frac{Y'(\tau)}{X'(\tau)}\right]^l \frac{[X'(\tau)]^2}{X(\tau)Y(\tau)Y'(\tau)}, \tag{3.37}$$

$$a_i(\tau) \equiv \mathrm{e}^{x^i(\tau)} = \mathrm{e}^{\gamma^i}\left\{\left|\frac{Y'(\tau)}{X'(\tau)}\right|^{(2-l)b^i_{(1)} - (1-l)b^i_{(2)}} \left|\frac{Y(\tau)}{X(\tau)}\right|^{(2+n)b^i_{(1)} - (1+n)b^i_{(2)}}\right\}^{\frac{2\beta}{l+n}}. \tag{3.38}$$

By γ^i for $i = 1, 2$ we denoted the following constants

$$\gamma^i = 2\beta\left\{\frac{\ln C}{n+l}\left[(n-l+4)b^i_{(1)} - (n-l+2)b^i_{(2)}\right] - \ln\sqrt{\left|\frac{a^{(2)}}{a^{(1)}}\right|}\left[b^i_{(2)} - b^i_{(1)}\right]\right\}. \tag{3.39}$$

We recall that $b^i_{(\mu)}$ are coordinates of the vector b_μ in the canonical basis. In the special case $l = 0$ one may also use by the similar manner the transformation (3.33),(3.34) and the result of integration of the equation (3.35) to write the metric. This transformation was used in [10] for integration of models with two curvatures.

Thus the method described allows to integrate the cosmological models if the corresponding generalized Emden-Fowler equation is integrable. Note that if the model with some vectors b_1 and b_2 is integrable in such a manner then any model with vectors αb_1 and αb_2 (α is an arbitrary non-zero constant) is also integrable as the parameters n, m and l do not change under such transformation of vectors. Taking into account the classes 1-4 (the class 5 does not arise for $n = 2$) and the additional to them class, which may be integrated by the method described, we obtain quite a large variety of the integrable 2-factor spaces cosmological models with 2 components.

4 Examples of the integrable models

Now we apply the method proposed in Section 3 to cosmological models on the manifold (2.29) with both Ricci-flat spaces M_1, M_2 and the 2-component perfect fluid source. Let us represent such model by Table 2

manifold/source	external space $M_1^{N_1}$	internal space $M_2^{N_2}$
1-st component of the perfect fluid	$h_1^{(1)}$	$h_2^{(1)}$
2-nd component of the perfect fluid	$h_1^{(2)}$	$h_2^{(2)}$

TABLE 2. Representation of the model on the manifold $M = R \times M_1 \times M_2$ with Ricci-flat spaces M_1, M_2 for the 2-component perfect fluid.

We recall that $N_i = \dim M_i$ and $h_i^{(\mu)}$ are constant parameters in the barotropic equation of state (2.9). The model is entirely defined by these 6 parameters. One easily shows [12] that the dominant energy condition applied to the stress-energy tensor (2.8) implies $0 \leq h_i^{(\mu)} \leq 2$. Usually rational values of the parameter $h_i^{(\mu)}$ are employed in cosmology, for instance, $h_i^{(\mu)} = (N_i - 1)/N_i$ - radiation, $h_i^{(\mu)} = 1$ - dust, $h_i^{(\mu)} = 0$ - Zeldovich (stiff) matter, $h_i^{(\mu)} = 2$ - false vacuum (Λ-term), $h_i^{(\mu)} = (D-1)/D$ - superradiation etc. On the other hand the most known cases, when the generalized Emden-Fowler equation (3.28) is integrable, arise for the rational parameters n, m and l [27]. So if one demands the rationality of the parameters n, m and l in the equation (3.28) corresponding to the model under the condition of rationality for parameters $h_i^{(\mu)}$, then due to the following relation

$$\alpha^2 = u_{12}^2 - u_{11}u_{22} = \frac{N_1 N_2}{N_1 + N_2 - 1}\left(h_1^{(1)}h_2^{(2)} - h_2^{(1)}h_1^{(2)}\right)^2, \tag{4.1}$$

the dimensions N_1, N_2 with integer value of the expression $R \equiv \sqrt{N_1 N_2(N_1 + N_2 - 1)}$ are singled out. For instance, the expression R is integer for the following dimensions: $(N_1, N_2) = (3,6), (2,8), (5,5), (7,8), (3,25), (N_1, 1)$. From the physical viewpoint the following cases may be of interest: $(2,1), (3,1), (3,6), (3,25)$.

Let us consider the models of the type represented in Table 2 leading to the Emden-Fowler equation (3.28) with

$$l = 0. \tag{4.2}$$

Due to the relation (3.32) arising for our models the equation is integrable for arbitrary parameter m and its exact solution has been written in [27]. It is worth to mention that other 2 integrable classes of the generalized Emden-Fowler equation (3.28), arising when $n = 0$ or $m = 0$, describe the same cosmological models. It follows easily from (3.29),(3.31) that if the model is such that $l = 0$ for $\delta = 1$ (or $\delta = -1$) then $n = 0$ for $\delta = -1$ (correspondingly, $\delta = 1$). It is easy to see also from (3.30),(3.31) that the condition $l = 0$ transforms to the condition $m = 0$ under the inverse numbering of the components. Thus, from 3 integrable classes, arising for $n = 0$, $m = 0$ and $l = 0$, correspondingly, of the generalized Emden-Fowler equation (3.28) for our models it is enough to study any one of them, let it be the class with $l = 0$. In this case (3.28) has the form

$$\frac{\mathrm{d}^2 Y}{\mathrm{d}X^2} = X^{-m-3} Y^m, \quad m = -\beta_1 - \beta_2. \tag{4.3}$$

By the transformation

$$Y = \frac{\tau}{\xi}, \quad X = \frac{1}{\xi} \tag{4.4}$$

it reduces to

$$\frac{\mathrm{d}^2 \tau}{\mathrm{d}\xi^2} = \tau^m, \tag{4.5}$$

which is easily integrable. Then the general solution of the equation (4.3) has the form

$$Y = \pm \frac{\tau}{F(\tau)}, \quad X = \pm \frac{1}{F(\tau)}, \tag{4.6}$$

where

$$F(\tau) = \int \left[\frac{2}{m+1} \tau^{m+1} + C_1 \right]^{-1/2} \mathrm{d}\tau + C_2, \quad m \neq -1, \tag{4.7}$$

$$= \int [2 \ln|\tau| + C_1]^{-1/2} \mathrm{d}\tau + C_2, \quad m = -1. \tag{4.8}$$

We suppose that both components of the perfect fluid have the positive mass-energy densities given by (2.11). It means $a^{(\mu)} = \kappa^2 A^{(\mu)} > 0$ for $\mu = 1, 2$, so $\varepsilon = \mathrm{sgn}\left[a^{(1)} a^{(2)}\right] = 1$. Then taking into account the formula (3.26), we must consider the general solution (4.6) on such interval of the variable τ where

$$G(\tau) \equiv -\frac{X(\tau) Y'(\tau)}{Y(\tau) X'(\tau)} = \frac{F(\tau)}{\tau F'(\tau)} - 1 > 0. \tag{4.9}$$

Finally using the results of Section 3 we obtain the following exact solution for the cosmological model represented by Table 2 in the special case $l = 0$:
the metric is given by the formula (3.36), where

$$f^2(\tau) = \frac{2|\beta| C^{-2\delta\alpha\beta}}{[A^{(1)}]^{\beta_2} [A^{(2)}]^{-\beta_1}} \left[\frac{F'(\tau)}{G(\tau)\tau^2} \right]^2 , \tag{4.10}$$

the scale factors of spaces M_1, M_2 have the form

$$a_i(\tau) = \left[\frac{A^{(1)}}{A^{(2)}} \right]^{\beta \left[u^i_{(2)} - u^i_{(1)} \right]} \left\{ [G(\tau)]^{u^i_{(2)} - 2u^i_{(1)}} \left[C^2 \tau^2 \right]^{\beta_1 u^i_{(2)} - \beta_2 u^i_{(1)}} \right\}^{\delta/\alpha} , \tag{4.11}$$

the mass-energy densities of the components read

$$\rho^{(1)}(\tau) = \frac{\left[A^{(1)}\right]^{\beta_2} \left[A^{(2)}\right]^{-\beta_1}}{[a_1(\tau)]^{2N_1} [a_2(\tau)]^{2N_2}} \left\{ [G(\tau)]^{u_{12} - 2u_{11}} \left[C^2 \tau^2 \right]^{\alpha^2 \beta} \right\}^{\delta/\alpha} , \tag{4.12}$$

$$\rho^{(2)}(\tau) = \frac{\left[A^{(1)}\right]^{\beta_2} \left[A^{(2)}\right]^{-\beta_1}}{[a_1(\tau)]^{2N_1} [a_2(\tau)]^{2N_2}} \left\{ [G(\tau)]^{u_{22} - 2u_{12}} \left[C^2 \tau^2 \right]^{\alpha^2 \beta} \right\}^{\delta/\alpha} . \tag{4.13}$$

The functions $F(\tau)$ and $G(\tau)$ are defined in (4.7)-(4.9); components $u^i_{(\mu)}$ of the vectors in the canonical basis are given in (2.23); the parameters $\alpha, \beta, \beta_1, \beta_2$ are defined in (3.12),(3.13); the values $u_{\mu\nu} =< u_\mu, u_\nu >$ for $\mu, \nu = 1, 2$ may be calculated from the parameters $N_i, h_i^{(\mu)}$ by the formula given in Table 1.

The following relation is valid for the densities (4.12)-(4.13)

$$\rho^{(2)}(\tau)/\rho^{(1)}(\tau) = G(\tau). \tag{4.14}$$

We recall the possible existence of the special solution (3.21).

Concluding the paper, we mention in Table 3 some special models interesting from our viewpoint with $l = 0$ for the dimensions $N_1 = 3$ and $N_2 = 6$. One easily shows that for these dimensions and given in Table 3 values of the parameters $h_i^{(\mu)}$ the parameter l given by (3.31) is equal to zero, so the special model is described by the exact solution (4.10)-(4.13). It follows from (2.22),(2.23) that for these dimensions the 2 perfect fluid components with the parameters 1-st: $h_1^{(1)} = 4/3, h_2^{(1)} = 1$, 2-nd: $h_1^{(2)} = 4, h_2^{(1)} = 5/3$ induce vectors, which coincide with the vectors induced by the curvatures of M_1^3 and M_2^3, correspondingly. Then adding such 2 components to the integrable vacuum model on $R \times M_1^3 \times M_2^6$ with 2 curvatures (see investigation of this model in [10]) provides us with the integrable model for the two non Ricci-flat spaces and the 2-component perfect fluid.

manifold/source	external space M_1^3	internal space M_2^6
1-st component of the perfect fluid	radiation $h_1^{(1)} = \frac{2}{3}$	radiation $h_2^{(1)} = \frac{5}{6}$
2-nd component of the perfect fluid	radiation $h_1^{(2)} = \frac{2}{3}$	Zeldovich matter $h_2^{(2)} = 0$
1-st component of the perfect fluid	dust $h_1^{(1)} = 1$	radiation $h_2^{(1)} = \frac{5}{6}$
2-nd component of the perfect fluid	radiation $h_1^{(2)} = \frac{2}{3}$	dust $h_2^{(2)} = 1$
1-st component of the perfect fluid	radiation $h_1^{(1)} = \frac{2}{3}$	radiation $h_2^{(1)} = \frac{5}{6}$
2-nd component of the perfect fluid	dust $h_1^{(2)} = 1$	radiation $h_2^{(2)} = \frac{5}{6}$
1-st component of the perfect fluid	radiation $h_1^{(1)} = \frac{2}{3}$	radiation $h_2^{(1)} = \frac{5}{6}$
2-nd component of the perfect fluid	false vacuum $h_1^{(2)} = 2$	false vacuum $h_2^{(2)} = 2$
1-st component of the perfect fluid	false vacuum $h_1^{(1)} = 2$	false vacuum $h_2^{(1)} = 2$
2-nd component of the perfect fluid	Zeldovich matter $h_1^{(2)} = 0$	false vacuum $h_2^{(2)} = 2$

TABLE 3. Examples of the integrable models for dimensions $N_1 = 3$ and $N_2 = 6$. The corresponding exact solutions are given by the formulas (4.10)-(4.13).

References

[1] Bleyer U, Liebscher D-E and Polnarev A G 1991 *Class. Quantum Grav.* **8** 477

[2] Bleyer U, Ivashchuk V D, Melnikov V N and Zhuk A I 1994 *Nucl. Phys.* **B429** 177

[3] Bogoyavlensky O I 1976 *Comm. Math. Phys.* **51** 201

[4] Börner G 1992 *The Early Universe. Facts and Fiction* (Berlin: Springer)

[5] Forgacs P and Horvath Z 1979 *Gen. Relativ. Grav.* **11** 205

[6] Chodos A and Detweiler S 1980 *Phys. Rev.* **D21** 2167

[7] Demiansky M and Polnarev A G 1990 *Phys. Rev. D* **41** 3003

[8] Gavrilov V R, Ivashchuk V D and Melnikov V N 1995 *J. Math. Phys.* **36** 5829

[9] Gavrilov V R, Melnikov V N and Novello M 1995 *Gravitation and Cosmology* **1** 149

[10] Gavrilov V R, Ivashchuk V D and Melnikov V N, 1996 *Class. Quantum Grav.* **13** 3039

[11] Gavrilov V R, Melnikov V N and Novello M 1996 *Gravitation and Cosmology* **4(8)** 325

[12] Gavrilov V R, Melnikov V N and Triay R Exact solutions in multidimensional cosmology with bulk and shear viscosity *Preprint* CPT-96/P.3396, Marseille, France; 1997 *Class. Quantum Grav.* **14** 2203.

[13] Gavrilov V R, Ivashchuk V D, Kasper U and Melnikov V N, 1997 *Gen. Relativ. Grav.* **29** 599

[14] Gibbons G W and Wiltshire D L 1987 *Nucl. Phys.* **B287** 717

[15] Gleiser M, Rajpoot S and Teylor J G 1985 *Ann. Phys.* (NY) **160** 299

[16] Ivashchuk V D and Melnikov V N 1988 *Nuovo Cimento* **B102** 131

[17] Ivashchuk V D and Melnikov V N 1989 *Phys. Lett.* **A135** 465

[18] Ivashchuk V D, Melnikov V N and Zhuk A I 1989 *Nuovo Cimento B* **104** 575

[19] Ivashchuk V D 1992 *Phys. Lett.* **A 170** 16

[20] Ivashchuk V D and Melnikov V N 1994 *Int. J. Mod. Phys.* **D3** 795

[21] Ivashchuk V D and Melnikov V N 1995 *Class. Quantum Grav.* **12** 809

[22] Koikawa T and Yoshimura M 1985 *Phys. Lett. B* **155** 137

[23] Lorenz-Petzold D 1984 *Phys. Lett. B* **149** 79

[24] Lorenz-Petzold D 1985 *Phys. Lett. B* **158** 110

[25] Melnikov V N 1994 *Multidimensional Classical and Quantum Cosmology and Gravitation: Exact Solutions and Variations of Constants* In: Cosmology and Gravitation. Ed. M.Novello. Editions Frontieres, Singapore, p. 147

[26] Melnikov V N, *Multidimensional Cosmology and Gravitation*, CBPF-MO-002/95, Rio de Janeiro, Brasil

[27] Polyanin A D and Zaitsev V F 1995 *Handbook on Exact Solutions for Ordinary Differential Equations* (Roca Raton, CRC Press)

[28] Rainer M 1996 *Gravitation and Cosmology* **1(5)** 27

[29] Rainer M and Zhuk A I 1996 *Phys. Rev. D* **54** 6186

[30] Rubakov V A, Shaposhnikov M E 1983 *Phys. Lett.B* **125**, 136

[31] Sahdev D 1984 *Phys. Rev. D* **30** 2495

[32] Szydlowski M 1988 *Gen Relativ. Grav.* **20** 221

[33] Szydlowski M and Pajdosz G 1989 *Class. Quant. Grav.* **6** 1391

[34] Toda M 1981 *Theory of Nonlinear Lattices* (Springer-Verlag, Berlin)

[35] Wesson P S and Ponce de Leon J 1994 *Gen. Rel. Gravit.* **26** 555

[36] Wiltshire D L 1987 *Phys. Rev. D* **36** 1634

[37] Zaitsev V F and Polyanin A D 1994 *Discrete-Groups Methods for Integrating Equations of Nonlinear Mechanics* (Roca Raton, CRC Press-Begel House)

World Spinors in Metric and Nonmetric Backgrounds

Yuval Ne'eman

1. Introduction: Spinors on Curved Spaces - They Do Exist!

It appears that the discovery and understanding of *spin* have been fraught with errors, beyond what we are used to in scientific research programs. First, there was, both on the part of Lorentz, in his criticism of Goudsmit and Uhlenbeck when they presented to him their idea, and on the part of Pauli, when Kronig brought him the same idea, the factor of 2 "error" in the magnitude of the electron's intrinsic magnetic moment - an error on the part of the critics, as was soon shown by Thomas, when he calculated the relativistic precession. Then, there was the Dirac equation, predicting the existence of an antiparticle to the electron - and Pauli's comment "what a pity that this beautiful equation should be so wrong!", commenting on the factor 1840 between the masses of the proton and the electron - whereas they should have been equal, according to the equation. This error too did not last long, as Anderson soon discovered the positron.

These lectures will deal with a third error, one which lasted much longer, from the 1930s to 1977.. Throughout these years, the gravity establishment was convinced (and many still wrongly believe) that curved space could not support *"world spinors"*, in contradistinction to the case of *world tensors*. What are *world tensors*? The General Theory of Relativity is built on two principles, namely *Covariance* and *Equivalence*. The two principles have group theoretical implications. "Covariance" means *invariance under the infinite group of local R^4 diffeomorphisms*. Although the theory appears to be expressed in terms of fields carrying linear (non-unitary) representations of $SL(4,R)$ or $GL(4,R)$, these are *world tensors*, i.e. *tensors under the entire group of local diffeomorphisms, in which the covariance group is realized nonlinearly over the linear representations of its linear subgroup*. This means that whereas transformations involving the subgroup $GL(4,R) \subset Diff(R^4)$ are represented linearly (the algebraic parameters multiplying the $GL(4,R)$ representation matrices), the transformations involving the quotient $[Diff(R^4)/GL(4,R)]$ are represented by *functions* multiplying the same constant matrices. The precise derivation of these functions is given explicitly in B. DeWitt's 1963 Les Houches lectures[1]. For example, an infinitesimal diffeomorphism $x^\mu \to x^\mu + \xi^\mu(x)$, transforms the world-tensor

$\Phi^m(x)$ (the index ^{m} is an abstraction of some tensor indices) by

$$\delta\Phi^m(x) = (\partial_\nu \xi^\mu)\{\mathcal{A}^\nu_\mu\}^m_n \Phi^n(x), \tag{1}$$

where the brackets denote the algebraic generator $\mathcal{A}^\nu_\mu$ of $SL(4,R)$ in the appropriate linear representation of that group (a matrix in n,m). In the cases where $\xi^\mu(x) = a^\mu_\nu x^\nu$, i.e. a linear $GL(4,R)$ transformation, we get $\delta\Phi^m(x) = a^\mu_\nu\{\mathcal{A}^\nu_\mu\}^m_n\Phi^n(x)$ i.e. the usual action on $GL(4,R)$ linear representation matrices. Had $\xi^\mu(x)$ contained more than the linear term, in a Taylor expansion, we would have the same matrices of the $GL(4,R)$ algebra multiplied by functions of x.

Note that when working in Quantum Field Theory in flat spacetime, we still use $SL(4,R)$ tensors, even though the "covariance" is now reduced to global Lorentz invariance. This is most appropriate for a smooth transition to curved spacetime, with $g_{\mu\nu}(x)$ a gravitational potential of a sort, replacing the set of constants in the Minkowski metric $\eta_{\mu\nu}$ and the coupling of matter to gravitation arising through the "minimal" coupling , provided by replacing ordinary partial derivatives by covariant derivatives. However, there are profound disadvantages too. The mid-fifties "discovery" that RQFT violates unitarity off-mass-shell was due to the fact that in the Yang-Mills Lagrangian, the field-potentials (or connections) are 4-component $SL(4,R)$ vectors of the defining representation, whereas the Hilbert space contains only 2 physical components, the transverse ones. This was resolved by Feynman, through the introduction of *ghost fields*, whose contributions cancel those of the unphysical components of the field. As to the Principle of Equivalence, it constrains the tangent manifold (or flat spacetime) to be Lorentz invariant. In the flat case, one may add the (mutually Abelian) translations, a fact which is reflected in the requirement that the particle Hilbert space be Poincaré covariant. The RQFT solution for bosons is thus to have the fields as representations (nonunitary - except for the scalar case) of $SL(4,R)$ and the particles as unitary irreducible representations of the extended (by CPT at least, so that the two components of a massless state sit in one unirrep) Poincaré group. The fermions get an altogether different treatment, however.

We start by recalling what spinors are. They are *single-valued representations of the double-covering of the rotation group:* for every matrix of $SO(N)$, there are two of $Spin(N) = \overline{SO}(N)$. The simplest example, for $SO(3)$, would consist in a rotation by 2π around the z-axis (or any axis), represented by the identity for $SO(3)$, but corresponding to -1 for 2π and $+1$ for 4π in $Spin(3) = SU(2) = \overline{SO}(3)$ (note also that in the two complex dimensions of $SU(2)$, both $diag(-1,-1)$ and $diag(1,1)$ matrices have $det = 1$). Every other $SO(3)$ rotation around the z axis can now be compounded (group-multiplied) with either of these two $2m\pi$ cycles ($m = 1, 2$), and we have the double covering of this subset. For $SO(2)$ - the line R^1 is an infinite covering of the circle S^2, which is isomorphic to $SO(2)$ or to $U(1)$. In $SO(3)$, we have three parameters (e.g. the Euler angles) and the covering's multiplicity is reduced to two. This is true for most $SO(N)$, although there are some richer cases, one of which we shall encounter in the sequel ($SO(3,3)$, which has a quadruple covering).

When the electron spin was discovered, it was at first treated nonrelativistically by Pauli, as an $SU(2) = \overline{SO}(3)$. Three years later, Dirac was pointing out its role and derivation in a relativistic equation. Cartan[2] then proved that the next "natural" inclusion, following the tensors, i.e. in the $SL(4, R)$ subgroup of General Covariance, *was not possible for a finite number of components*. Almost simultaneously, H. Weyl and Fock and Ivanenko showed how to perform the inclusion directly from the Lorentz group into the Diffeomorphisms, applying Darboux' method of *frames*, or *repères mobiles*. Here the linear action is that of local $SL(2, C)$ on the frame indices. The tetrad fields, through their one *world* index, carry the action of the Diffeomorphisms (nonlinearly over $SL(4, R)$ like a tensor) but also, by construction, provide the further nonlinear realization of the quotient $SL(4, R)/SL(2, C)$.

Cartan's point about the inexistence of world spinors with a finite number of components was erroneously taken as a complete negation (see examples quoted in ref. 3). Actually, a spinor can be described as the square-root of a tensor - and we are used to square-roots which require an infinite component description - the irrationals! Every time we write $\sqrt{2}$ we imply a number requiring an infinite series of integers for its description, whatever the base. Similarly, I proved[4] in 1977 that "taking the square-root of a tensor" in curved space is indeed possible, but Cartan's theorem requires it to be an infinite component spinor. With Dj. Šijački, we have since provided a more complete analysis[3], including a simplified proof of Cartan's result; we have also provided the classification and construction of the manifield representations[5]. In addition, we have provided a classification of the relevant Hilbert space representations of $\overline{SA}(4,R)$, the unimodular affine group[6].

Not only is this "natural", but we can even point[7] to a realization in nature: indeed, the hadrons, constituting 99.9% of atomic matter, are just such infinite component fields. In the next section, we first demonstrate the existence of the double-covering groups for the diffeomorphisms and their subgroups. We shall then describe these *world spinor manifields* and their equations and relate these to the Hilbert space states in flat and curved spacetime.

2. The Double-Covering of the Linear Groups

To the algebraic topologist, our problem rightly appears to have been answered long ago, with the realization that *the topology of a noncompact Lie group follows that of its maximal compact subgroup.* So much for the main current error, the erroneous assumption with respect to the non-existence of this double-covering. However, there is in this case, a second source of confusion, in the opposite direction, i.e. assuming the existence of the double-covering but disregarding Cartan's result, which limits it to infinite-component systems. The source of confusion here is generated by errors in the identification of the embedding map $GL(4,R) \rightarrow GL(4,C)$. $GL(4,C)$ does possess finite spinorial representations, and an error in the identification of the physical $GL(4,R)$ subgroup gives the impression that $\overline{GL}(4,R)$ does have finite spinorial representations. In ref. 3, we have proved that the embedding $\overline{GL}(4,R) \rightarrow GL(4,C)$ does not exist; neither can we embed $\overline{GL}(4,R)$ in a "double-covering of $GL(4,C)$", since there is no such thing for the latter, a simply-connected group. The forbidden embedding $\overline{GL}(4,R) \not\rightarrow GL(4,C)$ implies the inexistence of finite spinorial representations for $\overline{GL}(4,R)$.

A third source of confusion concerns the unitarity of the relevant spinor representations. In dealing with noncompact groups, it is customary to select infinite-dimensional unitary representations, where the *particle-states* are concerned. For both tensor or spinor *fields*, however, finite and nonunitary representations are used (of $GL(4,R)$ and $SL(2,C)$ respectively). We shall show that the correct answer for spinorial $\overline{GL}(4,R)$ consists in using the infinite unitary representations in a physical base in which they become nonunitary[5].

The proof of the existence of the double-covering of the linear groups can be deduced from the "Iwasawa decomposition". Any element of a connected

Lie group G with Lie algebra g_0 over R, can be decomposed into the product of three elements, each belonging to one (and only one) of the following analytic subgroups: the *maximal compact* subgroup K, an *Abelian* subgroup A and a *nilpotent* subgroup N (with "triangular" matrices, i.e. filled only above the main diagonal), with Lie algebras k_0,a_0 and n_0 respectively. The groups A and N are simply connected. We refer the reader to ref. 8 for the theorem which states that, with $g_0 = k_0 + a_0 + n_0$,the mapping $(k, a, n) \rightarrow kan \quad (k \in K, a \in A, n \in N)$ is an analytic diffeomorphism of the product manifold $K \times A \times N$ onto G.

Only K is not guaranteed to be simply-connected. There thus exists a *universal covering group* $\overline{K}_u$ of K, and therefore also a universal covering of G:

$$\overline{G}_u \simeq \overline{K}_u \times A \times N. \tag{2}$$

An analogous situation exists in the complex case[3]. It was by using the above construction (2), that I first constructed $\overline{SL}(3, R)$ and its representations[4].

For the linear groups $GL(n, R), SL(n, R)$, the K are respectively $O(n)$ and $SO(n)$ and their universal coverings $\overline{K}$ are thus $Pin(n)$ and $Spin(n)$. As a result, the *metalinear* groups of the universal-covering are given by $\overline{GL}(n, R) \equiv \{Pin(n)A(n, R)N(n, R)\}$ and $\overline{SL}(n, R) \equiv \{Spin(n)A(n, R)N(n, R)\}$ where the $A(n, R)$ and $N(n, R)$ are the relevant subgroups of the original (uncovered) corresponding linear group.

For *the group of diffeomorphisms*, Stewart[9] proved the decomposition

$$Diff(n, R) = GL(n, R) \times H \times R_n \tag{3}$$

where the subgroup H is contractible to a point. As a result, as $O(n)$ is the compact subgroup of $GL(n, R)$, one finds that $O(n)$ is a deformation retract of $Diff(n, R)$.

As a result, there exists a universal covering of the Diffeomorphism group

$$\overline{Diff}(n, R)_u \equiv \overline{GL}(n, R)_u \times H \times R^n. \tag{4}$$

For $n = 2$, the universal covering $\overline{SL}(2, R)_u$ has a countable infinity of coverings, as $K = SO(2), \overline{K} = U(1)$. For $\overline{SL}(n, R)_u, \quad n > 2$ these will be double-coverings. For $n = 3$, $K = SU(2)$. In the case $n = 4$, we have the homomorphism between $SO(3) \times SO(3)$ and $SO(4)$. Since

$$SO(3) \simeq SU(2)/Z_2 \tag{5}$$

where Z_2 is the two-element center $\{1, -1\}$, we have here

$$SO(4) \equiv [SU(2) \times SU(2)]/Z_2^d$$
$$Z_2^d : \{1, (-1)^{2j_1} = (-1)^{2j_2}\} \tag{6}$$

Z_2^d is thus the diagonal discrete group and where j_1 and j_2 are the Casimir labels of the two $SU(2)$ representations. The full $Z_2 \times Z_2$ group given by the representations

$$\{1, (-1)^{2j_1}\} \otimes \{1, (-1)^{2j_2}\} \tag{7}$$

is the *center* of $\overline{SO}(4) = SU(2) \times SU(2)$, which is thus the quadruple-covering of $SO(3) \times SO(3)$ and a double-covering of $SO(4)$. $SO(3) \times SO(3)$, $SO(4)$ and $\overline{SO}(4) = SU(2) \times SU(2)$ are thus the maximal compact subgroups of $SO(3,3)$, $SL(4,R)$ and $\overline{SL}(4,R)$ respectively. This also clarifies the case of the quadruple covering of $SO(3,3)$, where $SL(4,R)$ is on the one hand the double-covering of $SO(3,3)$ and on the other hand it is itself double-covered by $\overline{SL}(4,R)$.

3. Algebraic and physical role of the Generator Algebra $sl(n, R)$

We should acquire for the $sl(n,R)$ generators the kind of intuitive understanding that we have for angular momentum $so(3)$. It is simplest to make the acquaintance for the case $n = 3$. I assume the reader is familiar with the eight λ_i matrices of $su(3)$. Only three of these are *purely imaginary* - a necessary condition for the exponentiated group elements $g = exp(i\alpha^i\lambda_i)$ to be *real*, as required for $SL(n,R)$ by definition. These generate the $so(3)$ compact subgroup,

$$\{\lambda_2, -\lambda_5, \lambda_7\} : so(3) \subset sl(3,R) \tag{8}$$

For the rest, we thus have to multiply the matrices by an i, which makes them antihermitian; this is *noncompactness* and the theory of group representations tells us that for such representations to be unitary, their matrices have to be infinite. To get a physical insight, place the matrices between a *horizontal* coordinate vector $r(x,y,z)$ and a *(vertical) nabla* vector $\nabla(\partial_x, \partial_y, \partial_z)$. Multiplying through, we get

$$\begin{aligned} J_z &= r\ \lambda_2\nabla = xp_y - yp_x \\ J_x &= r\ \lambda_7\nabla = yp_z - zp_y \\ J_y &= r\ (-\lambda_5)\nabla = zp_x - xp_z \end{aligned} \tag{9}$$

We can now do the same r $\lambda_i\nabla$ for the 5 symmetric matrices, representing the deformations produced by ***shears***. These are given by the symmetric combinations $xp_y + yp_x, yp_z + zp_y, zp_x + xp_z$ and $(xp_x - yp_y)$, $(1/\sqrt{3})(xp_x + yp_y - 2zp_z)$.

Physically, we first consider non-relativistic gravitational quadrupoles $m(xy + yx), m(yz + zy)$ etc.. describing deformations, violating spherical symmetry. If we now take their time-derivatives[10], i.e. *pulsation rates*, and since $m\dot{x} = p_x$, we recover our five generators of shear. The $sl(3, R)$ charges and their currents are thus the three components of angular momentum and the five deformation pulsation rates $\frac{d}{dt}\int d^3xm(r^ir^j + r^jr^i)$. Together, they compose the hypermomentum tensor. The representations of $SL(3, R)$ and $\overline{SL}(3, R)$ are reduced over the compact subgroups $SO(3)$ or $SU(2)$ respectively; note that the generators of the quotient $SL(3, R)/SO(3)$ (the shears) form a spin $J = 2$ under the $SO(3)$ spin subalgebra. As such they appear appropriate for use in descriptions of the actions of the gravitational field.

4. Unirreps of the $\overline{SL}(n, R)$, for $n = 2, 3, 4$.

The case $n = 2$

The *uni*tary infinite–dimensional *irr*educible *rep*resentations (*unirreps*) of $\overline{SL}(2, R)$ were constructed and catalogued by Bargmann[11]. The two–dimensional linear group $SL(2, R)$ is special insofar as it has infinite coverings: The maximal compact subgroup $SO(2)$ is, regarded as a manifold, isomorphic to the circle S^1, which is infinitely often covered by the line.

Bargmann listed four classes of representations, defined by τ, the eigenvalue of the quadratic $SL(2, R)$ Casimir operator, negative here (the noncompact generators have been multiplied by an i), $-\tau^2 = -(\sigma_1)^2 + (\sigma_2)^2 - (\sigma_3)^2$ and by m, the (helicity–like) eigenvalue of the normal (compact) subalgebra generated by σ_2:

(i) *Principal series* $\mathcal{D}^{princ.}_{SL(2,R)}(\underline{m}, \tau)$:

$$1/4 \le \tau\,, \qquad \tau \in R\,;$$
$$\underline{m} = 0, \quad \{m\} = \{0, \pm 1, \pm 2, \pm 3, \cdots\}, \quad i.e. \{m\} = Z, or$$
$$\underline{m} = 1/2, \quad \{m\} = \{\pm\tfrac{1}{2}, \pm\tfrac{3}{2}, \pm\tfrac{5}{2}, \cdots\}, \quad i.e. \{m\} = Z/2\,,$$

(ii) *Supplementary series* $\mathcal{D}^{suppl.}_{SL(2,R)}(\underline{m}, \tau)$:

$$0 < \tau < \tfrac{1}{4}\,,$$

$$\underline{m} = 0, \quad \{m\} = \quad \{0, \pm 1, \pm 2, \pm 3, \cdots\}, \quad i.e. \{m\} = Z, or$$
$$\underline{m} = 1/2, \quad \{m\} = \quad \{\pm\tfrac{1}{2}, \pm\tfrac{3}{2}, \pm\tfrac{5}{2}, \cdots\}, \quad i.e. \{m\} = Z/2\,.$$

(iii) *Discrete series (mounting)* $\mathcal{D}^{disc.}_{\overline{SL}(2,R)}(\underline{m})$:

$$\tau = \underline{m}\,(1 - \underline{m})\,; \quad \{\underline{m}\} = \{\tfrac{1}{2}, 1, \tfrac{3}{2}, 2, \tfrac{5}{2}, \cdots\}\,,$$
$$\{m\} = \{\underline{m}, \underline{m} + 1, \underline{m} + 2, \cdots\},$$

(iv) *Discrete series (descending)* $\mathcal{D}^{disc.}_{\overline{SL}(2,R)}(\overline{m})$:

$$\tau = -\overline{m}\,(1 + \overline{m})\,; \quad \{\overline{m}\} = \{-\tfrac{1}{2}, -1, -\tfrac{3}{2}, -2, -\tfrac{5}{2}, \cdots\}\,,$$
$$\{m\} = \{\cdots, (\overline{m} - 2), \overline{m} - 1, \overline{m}\}$$

In these formulae, $\underline{m}$ and $\overline{m}$ are the minimal or maximal eigenvalues of m in a representation in which m is mounting or descending, respectively. The names of the various series refer to the values of τ.

Note that besides being a real form of $SU(1,1)$, the group $SL(2,R)$ is in itself the double–covering group of $SO(1,2)$. As a result, for instance, $\overline{SL}(2,R)$ is the quadruple covering of $SO(1,2)$. The fact that $SL(2,R)$ is itself a double covering of $SO(1,2)$ added to the confusion and strengthened the impression,throughout 1928–1977, that linear groups have no double covering

The representations with half–integer m in the supplementary series are thus two–valued in $SO(1,2)$ and *single–valued* in $SL(2,R)$, but they are not faithful representations of the double covering of that group itself. The representations of the double–covering group $\overline{SL}(2,R)$, displaying the structure $\overline{SL}(2,R)/Z_2 = SL(2,R)$, are of class (ii), i.e. are $\mathcal{D}^{suppl.}_{\overline{SL}(2,R)}(h,\tau)$, with $h = 1/4$. More generally, the representations of the multiple covering have $h = 1/(2c)$, where c denotes the order of the covering.

The case n=3

The representations of $\overline{SL}(3,R)$ were fully classified by Dj. Šijački[12].

There are four series of unirreps. Denoting by j the angular momentum eigenvalue corresponding to the maximal compact subgroup $\overline{SO}(3) = SU(2)$, and k an additional quantum number of this rank two group, by $\underline{j}$ and $\underline{k}$ the minimal values of J and K in a unirrep, and n the multiplicity, we have:

(i) *Principal series* $\mathcal{D}^{princ.}_{SL(3,R)}(\underline{j},\underline{k};\sigma_2,\delta_2)$:

$$\sigma_2,\delta_2 \in R; \quad \underline{k}=0,\tfrac{1}{2},1;$$
$$\underline{k}=0 \to \underline{j}=0, \{j^n\}=\{0,2^2,3,4^3,5^2,6^4,7^3,\cdots\},$$
$$\underline{k}=\tfrac{1}{2} \to \underline{j}=\tfrac{1}{2}, \{j^n\}=\{\tfrac{1}{2},\left(\tfrac{3}{2}\right)^2,\left(\tfrac{5}{2}\right)^3,\cdots\},$$
$$\underline{k}=0,1 \to \underline{j}=1, \{j^n\}=\{1,2,3^2,4^2,5^3,6^3,7^4,\cdots\},$$

(ii) *Supplementary series* $\mathcal{D}^{suppl.}_{SL(3,R)}(\underline{j},\underline{k};\sigma_2,\delta_1)$:

$$\sigma_2 \in R, 0<|\delta_1|\leq \tfrac{1}{2}, \underline{k}=\tfrac{1}{2} \quad or \quad 0<|\delta_1|<1, \underline{k}=0, \underline{j}=0,1,;$$
$$\underline{k}=0 \to \underline{j}=0, \{j^n\}=\{0,2^2,3,4^3,5^2,6^4,7^3,\cdots\},$$
$$\underline{k}=\tfrac{1}{2} \to \underline{j}=\tfrac{1}{2}, \{j^n\}=\{\tfrac{1}{2},\left(\tfrac{3}{2}\right)^2,\left(\tfrac{5}{2}\right)^3,\cdots\},$$
$$\underline{k}=0,1 \to \underline{j}=1, \{j^n\}=\{1,2,3^2,4^2,5^3,6^3,7^4,\cdots\},$$

(iii) *Discrete series* $\mathcal{D}^{disc.}_{SL(3,R)}(\underline{j};\sigma_2)$:

$$\sigma_2 \in R; \underline{j}=\underline{k}, \quad \underline{j}=\frac{3}{2},2,\frac{5}{2},3,\cdots$$
$$\{j^n\}=\{\underline{j},\underline{j}+1,(\underline{j}+2)^2,(\underline{j}+3)^2,(\underline{j}+4)^3,(\underline{j}+5)^3,\cdots\},$$

(iv) *Ladder series* : $\mathcal{D}^{ladd.}_{SL(3,R)}(\underline{j};\sigma_2)$:

$$\underline{j}=0,1 \to \sigma_2 \in R, \quad \underline{j}=\tfrac{1}{2} \to \sigma_2=0;$$
$$\{j^n\}=\{\underline{j},\underline{j}+2,\underline{j}+4,\underline{j}+6,\underline{j}+8,\cdots\}, \quad i.e. \quad \Delta j=2.$$

In terms of the unirreps of the maximal compact subgroup, the last sequence consists of those multiplicity–free representations which are known as the "ladder representations". These are the simplest - in the words of the mathematician B. Kostant, they represent "the last thing a representation can do before becoming finite". The unirrep $\underline{j}=1/2$ is particularly interesting; it fits the nucleon's Regge trajectory, namely a $|\Delta j|=2$ sequence starting with the nucleon.

In addition, there are two discrete quantum numbers ϵ and ϵ', in the principal, the supplementary, and the discrete series. For principal series, they are $\epsilon=\pm1$ and $\epsilon'=\pm1$. In the supplementary series, they are $\epsilon=+1$ (only) whereas $\epsilon'=\pm1$. For the discrete series, they are again both $\epsilon=\pm1$

and $\epsilon' = \pm 1$. They do not exist in the ladder series. These quantum numbers are related to the question of the occurence of even or odd values of k and j.

The case $n = 4$

The $\overline{SL}(4, R)$ unirreps reduce into infinite sums of finite-dimensional $\overline{SO}(4)$ representations. We list first the multiplicity-free set:

(i) *Principal series* $\mathcal{D}^{princ.}_{SL(4,R)}(\underline{m}, \underline{n}; e_2)$:

$$\underline{m} = \{0, 1\}, \underline{n} = 0;\ e_1 = 0,\ e_2 \in R;$$
$$\{(m, n)\} : m + n \cong \underline{m}(mod2)\,.$$

(ii) *Supplementary series* $\mathcal{D}^{suppl.}_{SL(4,R)}(0, 0; e_1)$:

$$0 \leq |e_1| \leq 1,\ e_2 = 0;$$
$$\{(m, n)\} : m + n \cong 0(mod2).$$

(iii) *Discrete series* $\mathcal{D}^{disc.}_{SL(4,R)}((\underline{j}, 0) or (0, \underline{j}))$:

$$e_1 = 1 - \underline{j}, e_2 = 0;$$
$$\{(m, n)\} : m + n \cong \underline{j}(mod2), \quad |m - n| \geq \underline{j},$$
$$\underline{j} = \tfrac{1}{2}, 1, \tfrac{3}{2},$$

(iv) *Ladder series* $\mathcal{D}^{ladd.}_{SL(4,R)}(\underline{j}, e_2)$:

$$\underline{j} = \{0, \tfrac{1}{2}\}, e_1 = 0, e_2 \in R,$$
$$m = n = j, \quad j \cong \underline{j}(mod1) for \underline{j} = \tfrac{1}{2}; \quad j \cong \underline{j}(mod2) for \underline{j} = 0.$$

The enumeration and classification of $\overline{SL}(4, R)$ unirreps are given in ref. 13. There are 16 classes and we do not list them here.

5. The Interplay between the Linear and Affine Groups: Fields and Particles

A. The algebras

As we explained in section 2, fields - whether ordinary tensors or our (infinite-component) *manifields*, both the fermionic *world spinors* and the bosonic *"infinitensors"*, - cannot be described by unitary representations.

The Lorentz group $\overline{SO}(1,3) = SL(2,C)$ is contained in $\overline{SL}(4,R)$ and therefore appears in its unitary representations within those of $\overline{SL}(4,R)$. These are the Gelfand-Neumark representations, exciting spin, and the action of the Lorentz boost on a particle with spin j will (in the simplest case) modify it to $j \pm 1$. This is unphysical: in the physical world, the particles correspond to representations of the Poincaré group, i.e. they have momenta, and the Lorentz boost acts orbitally and modifies the momentum, preserving the particle's spin throughout. This is ensured through the fact that Lagrangians are set so as to be hermitean; writing $\overline{\psi}(\gamma^\mu \partial_\mu + m)\psi$, one adds " + hermitean conjugate". Deriving the expressions for the intrinsic angular momentum (including the Lorentz boosts) one then finds $\int d^3x \overline{\psi} \sigma_{\mu\nu} \psi + h.c.$. ($\sigma_{\mu\nu} = \frac{1}{2}[\gamma_\mu, \gamma_\nu]$) The $\sigma_{\mu\nu}$ are a finite (and thus nonunitary) representation of $SO(1,3)$. Since σ_{ij} represents the compact $SO(3)$, it is hermitean, while σ_{i0} cannot be so. This boost thus cancels and physics makes use of the orbital piece only.

To achieve the same result in the infinite case, we make use of an *inner automorphism* $\mathcal{A}$ of the $sl(4,R)$ algebra.

Let Q_{ab}, $a,b = 0,1,2,3$ be the generators of the $sl(4,R)$ algebra. The $sl(n,R)$ commutation relations read

$$[Q_{ab}, Q_{cd}] = i g_{bc} Q_{ad} - i g_{ad} Q_{cb}, \tag{10}$$

where for the structure constants g_{ab} one may take the relevant invariant metric tensor: either the Euclidean $\delta_{ab} = diag(+1,+1,+1,...,+1)$ or the Minkowskian $\eta_{ab} = diag(+1,-1,-1,...,-1)$.

The metric tensor g_{ab} is $\overline{GL}(4,R)$-*covariant.*

Taking $g_{ab} = \eta_{ab}$, the antisymmetric generators

$$J_{ab} = Q_{[a,b]} \tag{11}$$

generate the physical Lorentz subgroup in n-dimensional Minkowski spacetime.

The traceless symmetric operators

$$T_{ab} = Q_{\{a,b\}} - \frac{1}{n}\eta_{ab} Q^c_c \tag{12}$$

act as "shears" and deform, though all the while preserving the 4-volume.

The trace

$$D = Q^a_a \tag{13}$$

is the dilation operator. The $gl(n, R)$ commutators are:

$$\begin{aligned}
[J_{ab}, J_{cd}] &= -i(\eta_{ac}J_{bd} - \eta_{ad}J_{bc} - \eta_{bc}J_{ad} + \eta_{bd}J_{ac}) \\
[J_{ab}, T_{cd}] &= -i(\eta_{ac}T_{bd} + \eta_{ad}T_{bc} - \eta_{bc}T_{ad} - \eta_{bd}J_{ac}) \\
[T_{ab}, T_{cd}] &= +i(\eta_{ac}J_{bd} + \eta_{ad}J_{bc} + \eta_{bc}J_{ad} + \eta_{bd}J_{ac}) \\
[D, J_{ab}] &= 0; \qquad [D, T_{ab}] = 0
\end{aligned} \tag{14}$$

Together with the translations P_a, $a = 0, 1, 2, 3$ these generate the general affine group $GA(n, R)$ and its double-covering. Without the dilations, this reduces to the group $\overline{SA}(n, R)$ with the additional commutators

$$\begin{aligned}
[Q_{ab}, P_c] &= -ig_{ac}P_b \\
[P_a, P_b] &= 0
\end{aligned} \tag{15}$$

B. Fields: the Deunitarizing Automorphism $\mathcal{A}$.

In what follows, we concentrate on the algebraic *simple* part $sl(n, R)$ and the group $\overline{SL}(n, R)$. Denoting the space-like co-ordinates by $i, j = 1, 2, 3$ we regroup the J_{ab} and T_{ab} in the following subsets:

J_{ij}	angular momentum
$K_i = J_{0i}$	Lorentz boosts
T_{ij}	shears
$N_i = T_{0i}$	
T_{00}	

The relevant subgroups are

$\overline{SO}(3)$	(spatial rotations), generated by the J_{ij}
$\overline{SO}(4)$	the maximal compact subgroup, generated by the $J_{ij}\&N_i$
$\overline{SO}(1,3)$	the Lorentz subgroup, generated by the $J_{ij}\&K_i$
$\overline{SL}(3,R)$	the"3volume"-preserving group, generated by the$J_{ij}\&T_{ij}$
R_+	a one-parameter subgroup generated by T_{00}.

The commutation relations for $sl(n, R)$ now read:

$$[J_{ij}, J_{kl}] = i(\delta_{ik}J_{jl} - \delta_{il}J_{jk} - \delta_{jk}J_{il} + \delta_{jl}J_{ik})$$

$$
\begin{aligned}
[J_{ij}, N_k] &= i(\delta_{ik}N_j - \delta_{jk}N_i) \\
[J_{ij}, K_k] &= i(\delta_{ik}K_j - \delta_{jk}K_i) \\
[N_i, N_j] &= iJ_{ij} \\
[K_i, K_j] &= -iJ_{ij} \\
[J_{ij}, T_{kl}] &= i(\delta_{ik}T_{jl} + \delta_{il}T_{jk} - \delta_{jk}T_{il} - \delta_{jl}T_{ik}) \\
[J_{ij}, T_{00}] &= 0 \\
[T_{ij}, T_{kl}] &= -i(\delta_{ik}J_{jl} + \delta_{il}J_{jk} + \delta_{jk}J_{il} + \delta_{jl}J_{ik}) \\
[K_i, N_i] &= -i(T_{ij} + \delta_{ij}T_{00}) \\
[K_i, T_{jk}] &= -i(\delta_{ij}N_k + \delta_{ik}N_j) \\
[N_i, T_{jk}] &= -i(\delta_{ij}K_k + \delta_{ik}K_j) \\
[K_i, T_{00}] &= -2iN_i \\
[N_i, T_{00}] &= -2iK_i \\
[T_{ij}, T_{00}] &= 0.
\end{aligned} \tag{16}
$$

The compact operators are J_{ij} and N_i, while the remaining ones K_i, T_{ij} and T_{00} are noncompact. We have proved the following theorem: for any $\overline{SL}(n,R)$, $n \geq 3$ group, there exists an inner automorphism ("deunitarizing automorphism") generated by

$$
\mathcal{A} = \exp(\frac{\pi}{4}T_{00}) \tag{17}
$$

which leaves the $R_+ \otimes \overline{SL}(n-1,R)$ subgroup intact, and which acts on the $\overline{SL}(4,R)$ generators in the following way

$$
\begin{aligned}
&\mathcal{A}J_{ij}\mathcal{A}^{-1} = J_{ij}, \qquad \mathcal{A}T_{ij}\mathcal{A}^{-1} = T_{ij}, \quad \mathcal{A}T_{00}\mathcal{A}^{-1} = T_{00}, \\
&\mathcal{A}N_j\mathcal{A}^{-1} = iK_j, \qquad \mathcal{A}K_j\mathcal{A}^{-1} = iN_j, \quad \mathcal{A}D\mathcal{A}^{-1} = D.
\end{aligned} \tag{18}
$$

The *deuniterizing automorphism* $\mathcal{A}$ allows us to start by applying it to the generators, i.e. identify iN_i as the noncompact boost and iK_i as the compact vector-shear. We now construct the unitary representations of that $\overline{SL}(n,R)_A$ group. This puts iK_i finite multiplets in the reduction over the $\overline{SO}(4)$ compact subgroup. Having constructed the representation, we now apply $\mathcal{A}^{-1}$ to the generator algebra, thus reidentifying N_i as the vector-shears and K_i as the Lorentz boosts. In this way, we identify the finite (unitary) representations of the abstract $\overline{SO}(n)$ compact subgroup with nonunitary representations of the physical Lorentz group (J_{ij}, K_i); moreover, the infinite (unitary) representations, which without our transformation would have represented the abstract $\overline{SO}(1,n-1)$ Lorentz group of (J_{ij}, K_i), now represent (nonunitarily) the compact (J_{ij}, N_i) generators. The non-Hermiticity of the intrinsic boost operator parts here therefore cancels their physical action, precisely as in finite tensors or spinors, the boosts thus acting kinetically only. In this way, we avoid a disease common to infinite-component wave equations and to spectrum generating groups.

C. Particles: Casimir Invariant and the Unirreps of SA(4,R)

In *flat spacetime*, particle states in the Hilbert space are described by the unirreps of the Poincaré group, since they are characterized by momenta. In *curved spacetime*, this role has to be played by the Affine group $SA(4,R)$ (or alternatively, by $A(4,R)$, if we adjoin the operation of scaling). However, particle states are created and annihilated by field operators; thus, *there is a constraining algebraic "interface" bridging problem, between the field description, via nonunitary representations of the homogeneous* $\overline{SL}(4,R)$ (or of its $\overline{SO}(1,3)$ subgroup, for spinors in flat spacetime) *and the particle description, applying the inhomogeneous Poincaré or* $SA(4,R)$ *groups.* That interface is produced by *the overlap between the unitarily-represented subgroup of the homogeneous group, acting on fields, and the stability subgroup (or "little group") of the inhomogeneous group acting on particles.* The compact subgroup of the homogeneous group survives as the only physical subgroup acting intrinsically (the noncompact part is limited to a kinematical action by the implicit or explicit application of $\mathcal{A}$). The stability subgroup in the inhomogeneous group is the only one acting linearly, in the *induced representations* introduced by Wigner; moreover, the rest of the inhomogeneous action is indeed "induced" over that linearly acting subgroup. In the flat spacetime case, we thus have - for massive states - spin $\overline{SO}(3)$ as the overlap; for the massless case, the little group is isomorphic to a Poincaré-like two-dimensional inhomogeneous Euclidean group of rigid motions $SO(2)\times T^{2'}$. The translations $T^{2'}$ are not physical translations, and one has to select representations for which their eigenvalues vanish $T^{2'}| >= 0$. The effective part of the little group is reduced to $SO(2)$, the helicity subgroup, which is then the overlap, here therefore only with a subgroup of the the homogeneous group's $SO(3)$ compact subgroup. We now turn to the situation in the curved spacetime case.

Such an analysis and the resulting classification was performed[6] some years ago. The first step in a Wigner-like analysis of the unirreps of $SA(4,R)$ is to identify its Casimir invariants. $SA(n,R$ has one invariant, $A(n,R)$ has none, but the invariant of $SL(4,R)$ is also invariant under $A(n,R)$ in some representations. The unique Casimir invariant of $SA(n,R)$ can be written equivalently in terms of a Cartan-Weyl basis of the related $gl(n,R)$,

$$(L^a{}_b)_\delta{}^\gamma = \delta^a_\delta \delta^\gamma_b, \tag{19}$$

or in terms of the traceless $Q^a{}_b$. It is given by

$$\begin{aligned} C(n) = \; & Sym\Big[\epsilon^{a_0\cdots a_{n-1}} P_{a_0}(L^{b_0}{}_{a_1} P_{b_0})(L^{c_0}{}_{a_2} L^{d_1}{}_{c_0} P_{d_1})\cdots \\ & (L^{k_0}{}_{a_{n-1}} L^{k_1}{}_{k_0}\cdots L^{k_{n-2}}{}_{k_{n-3}} P_{k_{n-2}})\Big], \end{aligned} \tag{20}$$

where Sym denotes the symmetrization of all generators. Eqn. (20) is equivalent to the determinant

$$C(n) = \det \begin{pmatrix} P_0 & L^b{}_0 P_b & L^c{}_0 L^d{}_c P_d & \cdots \\ P_1 & L^b{}_1 P_b & L^c{}_1 L^d{}_c P_d & \cdots \\ \vdots & \vdots & \vdots & \cdots \\ P_{n-1} & L^b{}_{n-1} P_b & L^c{}_{n-1}, \; L^d{}_c P_d & \cdots, \end{pmatrix} . \tag{20a}$$

or, in shorthand notation, to

$$C(n) = \det\left(P, LP, (L)^2 P, .., (L)^{n-1} P\right), \tag{20b}$$

thus involving powers of the basis L in the Lie algebra $gl(n,R)$ going from 0 to $n-1$. Hence the invariant $C(n)$ is a polynomial of degree n in the translations p and of degree $n(n-1)/2$ in the $sl(n,R)$ generators; altogether, it is thus of degree $n(n+1)/2$. The Casimir operator can also be represented as

$$C \equiv (n) = \det\left(P, QP, (Q)^2 P, \cdots, (Q)^{n-1} P\right). \tag{20c}$$

We can now perform our analysis. The main difference between this and the Poincaré case is that whenever in n-dimensional Poincaré, a massive rest state is chosen, namely $P_i| >= 0, P_0| >= m$, for the definition of the little group, the homogeneous group being pseudo-orthogonal, an $\overline{SO}(n-1)$ is projected out as the little group. In $\overline{SA}(n,R)$, however, there are $n-1$ additional generators preserving the defining direction, namely Q^0_i, each the sum of a rotation and a shear. It acts like an abstract translation, so that the little group is $\overline{SA}(n-1,R)'$, with $P_{i'} := Q_i^0$. This resembles the case of massless particles in flat spacetime, where the Poincaré little group is isomorphic to the group of rigid motions in Euclidean 2-space. Here we then have two subcases: if the eigenvalues of P' vanishes, the little group becomes $\overline{SL}(n-1,R)$, which overlaps with the unitarily-represented subgroup of the homogeneous $\overline{SL}(n,R)$: the noncompact operators of $\overline{SL}(n-1,R)$ are unchanged by $\mathcal{A}$ (eqn. (17) and the representations are infinite-dimensional and unitary. This happens, for instance, in matter manifields, i.e. their particle states will correspond to *Regge trajectories*, described by unitary infinite–dimensional representations of $\overline{SL}(3,R)$. The alternative, in which the abstract $P_{i'}$ momenta do not vanish amounts to the start of a new logical cycle, i.e. defining a little group for $\overline{SA}(n-1,R)'$, etc..

The group $\overline{SA}(n,R)$, acting on the space of momenta, has two orbits:

$$Orb_1 = \{0\}, \qquad Orb_2 = R^n - \{0\}. \tag{21}$$

For the null orbit,i.e. when we select states for which all n components of the momenta vanish, the Casimir operator vanishes, since it is a homogeneous symmetric polynomial of degree n in the momenta. In the second orbit — which, incidentally, is invariant under the entire $\overline{GL}(n,R)$ — for small values of the momenta, the invariance of the Casimir operator implies that the eigenvalues of the $\overline{SL}(n,R)$ homogeneous operators must grow fast.

These two orbits provide for a classification [ref. 6] of the unitary irreducible representations of $\overline{SA}(4,R)$. We have a hierarchy of stability subgroups over which the unirreps are constructed as *induced* representations. The eigenvalues of the four–vector P_μ either vanish, $p = 0$ (case I) and $C(4) = 0$ or it does not, $p \neq 0$ (case II) and $C(4) \sim p^4 = m^4$, where m denotes the particle's mass.

Case I: The little group is $\overline{SL}(4,R)$. The unirreps of this group have been classified[13]. they are rather unphysical in that the Lorentz subgroup will appear in unitary infinite representations, the unirreps of Gel'fand and Yaglom. These contain all spins, and the action of the Lorentz boost on a state with spin j connects it with those of spin $j+1$ and $j-1$. Such 'particles' are thus not characterized by definite spins, as phenomenologically required. This is the difficulty we have avoided for fields and manifields, by applying the *deunitarizing automorphism* $\mathcal{A}$.

Case II: The little group is $\overline{SA}(3,R)'$. This affine group consists of the semi-direct product of the spatial $\overline{SL}(3,R)$ with a 'fake' set of three 'translation' momenta p'. There are two subcases:

Case IIA: All three components $p' = 0$. The effective little group is then $\overline{SL}(3,R)$. The unirreps are induced over this subgroup, they can be reduced to infinite discrete sums of spins, fitting the hadron situation and also providing an interesting model for primordial fermion fields (in fact manifields). This picture has been studied in ref. 14-17. Note that $C(3') = 0$, and as a result $C(4) = 0$ as well, since the multiplier of p^4 is precisely the $C(3')$ Casimir invariant of the stability subgroup defined by p.

Case IIB: The fake momenta satisfy $p' \neq 0$. We can select a frame in which only p'^0 does not vanish, a fake energy-like component. $C(3') \sim (p'^0)^3 = (m')^3$, m' a mass-like eigenvalue. The new little group is $\overline{SA}(2,R)''$. Again, the "translations" are fake momenta p''. We can have two subcases:

Case IIB1: All components of $p'' = 0$ and $C(2'') = 0$. In that case, we get again both $C(3") = 0$ and $C(4) = 0$. The effective little group is $\overline{SL}(2,R)$ (i.e. the double-covering, in an infinitely covered group). We have seen that the unirreps have been classified by Bargmann and are useful in a variety of physical contexts.

Case IIB2: $p'' \neq 0$, $C(2'') \sim (p'')^2 = (m'')^2$. The little group is $SA(1,R)$, with one fake momentum p'''. Again we have two possibilities:

Case IIB2a: $p''' = 0$, $C(1''') = 0$. This is a scalar representation. As a result, $C(2'') = C(3') = C(4) = 0$.

Case IIB2b: $p''' \neq 0$, $C(1''') = q = m'''$. Note that here $C(2'') = (m'')^2 m'''$, $C(3') = (m')^3 (m'')^2 m'''$ and $C(4) = m^4 (m')^3 (m'')^2 m'''$.

To summarize, we have 5 classes of representations: I, IIA, IIB1, IIB2a, IIB2b, which are illustrated in the following skeleton,

$$
\begin{array}{lcll}
\overline{SA}(4,R) & \supset & \overline{SL}(4,R) & : \quad I \\
\uparrow & & & \\
\overline{SA}(3,R)' & \supset & \overline{SL}(3,R) & : \quad IIA \\
\uparrow & & & \\
\overline{SA}(2,R)'' & \supset & \overline{SL}(2,R) & : \quad IIB1 \\
\uparrow & & & \\
SA(1,R) & & & : \quad IIB2a\,,\ IIB2b
\end{array}
\tag{22}
$$

Moreover,

$$C(4) = 0, \quad \text{for I, IIA, IIB1, IIB2a;} \quad C(4) = m^4(m')^3(m'')^2m''' \text{ for IIB2b.} \tag{23}$$

At first sight, the Casimir invariant appears to constrain the masses and spins in a wrong manner, as in the Majorana infinite–dimensional equation: the higher the spin, the lower the mass; this is the opposite of what we observe in hadron phenomenology and of what is assumed in the Chew–Frautschi plot for a Regge trajectory. However, in the general case (including the most useful case IIA) the invariant vanishes, and the value of m^4 thus stays unconstrained in all but case IIB2b. Instead, constraints on the value of the masses may be derived dynamically. It is remarkable that an evaluation based on the 'chromogravity' approximation for QCD in the infrared region does reproduce the linear correlation between m^2 and the spin j.

6. Manifield Equations and Manibein Frames

In gravity, aside from providing for matter fields in affine and metric–affine theories, the new structures can be used for phenomenological studies and calculations in Einsteinian gravity, when protons or neutrons, for instance, are involved. In a fundamental Lagrangian, we would have quark fields and QCD gluons. ¿From these it would be almost impossible to calculate precise effects on their bound states, the protons and neutrons. In the past, the method used was to treat the protons and neutrons as Klein-Gordon particles, or at best, as phenomenological Dirac fields; this however is an incorrect prescription, since nucleons are not Dirac particles. Not only do they have large anomalous magnetic moments, but in addition, they possess a large sequence of excited states, for which there is no description in the Dirac equation. I shall present here the covariantly correct treatment, i.e. in terms of manifields.

Returning now to our algebraic constructs, we note that the simplest $\overline{SL}(4,R)$ manifields are the multiplicity–free ones. In particular, the two conjugate representations in the discrete series

$$\mathcal{D}^{disc.}_{\overline{SL}(4,R)}(\frac{1}{2},0) \oplus \mathcal{D}^{disc.}_{\overline{SL}(4,R)}(0,\frac{1}{2}) \tag{24}$$

contain, as representations of the "little group" $\overline{SL}(3,R)$, sequences of

$\mathcal{D}^{ladd.}_{\overline{SL}(3,R)}(\underline{j} = \frac{1}{2}; \sigma_2 = 0)$, the special spinorial unirrep we discussed in section 4. This reduces to an infinite sum of $SU(2)$ representations, $j = \frac{1}{2} \oplus \frac{5}{2} \oplus \frac{9}{2} \oplus \cdots$

The Lorentz group representation $(1/2, 0) \oplus (0, 1/2)$ appearing as the lowest $SO(4)$ submultiplet in eqn. (24) is the direct sum of two finite, chiral, and non–unitary representations, related by parity. This describes the Lorentz behavior of the Dirac field. In a region where the deformations are switched off, with no tensor field to relate the various $\Delta j = 2$ levels, this ground state decouples and becomes a true Dirac spinor field.

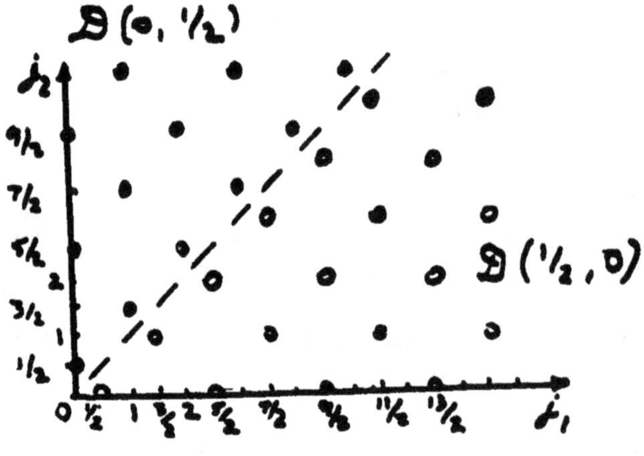

Figure 1: The basic spinor manifield $\mathcal{D}^{disc.}_{\overline{SL}(4,R)}(1/2, 0) \oplus \mathcal{D}^{disc.}_{\overline{SL}(4,R)}(0, 1/2)$, extending the Minkowski space Dirac spinor field of the same 'name' into $\overline{SL}(4, R)$. This reducible $\overline{SL}(4, R)$ representation is the direct sum of two infinite, $\mathcal{A}$–deunitarized irreducible representations, denoted by hollow and full circles respectively.

We now discuss various physical contexts which might involve these *spinorial manifields*. We take first the case in which the action of the linear group is defined on the frames, i.e. representing the anholonomic $\overline{SL}(4, R)_A$, the manifield is $\Psi^{\Xi}(x)$, where the anholonomic (upper case Greek) indices $\Xi, \Pi, \Theta = 1, 2, \ldots, \infty$ run over the countable infinity of the components of the manifield representations, namely, in the usual case, the quantum numbers of the representations of the $Spin(1,3) = \overline{SO}(1,3)$ - finite and non-unitary, by construction - isomorphic to the (finite) unitary representations of $\overline{SO}(4,R)$. For example, we may have a sequence $(0,0),(1,1),(2,2),\cdots$, i.e. single-valued representations, in the case of the *infinitensor* $D(0,0)$, or the sequence $(1/2,0),(3/2,1),(5/2,2)\cdots$, double-valued representations, for one of the two "ladders" in the spinorial "Dirac-manifield".

To realize the covariance group, we have to specify $\overline{SL}(4,R)_H$, the *holonomic* special linear subgroup of $\overline{Diff}(4,R)$; in addition, we shall have to correct for the action of the covariance group on the coordinate in the argument of the manifield.

If we lift the Lie generators Q^i_j of the linear connection Γ^j_i to the manifield representation, denoting it by $\rho(Q^i_j) = \{(Q^i_j)^\Pi_\Xi\}$, (for this holonomic case we use upper case Greek indices), we can introduce the exterior covariant derivative

$$D\Psi^\Pi(x) = \left[\delta^\Pi_\Xi d + \Gamma^b_a (Q^a_b)^\Pi_\Xi\right] \Psi^\Xi(x) . \tag{25}$$

The manifield representation is irreducible, because the shear generators (still in the holonomic case) $Q_{(ij)}$, i.e., the generators of the coset space $\overline{GL}(n,R)/\overline{SO}(n)$, connect all these various substates over the $|\Delta j| = 2$ intervals. Physically, it is the gravitational field (or the effective field of *"chromogravity"*, in the flat spacetime physics of the hadrons) which connects these substates, since they are coupled through the covariant derivative by means of the matrices $\{(Q^i_j)^\Pi_\Xi\}$. Note that to the extent that we are treating (in metric-affine gravity) the case of an anholonomic action of $\overline{SL}(4,R)_A$ on the frames, a similar covariant derivative can be defined, with the indices M, N replacing Ξ, Π.

The relationship between $\overline{SL}(4,R)_A$ and $\overline{SL}(4,R)_H$ is somewhat less straightforward than the one existing between the Lorentz groups in the anholonomic and holonomic physical realizations. Special Relativity does not allow a symmetry larger than the Lorentz group. Thus, the frames over which the anholonomic group is acting are not flat space frames as in GR. We shall however have occasion to use the anholonomic linear group and its double-covering in flat space; in such situations, the manifield breaks down and reduces to an infinite sum of Minkowski spacetime fields. Phenomenologically, $\overline{SL}(4,R)$ is then used as a Spectrum Generating Group[19]. There are, however, other physical situations in which the frame manifield remains irreducible – e.g. in models of quantum gravity in which the high-energy (above Planck mass) regime is non–Riemannian. In either case, we can introduce a local *manibein* frame[14] , by lifting the usual tetrad (one-form coefficient) fields $e^a_\mu(x)$ – 4×4 field–valued matrices relating Lorentz (the index) to $SL(4,R)_H$ in their $(1/2,1/2)$ 4-vector representations – so as to relate here $\overline{SL}(4,R)_A$ to $\overline{SL}(4,R)_H$ via the manifield representation. This defines an infinite–dimensional matrix $E^M{}_\Pi(x)$. As a result, we can define holonomic *world spinors*, $\Psi^\Xi(x)$,

$$\Psi^M(x) = E^M{}_\Xi(x)\Psi^\Xi(x), \qquad E^M{}_\Xi := E(e^a{}_\mu)^M{}_\Xi \tag{26}$$

where $\Psi^\Xi(x)$ (we can also solve (26) for it by inverting the frame $E^M{}_\Xi$) is now a spinorial manifield under the action of $\overline{SL}(4,R)_H$; with the action of the covariance group double–covering $\overline{Diff}(n,R)$ following the usual prescription of a non–linear realization over its linear subgroup. We have thereby obtained a field which will carry faithful representations of that covariance group's double–covering, fitting the *"world" spinor* description. Before dealing in detail with the transformation properties of both world spinors and affine–frame spinor manifields, we note that for a general world tensor field $\Phi^{\alpha\cdots\beta}_{\gamma\cdots\zeta}$, there are two ways of describing the transformations under $SL(4,R)$: (a) As a direct product of p covariant and q contravariant 4-vector "fundamental" representations, or, (b) after contractions over indices of complementary types, and symmetrization and antisymmetrization over indices of the same type, the tensor carries an *irreducible* representation of $SL(4,R)$. There is one single type of manifield, constructed as a ladder representation, which can be considered as a limiting case of some types of ordinary tensors – namely the totally symmetric contravariant $\Phi^{\alpha\cdots\beta}$ and its covariant analog – when the number of indices goes to infinity (for a spinor, Φ itself is a Lorentz spinor). In that case, option (a) is available. In the general case, however, we have to use option (b), i.e. we deal with the field as the carrier of an irreducible (non–unitary) representation. For an ordinary tensor field we have the infinitesimal variation of (1) (the *alias* piece), corrected by an *alibi* increment, i.e. together,

$$\delta\phi^m(x) = -i\partial_\mu\xi^\nu(x)\,[\rho(L^\mu{}_\nu)]^m{}_n\phi^n((\sigma^\mu{}_\nu)^{-1}x) + \xi^\mu\partial_\mu\phi^m(x) \tag{27}$$

where $\xi^\nu(x) = \delta x^\nu$, the indices m,n denote components of the field representation, $\rho(L)$ is the appropriate matrix representation and $\sigma(L^{-1})$ is the

4×4 matrix representation of the inverted action of that generator. The $SL(4,R)_H$ matrix is multiplied by a parameter which is the relevant component of the gradient of x^μ, resulting from the Jacobian determinant and producing the non-linear action of the $Diff(4,R)/SL(4,R)_H$ quotient.

The infinitesimal variation of the world spinor manifield under the action of the diffeomorphism $x \to x' = x + \xi$ will be very similar, except that the action on the spinorial manifield will be realized through the infinite matrices of $\overline{SL}(4,R)_H$,

$$\delta\Psi^\Xi(x) = -i\partial_\mu\xi^\nu(x)\left[\rho(\overline{L}^\mu{}_\nu\right]^\Xi{}_\Omega\Psi^\Omega((\sigma^\mu{}_\nu)^{-1}x) + \xi^\mu\partial_\mu\Psi^\Xi(x) \qquad (28)$$

We note that the generators of $\overline{SL}(4,R)_H$ are given, in terms of the anholonomic ones by

$$(\overline{L}^\mu{}_\nu)^\Xi{}_\Omega = H^\Xi{}_M(x)(\overline{Q}^a{}_b)^M{}_N E^N{}_\Omega(x) \qquad (29)$$

where $H = E^{-1}$ is the inverted manibein. The "infinitesimal" parallel–transport of a manifield under a one–parameter subgroup of $\mathcal{T}^n \approx Diff(n,R)$ with generator ξ is provided by the $GL(n,R)$-gauge covariant Lie derivative

$$\pounds_\xi\Psi^M(x) = \xi\rfloor D\Psi^M(x) + D\xi\rfloor\Psi^M(x)\,. \qquad (30)$$

Holonomically, i.e., for a world spinor $\Psi^\Xi(x)$, Eq. (30) becomes the usual covariant Lie derivative $\ell_\xi = \xi\rfloor d + d\xi\rfloor$.

Fermionic holonomic manifields (with half–integer Lorentz subgroup spin) are *world spinors*, to be clearly distinguished from *bosonic* ones. The bosonic ones contain the same algebraic information as ordinary tensor fields, in as much as they belong to $Diff(n,R)$. They are thus better known as *infinitensors.* The fermionic world spinors, on the other hand, display the action of the covering group $\overline{Diff}(n,R)$ *faithfully* and are therefore more general.

In order to introduce covariant differentiation for world spinors, the linear connection has to be lifted to the manifield and then written holonomically. We find the following inhomogeneous relation

$$\Gamma'^{\,b}_a(L^a{}_b)_M^N = H^\Pi_M\,\Gamma_\mu{}^\nu(L^\mu{}_\nu)_\Pi^{Xi}\,E_\Xi^N - H^\Pi_M dE_\Pi^N \qquad (31)$$

between the anholonomic (Latin indices) and the holonomic (Greek indices) connection for a manifield with respect to a manibein transformation.

In the minimal coupling prescription, the coupling to the connection is then achieved by means of the covariant manifield derivative $(D_\mu := \partial_\mu \rfloor D)$)

$$D_\mu \Psi^{\Xi}(x) = [\delta^{\Omega}_{\Xi}\partial_\mu + \Gamma_{\mu\nu}{}^{\rho}(L^{\nu}{}_{\rho})_{\Omega}{}^{\Xi}]\Psi^{\Omega}(x) \ . \tag{32}$$

Then, in first order formalism, the world spinor matter Lagrangian reads

$$L_{mat} = L_{mat}\Big(E^{L}{}_{\Lambda}(x), \Psi^{\Xi}(x), D_\mu \Psi^{\Xi}(x)\Big) \ . \tag{33}$$

Several theorems constrain the construction of manifield equations, aside from the use of $\mathcal{A}$. First,*in $\overline{SL}(4,R)$ or $\overline{Diff}(4,R)$ covariant equations, the manifield cannot correspond to a multiplicity-free representation.* In a curved space and covariant equation of motion, the 'γ-type matrices' X^a span a four-vector representation $(1/2, 1/2)$ under $sl(4,R)$, i.e. $[Q^a{}_b, X^c] = \delta^c_b X^a$. Taking the commutator between states connected by the X^a, we find that for the left hand side not to vanish, $Q^a{}_b$ has to act (non-trivially) as a $(0,0)$ transition, i.e. reproduce the same state, when it acts on a carrier space of $sl(4,R)$. This cannot happen by definition in the *multiplicity-free* representations. Note that if we start - for the sake of simplicity - with a multiplicity-free manifield *in flat spacetime*, we thus know that the effect of the manibeins will have to transform the manifield into a world spinor which will no more be multiplicity-free. This will be a convenient way of applying the formalism. For this purpose, we have to know more about *the flat spacetime limit.* Here we have another theorem. It states that *in the absence of the gravitational field, a Lorentz invariant equation decouples the states of $\mathcal{D}^{disc.}_{\overline{SL}(4,R)}(1/2,0) \oplus \mathcal{D}^{disc.}_{\overline{SL}(4,R)}(0,1/2)$ lying outside of the main diagonal of Fig. 1.* The infinite γ-like matrices X^a are still a Lorentz four-vector, i.e. a $(1/2,1/2)$ representation. Such operators cannot connect a state to any other state, except between both sides of the main diagonal "alley".

To construct a holonomic equation, i.e. one describing a world spinor, we thus first construct a Lorentz-invariant (i.e. flat space) equation, e.g. for the manifield $\mathcal{D}^{disc.}_{\overline{SL}(4,R)}(1/2,0) \oplus \mathcal{D}^{disc.}_{\overline{SL}(4,R)}(0,1/2)$ (while applying the deunitarizing automorphism) $[i\, X^\mu \partial_\nu - m(p)]\, \Psi = 0$

The anholonomic X^a are constructed in the following way: first we embed $\overline{SL}(4,R)$ in $\overline{SL}(5,R)$ and then select a pair of parity conjugate principal series representations which are contained in the $SL(4,R)$ reduction of our spinorial representations. Let the generators of $\overline{SL}(5,R)$ be $\overline{L}_A{}^B$, $A, B = 0, \cdots 4$. We define $X^a = \overline{L}^{a]}_{[4}$, $a = 0,1,2,3$, which yields an $SL(4,R)$ four-vector. Note that the X^a operators constructed in this way yield, upon commutation, the $SL(2,C)$ generators, generalizing a property of Dirac's γ-matrices. To go over to a world spinor field, we now use "manibeins", etc..

Note that in the conventional transition between holonomic and anholonomic indices for tensors, mediated by the vielbein coefficients $e_\mu{}^a$ we have $g_{\mu\nu} = e_\mu{}^a e_\nu{}^b \eta_{ab}$, with η_{ab} the Minkowski metric. We can construct an analog in the infinite representation of the world spinor. Denoting by X^0 the constant γ^0-like matrix in the X^μ set, we can form the Dirac-type adjoint $\overline{\Psi} := \Psi^\dagger X^0$ and find similarly from the scalar product of world spinors,

$$\overline{\Psi}(x)\,\Psi(x) := \Psi^{\dagger\Xi}(x)\,E_\Xi{}^M E_\Pi{}^N (X^0)_{MN}\,\Psi^\Pi(x) =: \Psi^{\dagger\Xi}(x)\,G_{\Xi\Pi}(x)\Psi^\Pi(x)\,, \tag{34}$$

The symmetric infinite-component tensor $G_{\Xi\Pi}(x)$ is a functional of the gravitational field realizing the metric $g_{\mu\nu}$ on the world spinor components. In Riemann–Cartan or in Riemannian spacetime, either reached directly, or through spontaneous symmetry breakdown[20–22], we expect, in accordance with the principle of equivalence, to get a metricity condition, perhaps (in the SSB case) a "weak" equation for "low energy states":

$$< \ell.e.s.|D_\mu G_{\Xi\Pi}(x)|\ell.e.s. > \cong 0\,. \tag{35}$$

7. An Application: a Gravity-like Component in QCD ("Chromogravity")

Some resemblances between Strong Interactions and Gravity were noticed by Salam and others in the sixties, leading them to the hypothesis of "Strong Gravity" as the theory of the Strong Nuclear Interaction. As a matter of fact, even the present (1997) most popular candidate for a theory of Quantum Gravity, namely the Quantum Superstring (or beyond it, "M theory") in the context of an "ultimate" theory, was first identified as the theory of hadron dynamics, i.e. of the Strong Interactions. With the advent of QCD, however, "naive" strong gravity had to be abandoned. For a while, new models were tried, in which QCD was "extended" so as to include elements of "strong gravity" *in addition* to QCD itself (example: an $sl(6,C)$ gauge theory).

An alternative approach has consisted in extracting a kind of "strong gravity" from within QCD itself. With Dj. Šijački, we have pioneered this idea through an algebraic derivation[6,7]. We have "sharpened" the derivation in two recent publications[23,24]. Similar programs, though utilizing different techniques, were launched by F.A. Lunev, D.Z. Freedman, K. Johnson, D. Singleton, V. Brindejonc and G. Cohen-Tannoudji and others (see listings in ref. 24).

In the first study[23], we provided a mathematical derivation demonstrating *the emergence of "effective" diffeomorphisms as a subset of the color-$SU(3)$ local gauge transformations in inter-hadron interactions.* That proof, however, involved an integration by parts, with the possibility of complications as a result of surface terms. In the second article[24], we showed that the proof can also be derived directly, using an expansion in terms of distances. Replacing the low-frequency limit by the (conjugate) large-distance limit results in avoiding integrations and surface terms.

We had originally suggested[6] that the spacetime-geometrical nature of QCD confinement – as discussed in *ad hoc* theories such as the "Bag Model" – *might induce* Regge systematics based on $\overline{SL}(4, R)$ or $\overline{SA}(4, R)$, with relevant stability subgroup $\overline{SL}(3, R)$, preserving a fixed volume. This would also fit with observations from the phenomenology and with an algebraic description of gravitational-like quadrupolar pulsations[10], even though unrelated to gravity. The approach was further strengthened by field theory covariance considerations, also involving an $\overline{SL}(4, R)$, this one related to true (gravity-extendable) covariance. Between the two groups, a transformation is involved, related to the one between current and constituent quarks[17].

We have chosen to work with variables displaying gauge variations, thus providing for a source of (effective) diffeomorphisms. As a result, (a) an $\overline{SL}(3, R)$ invariance then becomes implicit, as the stability subgroup of $\overline{SA}(4, R)$ – to the extent that the effective "chromogravity" might take on *affine* features (hadron spectroscopy does display a good fit between this dynamical algorithm and phenomenology, with respect to both classification and energy-spacings in Regge sequences); (b) the model supplies a confining mechanism, namely p^{-4} propagators, resulting from a Riemannian constraint relating the (pseudo) connection to the (pseudo) metric: (c) there is a smooth transition to the ("semi-soft") domain of hadron high-energies, in which the same color-neutral gluon combinations have provided a model for diffractive scattering [25,26]; (d) an understanding of the relationship between the quark current-to-constituent mass-growth and the slope of the Regge sequence, in terms of chromo-graviton self-energies; and (e) a dynamical derivation of the

$J = 2^+, 0^+$ ground state in even-even nuclei, the algebraic foundation for the successful "IBM" $SU(6)$ spectrum generating group in nuclei[27,28]. This appears as a long-range excitation of the chromo-gravitational quanta, somewhat similar to a Van der Waals effect in QED.

We had originally[7] conjectured (a) that color-neutral gluon exchange forces make up an important component of inter-hadron interactions, or of *"soft* QCD" interactions; (b) that this component produces *a longer-range force*, with many of the characteristics of gravity, including (and "protected" by) the basic mathematical criterion, namely *invariance under diffeomorphisms*; (c) that the simplest such n-gluon operator product, that of a two-gluon system

$$G_{\mu\nu}(x) = (\kappa)^{-2}\; k_{ab}\; B^a_\mu(x)\; B^b_\nu(x) \tag{36}$$

– where κ has the dimensions of mass, μ, ν, $\cdots$ are Lorentz 4-vector indices, a, b, $\cdots$ are $SU(3)$ adjoint representation (octet) indices, k_{ab} is the Cartan metric for the $SU(3)$ octet, B^a_μ is a gluon field – fulfills the role of an effective (pseudo) metric, with respect to these (pseudo) diffeomorphisms, in the same manner that the physical metric (through its Christoffel connection) "gauges" the true diffeomorphisms.

The proof of this conjecture then follows a three-step argumentation. The first step consists in the definition of an "IR limit" and the demonstration of *the presence of effective ("chromo") diffeomorphisms, as a class of QCD gauge functions*, in the color-gauge transformations of the above $G_{\mu\nu}$.

The gluon color-$SU(3)$ gauge field transforms under an infinitesimal local $SU(3)$ variation according to

$$\delta_\epsilon B^a_\mu = \partial_\mu \epsilon^a + g B^b_\mu \left\{\frac{\lambda_b}{2}\right\}^a_c \epsilon^c = \partial_\mu \epsilon^a + ig f^a_{\ bc} B^b_\mu \epsilon^c \tag{37}$$

(using the adjoint representation $\left\{\frac{\lambda_b}{2}\right\}^a_c = -i f_b{}^a{}_c = i f^a_{\ bc}$; g being the QCD coupling constant). The gauge field operator is then expanded around a constant vacuum solution (pure gauge) of the instanton type,

$$B^a_\mu = N^a_\mu + A^a_\mu, \tag{37a}$$

$$\partial_\mu N^a_\nu - \partial_\nu N^a_\mu - ig f^a_{\ bc}\; N^b_\mu\; N^c_\nu \to 0. \tag{37b}$$

By further defining a constant (flat) Euclidean metric, one can now replace κ in (36) by the "flat" density,

$$G_{\mu\nu} = \frac{g_{ab}\, B^a_\mu\, B^b_\nu}{[\det(g_{ab}\, N^a_\mu\, N^b_\nu)]^{1/4}} \tag{38}$$

thus yielding a non-singular dimensionless Riemannian "metric" for either signature. Its color-$SU(3)$ infinitesimal gauge variation is given by

$$\begin{aligned} \delta_\epsilon G_{\mu\nu} &= \delta_\epsilon \left\{ k_{ab}(N^a_\mu + A^a_\mu)(N^b_\nu + A^b_\nu) \right\} \\ &= k_{ab}(\partial_\mu \epsilon^a\, N^b_\nu + N^a_\mu\, \partial_\nu \epsilon^b\, + \partial_\mu \epsilon^a\, A^b_\nu + A^a_\mu\, \partial_\nu \epsilon^b) \\ &+\ ig\, k_{ab} \left\{ f^a_{cd}\, B^c_\mu\, \epsilon^d\, B^b_\nu + f^b_{cd}\, B^a_\mu\, B^c_\nu\, \epsilon^d \right\}. \end{aligned} \tag{39}$$

The last bracket vanishes, since it represents the homogeneous $SU(3)$ transformation of the $SU(3)$ scalar expression in (36), or, more technically, due to the total antisymmetry of f_{abc} in a compact group. It is at this stage that our first proof[23] involved an integration by parts, replacing the terms of type $A^a_\mu\ \partial_\nu \epsilon^b$ by terms $-\partial_\nu A^a_\mu \epsilon^b$, and defining the "IR" region through $\partial_\nu A^a_\mu = 0$. Although it is plausible to assume that the only surface term in integrating $B^a_\mu\ \partial_\nu \epsilon^b$ would come from N^a_μ rather than from A^a_μ, justifying that proof, it is good to know that this argument can be replaced by a more heuristic approach, foregoing the integration altogether.

At this stage, one of two sets of background conditions is assumed:

(a) a Euclidean metric, to adapt to quantum solutions (such as instantons), in which case one discusses the limit of large $|x|$;

(b) Alternatively, in Minkowski spacetime, one uses a non-relativistic picture. Working in polar coordinates (r, θ, ϕ, t), one expands in powers of r, having fixed $t = 0$, *taking the limit of large r – which again is the same as large $|x|$*. Since large r will also imply large $|x|$ for the Euclidean case, one is mostly using that definition for the limit at which the gauge field is expanded, thus covering both applications as large $|x|$ limits . Expanding B^a_μ in powers of r, one now writes,

$$B \simeq ...c_{-2}(\theta, \phi) r^{-2} + c_{-1} r^{-1} + N + c_1 r + c_2 r^2 \tag{40}$$

One now projects, out of this summation, states (or sometimes values of θ, ϕ) for which the c_1 and c_2 coefficients are small for large r, i.e. letting the N term dominate. *This fits with a picture in which the linear or higher terms correspond to a non-vanishing color source and generate confinement (possibly with flux-tube like departures from spherical symmetry), whereas the color-neutral hadrons do not confine (and fit spherical symmetry).* The above inter-hadron $\tilde{B}$ gauge field is thus the *effective long range gauge field, in which the confining features have already been cancelled between gluons, in making up a color-free set.* As to the non-constant $\tilde{A}^a_\mu$ *residual component, it thus involves, in this limit, either periodic terms or only negative powers of* r.

Returning to our proof, dealing with the non-perturbative (Minkowski "IR", or "soft") region, the expansion of the gauge field operator in eqs. (37) should now be redefined for large r (or large $|x|$): the constant global vacuum solution N^a_μ, is now defined by its vanishing field-strength *in the long range limit*, i.e. the equality sign should be replaced by an arrow for that limit! At large distances r, the N^a_μ field is thus required to approach a constant value, so that we have

$$k_{ab}\, N^a_\mu\, \partial_\nu \epsilon^b \to \partial_\nu (k_{ab}\, N^a_\mu\, \epsilon^b) \tag{41}$$

In the Euclidean case, the vacuum solution $N^a_\mu(x)$ is then indeed of the instanton type, and can be written, for instance, as

$$N_\mu(x) = N^a_\mu(x)\frac{\lambda_a}{2} = \frac{|x|^2}{|x|^2+\tau^2}[U^{-1}(x)\partial_\mu U(x)], \tag{42}$$

Only $\tilde{A}^a_\mu(x)$ preserves an x-dependence in the above limit and we have seen that it consists in negative powers of x. Thus, $\frac{\partial}{\partial r}\tilde{B}^a_\mu \simeq 0$ and as a result, in the Euclidean case – and effectively in Minkowski spacetime too (for spherical symmetric situations) one has the defining constraint,

$$\partial_\nu \tilde{B}^a_\mu \simeq 0. \tag{43}$$

Fourier-transforming, in Minkowski spacetime, to momentum space, one has a summation $\int dk^\nu \tilde{B}^a_\mu \simeq 0$, which, when applied to gluon Hilbert space "on mass shell" states, projects out those boundary conditions in which only the $k^\mu = 0$ states contribute, i.e. the low frequency (IR) gluon regime.

We can now go back to eq. (39). From the analysis of the r-expansion constituents, *we now conclude that the terms involving* N^a_μ in (39) *dominate those involving* $\tilde{A}^a_\mu$ *at large distances.* This is the basic term where we no more depend on the integration by parts.

With N^a_μ, N^b_ν, representing – in the $r \to \infty$ limit – constant fields, one now rewrites the terms in which they appear as a new infinitesimal variation,

$$\xi_\mu = k_{ab}\epsilon^a N^b_\mu \tag{44a}$$

We shall return and complete the definition of this "IR limit". Meanwhile, as a result, one can write in that limit,

$$\delta_\xi G_{\mu\nu} = \partial_\mu \xi_\nu + \partial_\nu \xi_\mu = \partial_\mu(\xi^\sigma G_{\sigma\nu}) + \partial_\nu(\xi^\sigma G_{\mu\sigma}) \tag{44b}$$

where we have changed over to the ξ^σ variable and where *we can reidentify δ_ξ as a variation under a formal diffeomorphism of the R^4 manifold.* Eq. (44b) simulates the infinitesimal variation of a "world tensor" $G_{\mu\nu}$ under Einstein's covariance group, $x^\sigma \to x^\sigma + \xi^\sigma$. ξ^σ thus has to be defined as a contravariant vector. Note that $G_{\mu\nu}$ in (36) is invertible, thanks to the constant part N^a_μ, in the long range limit, and using a Taylor expansion we can evaluate the inverse $G^{\mu\nu}(x)$. As the μ, ν indices are "true" Lorentz indices, acted upon by the physical Lorentz group (or by $SO(4)$ in the Euclidean case), the manifold has to be pseudo-Riemannian or Riemannian: only these manifolds – with or without torsion – have tangents with pseudo-orthogonal or orthogonal symmetry. Thus

$$D_\sigma G_{\mu\nu} = 0. \tag{45}$$

We shall see in the sequel that this result may explain *color confinement*, due to the emergence of p^{-4} propagators. To complete this proof, one now evaluates the commutator of two such variations,

$$[\delta_{\xi_1}, \delta_{\xi_2}]\, G_{\mu\nu} = \delta_{\xi_3} G_{\mu\nu}\ , \tag{46a}$$

and verifies that

$$\xi_{3\mu} := (\partial_\nu \xi_{1\mu})\xi_2^\nu + (\partial_\mu \xi_{1\nu})\xi_2^\nu - (\partial_\nu \xi_{2\mu})\xi_1^\nu - (\partial_\mu \xi_{2\nu})\xi_1^\nu \tag{46b}$$

The definition of this "IR limit", which is based on the vanishing of the gluon 4-momenta, is now extended so as to include similar terms with vanishing momenta in all many-gluon zero-color operator products. This can be taken as an operational definition, sufficient for general purposes.

To draw an easier physical picture for the following considerations, place them in the context of Minkowski spacetime. One can then write a generic

'IR state', carrying 4-momentum k, as follows:

$$|\phi_{IR},\ k\rangle = \sum_{m=1}^{\infty} f_m(k_1, k_2, \ldots, k_m)\delta_{k,k_1+k_2+\cdots+k_m}|k_1 k_2 \ldots k_m\rangle \tag{47}$$

where $|k_1 k_2 \ldots k_m\rangle$ represents a state of m soft gluons ($k_i \approx 0,\ i = 1, 2, \ldots m$). One can now verify the effect of taking the variation (39) (or (44b)) – as a typical result in this formalism – between such states. For the residual component in (39) one writes

$$\langle\phi'_{IR}, k'|g_{ab}(\epsilon^a\partial_\mu\tilde{A}^b_\nu + \partial_\nu\tilde{A}^a_\mu\epsilon^b)|\phi_{IR},\ k\rangle, \tag{48}$$

Our conditions will indeed make the terms in N dominate over those in A. As a result, we can change to the ξ^σ variables of (46b) and reidentify δ_ξ as a variation under a formal R^4 diffeomorphism. To identify the subset of color-$SU(3)$ gauge functions which produce appropriate N^a_μ and fit our IR condition, we apply (44) to the variation in (37).

$$\begin{aligned}\partial_\nu(\delta_\epsilon\tilde{B}_\mu)^a &= \partial_\mu\partial_\nu\epsilon^a + igf^a_{\ bc}\partial_\nu B^b_\mu\epsilon^c + igf^a_{\ bc}(\tilde{A}^b_\mu + N^b_\mu)\partial_\nu\epsilon^c \\ &\rightarrow \partial_\mu\partial_\nu\epsilon^a + igf^a_{\ bc}N^b_\mu\partial_\nu\epsilon^c = D^{(N)}_\mu(\partial_\nu\epsilon^a),\end{aligned} \tag{49a}$$

where one drops the $\partial_\nu\tilde{B}^b_\mu$ term because of (43) and the term in $\tilde{A}^b_\mu$ since it is diminishing fast, as compared to the term in N^b_μ. Thus, the condition amounts to the vanishing of the long-range covariant derivative, i.e.

$$\partial_\mu(\partial_\nu\epsilon^a) = -igf^a_{\ bc}N^b_\mu(\partial_{nu}\epsilon^c). \tag{49b}$$

To solve this conditions, we make use of a similarity transformation that diagonalizes (in an 8×8 color-represenration space) the $f^a_{\ bc}N^b_\mu$ matrix: $f^a_{\ bc}N^b_\mu \rightarrow U^a_d f^d_{\ be}N^b_\mu U^{-1}e_c = (UfN_\mu U^{-1})_{(a)}\delta^a_c$, and $\epsilon^a \rightarrow U^a_b\epsilon^b$. (49b) is solved by

$$\partial_\nu(U\epsilon)^a = e^{-ig(UfN_\mu U^{-1})_{(a)}x^\mu}, \tag{49c}$$

and finally,

$$(U\epsilon)^a = \frac{1}{-ig(UfN_\mu U^{-1})_{(a)}}\left(e^{-ig(UfN_\mu U^{-1})_{(a)}x^\mu} - 1\right), \tag{49d}$$

where the integration constant is fixed by requirçing $(U\epsilon)^a < 1$ (consistency of the calculation). This is indeed the long-distance limiting value for the class of gauge functions yielding "pure gauge" fields of the instanton type, i.e. fields tending to a constant value N_μ^a at large distance.

The second step in the proof consists in showing that *the infinite algebra of chromo-diffeomorphisms is realized within the set of all multi-gluon color-neutral operator products, thus generalizing the two-gluons construction.* I refer the reader to refs.[23,24] for the details.

The set of all possible color-singlet configurations of gluon fields is rearranged by lumping together contributions from n-gluon, $n = 2, 3, \ldots, \infty$ and with the same Lorentz quantum numbers. The corresponding color-singlet n-gluon field operator has the following form

$$G^{(n)}_{\mu_1\mu_2\cdots\mu_n} = d^{(n)}_{a_1a_2\cdots a_n} \tilde{B}^{a_1}_{\mu_1} \tilde{B}^{a_2}_{\mu_2} \cdots \tilde{B}^{a_n}_{\mu_n} \tag{50a}$$

the set of all $G^{(n)}_{\mu_1\mu_2\cdots\mu_n}$ operators, $n = 1, 2, \ldots$, forms a basis of a vector space of colorless purely gluonic configurations. These field operators are also all functionally independent.

In the QCD variation of the $G^{(n)}_{\mu_1\mu_2\cdots\mu_n}$ field, the terms in $N^{a_i}_{\mu_i}$, $i = 1, 2, \ldots n$, again dominate over the corresponding terms in $\tilde{A}^{a_i}_{\mu_i}$, yielding variations which can be cast in a diffeomorphic mold,

$$\delta_\epsilon G^{(n)}_{\mu_1\mu_2\cdots\mu_n} = \partial_{\{\mu_1}\xi^{(n-1)}_{\mu_2\mu_3\cdots\mu_n\}} \equiv \delta_\xi G^{(n)}_{\mu_1\mu_2\cdots\mu_n}, \tag{51a}$$

where $\{\mu_1\mu_2\cdots\mu_n\}$ includes symmetrization of indices, and

$$\xi^{(n-1)}_{\mu_1\mu_2\cdots\mu_{n-1}} \equiv d^{(n)}_{a_1a_2\cdots a_n} N^{a_1}_{\mu_1} N^{a_2}_{\mu_2} \cdots N^{a_{n-1}}_{\mu_{n-1}} \epsilon^{a_n} \tag{51b}$$

generalizing the results as derived for $G^{(2)}_{\mu\nu} = G_{\mu\nu}[det(k_{ab}N^a_\mu N^b_\nu)]^{1/4}$.

A subsequent application of two $SU(3)$-induced variations implies

$$[\delta_{\epsilon_1}, \delta_{\epsilon_2}]G^{(n)}_{\mu_1\mu_2\cdots\mu_n} = \delta_{\epsilon_3} G^{(n)}_{\mu_1\mu_2\cdots\mu_n} \quad i.e. \quad [\delta_{\xi_1}, \delta_{\xi_2}]G^{(n)}_{\mu_1\mu_2\cdots\mu_n} = \delta_{\xi_3} G^{(n)}_{\mu_1\mu_2\cdots\mu_n} \tag{52}$$

generalizing the $n = 2$ case in (46), i.e. an infinitesimal nonlinear realization of the $Diff(4, R)$ group in the space of fields $\left\{G^{(n)}_{\mu_1\mu_2\cdots\mu_n} \,\middle|\, n = 2, 3, \ldots\right\}$.

The third and last step in the proof[23,24] consists in *the construction of the operators of the chromo-diffeomorphism algebra.* The resulting generators are given by

$$L^{(m)\rho}_{\nu_1\nu_2\cdots\nu_{m+1}} = d^{(m+2)}_{a_1a_2\cdots a_{m+2}} \tilde{B}^{a_1}_{\nu_1} \tilde{B}^{a_2}_{\nu_2} \cdots \tilde{B}^{a_{m+1}}_{\nu_{m+1}} \frac{\delta}{\delta(g_{a_{m+2}b}\tilde{B}^b_\rho)}, \tag{53}$$

acting on the $G^{(n)}$ fields of (50a), with the latter making up an ∞-dimensional vector space $V(G^{(2)}, G^{(3)}, \ldots)$. The $L^{(n)}$ form a Z_+-graded algebra, with the grading given by a dilation-like operator, counting the number of gluon fields in a multi-gluon configuration. The bracketing of $L^{(l)}$ and $L^{(m)}$ preserves that grading

$$[L^{(l)}, L^{(m)}] \subset L^{(l+m)}, \tag{54}$$

the $L^{(m)\rho}_{\nu_1\nu_2\cdots\nu_{m+1}}, m = 0, 1, 2, \ldots$ spanning the $diff_0(4, R)$ algebra of homogeneous diffeomorphisms.

8. Color Confinement and Regge Excitation Sequences in Chromogravity

The effective Lagrangian for this (IR) region in QCD can be written as $L = L_M(\Psi, \partial\Psi, e, \Gamma) + L_g(e, \partial e, \Gamma, \partial\Gamma)$. L_M is the matter Lagrangian and is given in terms of a matter manifield Ψ and its gradient $\partial\Psi$, and of the "pseudo-tetrad" $e^A{}_\mu(x)$ and "pseudo-connection" $\Gamma^A{}_{B\mu}(x)$ derived from $G_{\mu\nu}(x)$. L_g is the pseudo-gravitational Lagrangian. This L is written in a Palatini "first order" formalism, with formally independent tetrad (or metric) and connection. The manifield $\Psi(x)$ is a "$\mathcal{A}$-deunitarized" $[(\frac{1}{2}, 0) \oplus (0, \frac{1}{2})]$ of $\overline{SL}(4, R)$.

In the absence of gravity (here pseudo-gravity), the matter Lagrangian would be $L_M = \overline{\Psi} i X^\mu \partial_\mu \Psi$, invariant under global $\overline{SL}(4, R)$, as would be true for any tensor field by construction.

(Pseudo)- gravity enters through the replacement $\partial_\mu \rightarrow \hat{D}_A$, where the index "$A$" denotes a local frame: $\hat{D}_\mu = \partial_\mu - \Gamma^A_{B\mu} Q^B_A$, $\hat{D}_A = \beta_A{}^\mu \hat{D}_\mu$ with Γ the connection and $\beta_A{}^\mu \cdot e_\mu{}^B = \delta_A{}^B$, e the pseudo-gravity tetrad; $Q_A{}^B$ is the $sl(4, R)$ algebraic generator in the tangent frame. We use $\hat{D}$ for the full covariant derivative with $sl(4, R)$ connection. $e^A{}_\mu(x)$ and $\Gamma^A{}_{B\mu}(x)$ can be taken as gauge fields for $\overline{SA}(4, R)$: $\delta_{(\epsilon,\alpha)}\Psi = [-\epsilon^A(x)\partial_A - \alpha^A{}_B(x) Q_A{}^B]\Psi$. As in gravity, the corresponding field strengths are the *torsion* and the *(generalized) curvature* (i.e. with both symmetric (AB) and antisymmetric [AB] pairs)

$$\begin{aligned} \hat{R}^A{}_{\mu\nu} &= \partial_\mu e^A_\nu + \Gamma^A_{B\mu} e^B_\nu - (\mu \leftrightarrow \nu) \\ \hat{R}^A{}_{B\mu\nu} &= \partial_\mu \Gamma^A_{B\nu} + \Gamma^C_{B\mu} \Gamma^A_{C\nu} - (\mu \leftrightarrow \nu) \end{aligned} \tag{55}$$

The Noether currents resulting from this $\overline{SA}(4, R)$ invariance are the energy-momentum and hypermomentum,

$$\begin{aligned} \Theta^{\mu}_{A} &= \frac{1}{\hat{e}}\frac{\delta L_M}{\delta e^A_\mu}, \qquad \hat{e} \equiv det(e^A_\mu), \\ \Upsilon^{B}_{A\mu} &= \frac{1}{\hat{e}}\frac{\delta L_M}{\delta \Gamma^A_{B\mu}}, \end{aligned} \tag{56}$$

with the symmetric (AB) pairs in $(\Upsilon)^B_A$ denoting shear currents and the antisymmetric pairs $[AB]$ representing angular momentum.

We apply a more general approach, provided by Metric-Affine Gravity. Variation with respect to e^A_μ and $\Gamma^A_{B\mu}$ yields the equations,

$$\begin{aligned} \Pi^{\mu\nu}_{A} &= \frac{\partial L_g}{\partial \partial_\nu e^A_\mu} = \frac{\partial L_g}{\partial \hat{R}^A{}_{\nu\mu}}, \\ E_A{}^{\mu} &= e_A{}^{\mu} Lg - \hat{R}^B{}_{A\nu}\Pi_B{}^{\nu\mu} - \hat{R}^B{}_{CA\nu}\Pi^C{}_B{}^{\nu\mu}, \\ \hat{D}_\nu \Pi^{\mu\nu}_A - E^{\mu}_A &= \hat{e}\Theta^{\mu}_A, \end{aligned} \tag{57a}$$

and

$$\begin{aligned} \Pi^A{}_B{}^{\mu\nu} &= \frac{\partial L_g}{\partial \partial_\nu \Gamma^B{}_{A\mu}} = \frac{\partial L_g}{\partial \hat{R}^B{}_{A\nu\mu}}, \\ E^A{}_B{}^{\mu} &= e^A{}_\nu \Pi_B{}^{\nu\mu}. \\ \hat{D}_\nu \Pi^A{}_B{}^{\mu\nu} - E^A{}_B{}^{\mu} &= \hat{e}\Upsilon^A{}_B{}^{\mu}, \end{aligned} \tag{57b}$$

where the Π are the canonically conjugated momenta and the E represent the gravitational contributions to the momentum and hypermomentum current tensor densities,

All expressions can be rewritten holonomically, i.e. with "curved space" indices μ, ν, $\cdots$ only, i.e. in term of the fields

$$\begin{aligned} G_{\mu\nu} &= k_{AB} e^A{}_\mu e^B{}_\nu, \\ \Gamma^\sigma{}_{\mu\nu} &= \beta_A{}^\sigma e^B{}_\nu \Gamma^A{}_{B\mu}, \\ \hat{e} &= \sqrt{-det(G_{\mu\nu})}. \end{aligned} \tag{58}$$

For our considerations, it is more convenient to use as independent dynamical variables an equivalent set, namely the pseudo-metric $G_{\mu\nu}$, Cartan's torsion tensor $S_{\mu\nu}{}^\sigma = \Gamma^\sigma{}_{[\mu\nu]}$, and the nonmetricity tensor $Q_{\mu\nu\sigma} = \hat{D}_\mu G_{\nu\sigma}$. The linear connection can be expressed in terms of these variables as

$$\Gamma^\sigma{}_{\mu\nu} = \left\{ {\sigma \atop \mu\ \nu} \right\} + \frac{1}{2}\left(S_{\mu\nu}{}^\sigma - S_\nu{}^\sigma{}_\mu + S^\sigma{}_{\mu\nu}\right) + \frac{1}{2}\left(Q_{\mu\nu}{}^\sigma - Q_\nu{}^\sigma{}_\mu + Q^\sigma{}_{\mu\nu}\right), \tag{59}$$

$\left\{ {\sigma \atop \mu\ \nu} \right\}$ is the Christoffel symbol. The 64 components of Γ are replaced by 24 and 40 in S and Q respectively; the new variables are tensor quantities.

The most general first order gauge Lagrangian for Affine Gravity can thus be written as

$$L_g = L_G(G, \partial G) + L_S(S, \partial S) + L_Q(Q, \partial Q). \qquad (60)$$

The simplest such Lagrangian is the $GL(4,R)$ or $SL(4,R)$ scalar curvature tensor. It differs from Einstein's in that the original curvature 2-form $R_{\mu\nu}{}^{A}{}_{B}$ (prior to contraction) has all 16 (or 15 for $SL(4,R)$) combinations, whereas Einstein's has only the 6 anti-symmetric $[AB]$. Since $R = d\Gamma + \frac{1}{2}[\Gamma,\Gamma]$, we see from (59,60) that this will contain terms such as $\{\ \}^2$, S^2 and Q^2. Variation of the total Lagrangian $L = L_G + L_S + L_Q + L_M$ with respect to G, S and Q yields the equations of motion,

$$\begin{aligned} -\frac{\delta L_G}{\delta G_{\mu\nu}} &= \frac{\delta L_M}{\delta G_{\mu\nu}} \equiv \sqrt{-G}\Theta^{\mu\nu}, \\ -\frac{\delta L_S}{\delta S_{\mu\nu}{}^{\sigma}} &= \frac{\delta L_M}{\delta S_{\mu\nu}{}^{\sigma}} \equiv \sqrt{-G}(\Sigma_{\sigma}{}^{\nu\mu} - \Sigma^{\nu}{}_{\sigma}{}^{\mu} + \Sigma^{\mu\nu}{}_{\sigma}), \\ -\frac{\delta L_Q}{\delta Q_{\mu\nu}{}^{\sigma}} &= \frac{\delta L_M}{\delta Q_{\mu\nu}{}^{\sigma}} \equiv \sqrt{-G}(\Delta_{\sigma}{}^{\nu\mu} - \Delta^{\nu}{}_{\sigma}{}^{\mu} + \Delta^{\mu\nu}{}_{\sigma}), \end{aligned} \qquad (61)$$

$\Theta^{\mu\nu}$ is the symmetrized energy-momentum tensor, $\Sigma_{\sigma}{}^{\nu\mu}$ and $\Delta_{\sigma}{}^{\nu\mu}$ are the spin and shear currents, antisymmetric and symmetric in $\mu \leftrightarrow \nu$ respectively.

We can now apply this formalism to that sector in the IR region of QCD. Due to (45), we have only one propagating field, namely $G_{\mu\nu}$ We are thus led to a situation analogous to the Einstein-Cartan theory, in which spinor fields contribute to the torsion currents, but there is no propagating torsion piece of the connection (59), independent of $G_{\mu\nu}$. Torsion, and here Non-Metricity as well, exist pointwise in the matter distribution, but do not propagate from it through the vacuum as a free wave or via any interaction of non-vanishing range. The pseudo-gravity Lagrangian (60) therefore does not contain ∂S and ∂Q. The last two of equations (61) thus become purely algebraic equations, relating torsion to spin and non-metricity to shear, with some proportionality factors given by the couplings in L_S and L_Q:

$$S_{\mu\nu}{}^{\sigma} \sim \Sigma_{\mu\nu}{}^{\sigma} - \Sigma_{\mu}{}^{\sigma}{}_{\nu} + \Sigma^{\sigma}{}_{\nu\mu}, \quad Q_{\mu\nu}{}^{\sigma} \sim \Delta_{\mu\nu}{}^{\sigma} - \Delta_{\mu}{}^{\sigma}{}_{\nu} + \Delta^{\sigma}{}_{\nu\mu}. \qquad (62)$$

To simplify the expressions we replace S and Q by linear combinations $\tilde{S}_{\mu\nu}{}^{\sigma}$ (the contortion tensor) and $\tilde{Q}_{\mu\nu}{}^{\sigma}$ proportional to $\Sigma_{\mu\nu}{}^{\sigma}$ and $\Delta_{\mu\nu}{}^{\sigma}$. Thus,

$$L_S + L_Q = l_S^{-2}\, \Sigma_{\mu\nu}{}^{\sigma}\Sigma^{\mu\nu}{}_{\sigma} + l_Q^{-2}\, \Delta_{\mu\nu}{}^{\sigma}\Delta^{\mu\nu}{}_{\sigma}. \qquad (63)$$

As a matter of fact, selecting for L_g the generalized $\overline{SL}(4,R)$ scalar curvature, and substituting as in (62) generates (63). This then becomes an effective addition to the matter energy momentum tensor - once we segregate the pure Riemannian part of the Ricci tensor on the left hand side of (57) and thus move (63) to the right hand side.

The effective action for this IR (zero-colour) hadron sector of QCD, written as a pseudo-gravitational theory, with matter in $\overline{SL}(4,R)$ manifields, then becomes

$$I = \int d^4x\sqrt{-G}\{-aR_{\mu\nu}R^{\mu\nu}+bR^2-cl_G^{-2}R+l_S^{-2}\,\Sigma_{\mu\nu}{}^{\sigma}\Sigma^{\mu\nu}{}_{\sigma}+l_Q^{-2}\,\Delta_{\mu\nu}{}^{\sigma}\Delta^{\mu\nu}{}_{\sigma}+L_M\}. \tag{64}$$

The first three terms constitute the Stelle Lagrangian[29] and provide the p^{-4} propagators. The fourth and fifth terms in (64) are spin-spin and shear-shear contact interaction terms. The a, b, c are dimensionless constants; l_G, l_S and l_Q, have the dimensions of lengths; we estimate them to be of hadron size $\sim 1\ GeV$. Following Stelle we obtain the field equation (; denotes D, i.e. purely with the $\{\ \}$),

$$\begin{aligned}
&aR_{\mu\nu}{}^{;\eta}{}_{;\eta} - (a-2b)R_{;\mu;\nu} + (\tfrac{a}{2}-2b)(R_{\mu\nu}-\tfrac{1}{2}RG_{\mu\nu})R^{;\eta}{}_{;\eta} - \\
&-2aR^{\eta\lambda}R_{\mu\eta\nu\lambda} + 2bRR_{\mu\nu} + \tfrac{1}{2}(R_{\mu\nu}-\tfrac{1}{2}RG_{\mu\nu})(aR^{\eta\lambda}R_{\eta\lambda}-bR^2) - \\
&-cl_G^{-2}(R_{\mu\nu}-\tfrac{1}{2}RG_{\mu\nu}) - \tfrac{1}{2}\Sigma_{\rho\eta}{}^{\lambda}\Sigma^{\rho\eta}{}_{\lambda}G_{\mu\nu} - \tfrac{1}{2}\Delta_{\rho\eta}{}^{\lambda}\Delta^{\rho\eta}{}_{\lambda}G_{\mu\nu} = \tfrac{1}{2}\Theta_{\mu\nu}.
\end{aligned} \tag{65}$$

It is the IR region of QCD for which we are using pseudo-gravity. This is the *strong coupling limit of QCD*, corresponding to low energies and "large" distances. Thus, the corresponding sector in a gravity-like theory has to be its high energy limit; for true gravity, we know that the weak Newtonian coupling of macroscopic physics reaches in the Planck region strong interaction strength. Thus, our correspondence is between the low energy IR sector of QCD and *the UV strong coupling sector of pseudo-gravity*. In eq.(65), this implies dominance by the $R_{\mu\nu}R^{\mu\nu}$ and R^2 terms and we can neglect the R term (Riemannian, with only $\{\ \}$ as connection). We linearize our theory in terms of $H_{\mu\nu}(x) = G_{\mu\nu}(x) - \eta_{\mu\nu}$, where $\eta_{\mu\nu}$ is the Minkowski metric. Taking just the homogeneous part, as required for the evaluation of the propagator, we get for the $H_{\mu\nu}$ field the equation of motion

$$(\frac{a}{4}\,\Box^2 - \frac{1}{2}l_S^{-2}\,\Sigma_{\rho\eta}{}^{\lambda}\Sigma^{\rho\eta}{}_{\lambda} - \frac{1}{2}l_Q^{-2}\,\Delta_{\rho\eta}{}^{\lambda}\Delta^{\rho\eta}{}_{\lambda})H_{\mu\nu} = 0, \tag{66a}$$

which becomes in momentum space

$$(\frac{a}{4}\,(p^2)^2 - \frac{1}{2}l_S^{-2}f_S\ M_\eta{}^\lambda M^\eta{}_\lambda - \frac{1}{2}l_Q^{-2}f_Q\ T_\eta{}^\lambda T^\eta{}_\lambda)H_{\mu\nu}(p) = 0. \qquad (66b)$$

For pseudo-gravity, we may regard (66) as *the dynamical equations above the theory's "vacuum", as represented by hadron matter itself.* The equations represent the excitations produced over that ground state by the pseudo-gravity potential. In these expressions, we have factored out the $\overline{SL}(4,R)$ group factors (the bilinear forms in the algebra's generators). The factors f_S and f_Q represent the residual part of the configuration space integrals. Equation (66a) is like an equation for the $H_{\mu\nu}(x)$ field in an external field of hadronic matter. We saw in section 5 that the rest frame (stability) "little" group is $\overline{SL}(3,R) \subset \overline{SL}(4,R)$. Taking a hadron's rest frame $(i,\ j = 1,\ 2,\ 3)$

$$\begin{array}{rcll} M_\eta{}^\lambda M^\eta{}_\lambda & \rightarrow & M_i{}^j M^i{}_j & \rightarrow J(J+1), \\ T_\eta{}^\lambda T^\eta{}_\lambda & \rightarrow & T_i{}^j T^i{}_j & \rightarrow M_i{}^j M^i{}_j - C^2_{sl(3,R)} \rightarrow J(J+1) - C^2_{sl(3,R)}, \end{array} \qquad (67)$$

where C^2 is the $sl(3,R)$ quadratic invariant.

As a result, we find that in a rest frame, all hadronic states belonging to a single $\overline{SL}(3,R)$ (unitary) irreducible representation (i.e. one value of $C^2_{sl(3,R)}$) lay on a single trajectory in the Chew-Frautschi plane, i.e.

$$\begin{array}{rcl} (J+\frac{1}{2})^2 & = & (\alpha' m^2)^2 + \alpha_0^2, \\ (\alpha')^2 & = & [\frac{2}{a}(l_S^{-2}f_S + l_Q^{-2}f_Q)]^{-1}, \\ \alpha_0^2 & = & \frac{1}{4} + \frac{l_Q^{-2}f_Q}{l_S^{-2}f_S + l_Q^{-2}f_Q}C^2_{sl(3,R)}. \end{array} \qquad (68)$$

α' is the (asymptotic) trajectory slope. Reality of α' requires $f_S/a > 0$ and $f_Q/a > 0$. For $J = 0$ eq. (68) implies: $(m^2_{J=0})^2 = -\frac{2}{a}l_Q^{-2}f_Q C^2_{sl(3,R)}$. Thus eq. (68) is meaningful only if $C^2_{sl(3,R)} < 0$. This is indeed the case for the (relevant) unitary irreducible $\overline{SL}(3,R)$ representations Neglecting a slight bending at small m^2,i.e. the α_0^2 term, we finally obtain the linear Regge trajectory

$$J = \alpha' m^2 - \frac{1}{2}. \qquad (69)$$

As derived here from QCD-generated pseudo-gravity, the $J \sim M^2$ linear Regge trajectories embody the spin-2 nature of the effective di-gluon, dominating the IR sector of hadron interactions. The $(M^2)^2$ (like the p^{-4} propagators indicating confinement) results from the typical $(\text{curvature})^2$ Lagrangian of a gauge theory, when the connection Γ is replaced by the $\sim G^{-1}\partial G$ affine connection $\{\ \}$ of a Riemannian spin-2 theory. The J^2 results from the direct substitution, in the linear curvature scalar (characterizing spin-2 gauge theories) of non-propagating torsion by spin, through the algebraic Cartan equation of a Riemannian theory (higher powers of the curvature will yield corrections proportional to J^4 etc ...). This then fixes $J^2 \sim (M^2)^2$. Long ago, we noted[10] the existence of a link between Regge trajectories and what we then thought was plain gravity; we can now conclude that these moments of inertia become relevant to strong interactions because QCD emulates gravity in this sector. In nuclei, the missing dipolar excitations, the $2^+ + 0^+$ ground state of the IBM symmetry[27,28] , the quadrupolar nature of the $SL(3,R)$, $SU(3)$ and $Eucl(3)$ sequences[30] - all of these features again characterize the action of a gravity-like spin-2 effective gauge field. Overall, the evidence for the existence of such an effective component in QCD seems overwhelming.

Note that this theory provides a derivation of the observed linear relation in hadron Regge trajectories between the spins and squared masses, a relation which also emerges in a different context – perhaps related through gravity/chromogravity parallelism – namely for the Kerr metric in General Relativity, as pointed out by A. Salam.

The last two sections have served as one illustration of the potentialities of this formalism, based on post-Riemannian geometries. We have recently presented another example[31], in which the spacetime geometry was given by a solution of Metric-Affine Gravity, and test matter, in the guise of a spinor manifield, was exposed to its action.

References

[1] B.S. DeWitt, in *Relativity, Groups and Topology*, Les Houches Lectures 1963, C. and B. DeWitt eds., Blackie and son, London (1964) pp. 587-820.

[2] E. Cartan, *Leçons sur la Theorie des Spineurs*, Hermann and Co. Eds., Paris (1938), article 177.

[3] Y. Ne'eman and Dj. Šijački, *Intern. J. Mod. Phys.* **A2** (1987) 1655.

[4] Y. Ne'eman, *Proc. 8th Intern. Conf. on Gen. Rel. Grav.*, M.A. McKiernan, ed. (Un. of Waterloo, Canada 1977) p. 262; *ibid. Proc. Nat. Acad. Sci. USA* **74** (1977) 4157; *ibid. Ann. Inst. Henri Poincaré* **A28** (1978) 369.

[5] Dj. Šijački and Y. Ne'eman, *J. Math. Phys.* **26** (1985) 2457.

[6] J. Lemke, Y. Ne'eman and J. Pecina-Cruz, *J. Math. Phys.* **33** (1992) 2656.

[7] Dj. Šijački and Y. Ne'eman, *Phys. Lett.* **B247** (1990) 571.

[8] Y. Ne'eman and Dj. Šijački, *Phys. Lett.* **B276** (1992) 173.

[9] T.E. Stewart, *Proc. Am. Math. Soc.* **11** (1960) 559.

[10] Y. Dothan, M. Gell-Mann and Y. Ne'eman, *Phys. Lett.* **17** (1965) 148.

[11] V. Bargmann, *Ann. Math.* (Princeton) **48** (1947) 568.

[12] Dj. Šijački, *J. Math. Phys.* **16** (1975) 298.

[13] Dj. Šijački, in *Spinors in Physics and Geometry*, A. Trautman and G. Furlan eds., World Scientific Pub., Singapore (1987) p. 637.

[14] Y. Ne'eman and Dj. Šijački, *Phys. Lett.* **B157** (1985) 267.

[15] Y. Ne'eman and Dj. Šijački, *Phys. Lett.* **B157** (1985) 275.

[16] Y. Ne'eman and Dj. Šijački, *Phys. Rev.* **D37** (1988) 3267.

[17] Y. Ne'eman and Dj. Šijački, *Phys. Rev.* **D47** (1993) 4133.

[18] F.W. Hehl, J.D. McCrea, E.W. Mielke and Y. Ne'eman, *Phys. Reports* **258** (1995) 1-171.

[19] A. Barut, A. Bohm and Y. Ne'eman, eds., *Dynamical Groups and Spectrum Generating Algebras*, World Scientific Pub., Singapore (1988).

[20] Y. Ne'eman and Dj. Šijački, *Phys. Lett.* **B200** (1988) 489.

[21] C.-Y. Lee and Y. Ne'eman, *Phys. Lett.* **B233** (1989) 286.

[22] C.-Y. Lee and Y. Ne'eman, *Phys. Lett.* **B242** (1990) 59.

[23] Y. Ne'eman and Dj. Šijački, *Inter. J. Mod. Phys.* **A30** (1995) 4399.

[24] Y. Ne'eman and Dj. Šijački, *Mod. Phys. Lett.* **A11** (1996) 217.

[25] L.V. Gribov, E.M. Levin and M.G. Ryskin, *Phys. Rep.* **100** (1983) 1.

[26] E.M. Levin, in *QCD 20 Years Later*, P. M. Zerwas and H. A. Kastrup, eds. (World Scintific, 1993).

[27] A. Arima and F. Iachello, *Phys. Rev. Lett.* **35** (1975) 1069.

[28] Dj. Šijački and Y. Ne'eman, *Phys. Lett.* **B250** (1990) 1.

[29] K. S. Stelle, *Phys. Rev.* **D16** (1977) 953.; *ibid. GRG* **9** (1978) 353.

[30] J.P. Draayer and K.J. Weeks, *Phys. Rev. Lett.* **51**, (1983) 1422.

[31] Y. Ne'eman and F.W. Hehl, *Class. Quantum Grav.* **14** (1997) A2

PHYSICAL EFFECTS AND MEASURABILITY OF SPIN-TORSION INTERACTION

by Petr I. Pronin

Department of Theoretical Physics,
Physics Faculty, Moscow State University,
117234 Moscow, Russian Federation

CONTENTS

1 INTRODUCTION

Gravitational theory with torsion has a long history. The first steps in this subject have been made by Cartan (1922-1925) who remarked the independence of metrical structure and connection. Cartan proposed to consider as a model of space-time a smooth four dimensional manifold that is provided by metric and non-symmetrical affine connection. The antisymmetric part of the last was called the torsion of space-time. However at that time the Cartan's approach to geometry and gravity was not developed because the spin of particles has not yet been known.

The Cartan's ideas were regenerate in the end of fourtieth years. Stueckelberg (1948) proposed to introduce the torsion to coordinate theoretical predictions on superfine structure of hudrogen. He also introduced spacetime torsion to explane spin-spin interaction of elementary particles. Ather that Einstein (1955) used the non-symmetrical connection in the unified field model. Independently the analogous description of unified model was proposed by Schrödinger (1944). The equations of Einstein-Schrödinger theory give at the first approximation The Maxwell's equation in general relativity.

The crucial stage in the development and foundation of gravitational theory with torsion was the gauge treatment of gravitational interaction proposed by Utiyama (1956), Sciama (1958), Kibble (1961), Brodskii, Ivanenko and Sokolik (1962) and other authors. The gauge theory of Poincaré group allowes in explicit manner to base the introduction of torsion and to prove that the source of torsion is the spin of particles.

At the present time the papers on the gauge treatment of gravity are the considerable part of all gravitational investigations. The main attention is payed toe the physical foundation of torsion nature, exact solutions problem, cosmological models, investigation of equatios of motion for classical bodies, discussion of quantum approach and investigations on the renormalizability problem.

One of the most important subject of these investigations is the prediction of specific physical effects caused by spin-torsion interaction. We devoited our lecturec namely this subject. To make our consideration more clean we spent the time to introduce the main theoretical background in the Sect. 2.

Since the structure of gravitational theries with torsion is more reach than the structure of general relativity models based on the Riemannian geometry we consider independently the physical effects in so-called Einstein-Cartan theory with the Lagrangian of Hilbert-Einstein type (Sect. 3) and physical effects in quadratic model which are characterised by spreading of torsion field (Sect. 4). We focused our attention in these sections on the influence of torsion on the well known phenomena, namely: particle motion and spreading of electromagnetic waves.

The separate place ocupied by the quantum effects and problems measurability in quantum experiments. This problem is discussed in the last section.

We hope that all defects and incompleteness of our discussion will be compensated by the references and details of calculations that can be found in original papers.

2 THEORETICAL BACKGROUND

We are mainly interested in the formulation of physical aspects of the gravity theory with torsion. Namely, along with the analysis of kinematical problems the construction of dynamical scheme is an important issue.

2.2 Kinematics of gravity theory with torsion: the Riemann-Cartan geometry

The mathematical formalism for the manifold with torsion was developed to a considerable extent already by Cartan. The modern approach use the methods of the fibre bundle theory.

The Riemann-Cartan spacetime (denoted U_4) is defined as the four-dimensional smooth manifold on which two independent structures are introduced:

the pseudo-Riemannian metric

$$g_{\mu\nu}(x) \tag{1}$$

and affine connection

$$\tilde{\Gamma}^{\alpha}_{\mu\nu}(x) \tag{2}$$

These are restricted by the metricity condition

$$\nabla_{\lambda} g_{\mu\nu} = \partial_{\lambda} g_{\mu\nu} - \tilde{\Gamma}^{\alpha}_{\mu\lambda} g_{\alpha\nu} - \tilde{\Gamma}^{\alpha}_{\nu\lambda} g_{\mu\alpha} = 0. \tag{3}$$

The last equation can be solved, so that the connection is expressed in the form of a sum

$$\tilde{\Gamma}^{\lambda}{}_{\mu\nu} = \Gamma^{\lambda}{}_{\mu\nu} + K^{\lambda}{}_{\mu\nu} \tag{4}$$

of the Riemann connection - the Christoffel symbols

$$\Gamma^{\lambda}{}_{\mu\nu} = \frac{1}{2} g^{\lambda\sigma}(g_{\mu\sigma,\nu} + g_{\nu\sigma,\mu} - g_{\mu\nu,\sigma}) \tag{5}$$

and the non-Riemann part - the contorsion tensor

$$K^{\lambda}{}_{\mu\nu} = Q_{\mu\nu}{}^{\lambda} + Q_{\nu\mu}{}^{\lambda} + Q^{\lambda}{}_{\mu\nu} \tag{6}$$

As we already mentioned, the torsion tensor is defined as the antisymmetric part of connection

$$Q^{\lambda}{}_{\mu\nu} = \tilde{\Gamma}^{\lambda}_{[\mu\nu]} \equiv \frac{1}{2}(\tilde{\Gamma}^{\lambda}{}_{\mu\nu} - \tilde{\Gamma}^{\lambda}{}_{\nu\mu}) \tag{7}$$

Torsion has 24 independent components and it can be decomposed into three irreducible parts

$$Q^{\lambda}{}_{\mu\nu} = \bar{Q}^{\lambda}{}_{\mu\nu} + \frac{2}{3}\delta^{\lambda}_{[\nu} Q_{\mu]} + \frac{1}{3}\varepsilon^{\lambda}_{\ \mu\nu\beta}\check{Q}^{\beta}, \tag{8}$$

where the torsion trace is

$$Q_\mu \equiv Q^\lambda{}_{\mu\lambda} \tag{9}$$

the axial (or pseudo) trace is

$$\check{Q}_\alpha \equiv \frac{1}{3!}\,\varepsilon_{\alpha\mu\nu\lambda}Q^{\mu\nu\lambda} \tag{10}$$

and the first term in (8) is the so-called traceless part of torsion which satisfies

$$\bar{Q}^\lambda{}_{\lambda\mu\nu} = 0, \quad \bar{Q}_{\lambda\mu\nu} + \bar{Q}_{\mu\nu\lambda} + \bar{Q}_{\nu\lambda\mu} = 0 \tag{11}$$

Hereafter the Riemann-Cartan connection and all the geometrical objects and operators defined with (4) (e.g. the curvature, covariant derivatives etc) are denoted by the tilde. The same objects without tilde are constructed from the standard Riemannian connection.

The curvature tensor in U_4 is introduced as usually,

$$\tilde{R}^\alpha{}_{\beta\mu\nu} = \partial_\mu\tilde{\Gamma}^\alpha{}_{\beta\nu} - \partial_\nu\tilde{\Gamma}^\alpha{}_{\beta\mu} + \tilde{\Gamma}^\alpha{}_{\rho\mu}\tilde{\Gamma}^\rho{}_{\beta\nu} - \tilde{\Gamma}^\alpha{}_{\rho\nu}\tilde{\Gamma}^\rho{}_{\beta\mu}, \tag{12}$$

and its contractions are the Ricci tensor $\tilde{R}_{\mu\nu} = \tilde{R}^\lambda_{\mu\lambda\nu}$ and the curvature scalar $\tilde{R} = g^{\mu\nu}\tilde{R}_{\mu\nu}$. As compared to the Riemannian case the structure of curvature is more complicated: the Riemann-Cartan curvature tensor has 36 components since the skew symmetry holds for both pairs of indices

$$\tilde{R}_{\alpha\beta\mu\nu} = \tilde{R}_{[\alpha\beta]\mu\nu} = \tilde{R}_{\alpha\beta[\mu\nu]} \tag{13}$$

The curvature can be decomposed into six irreducible parts

$$\tilde{R}_{\alpha\beta\mu\nu} = C_{\alpha\beta\mu\nu} + T_{\alpha\beta\mu\nu} + S_{\alpha\beta\mu\nu} + A_{\alpha\beta\mu\nu} + B_{\alpha\beta\mu\nu} + D_{\alpha\beta\mu\nu} \tag{14}$$

Here the first three terms are the direct analogues of the standard Riemannian curvature: $C_{\alpha\beta\mu\nu}$ is the traceless tensor, $C^\alpha_{\mu\alpha\nu} = 0$, with the symmetry properties of the Weyl tensor

$$C_{\alpha\beta\mu\nu} = C_{\mu\nu\alpha\beta}, \quad C_{\alpha\beta\mu\nu} + C_{\alpha\mu\nu\beta} + C_{\alpha\nu\beta\mu} = 0. \tag{15}$$

Next

$$T_{\alpha\beta\mu\nu} = \frac{1}{2}(g_{\alpha\mu}C_{\beta\nu} - g_{\alpha\nu}C_{\beta\mu} - g_{\beta\mu}C_{\alpha\nu} + g_{\beta\nu}C_{\alpha\mu}) \tag{16}$$

and

$$S_{\alpha\beta\mu\nu} = \frac{1}{12}\tilde{R}(g_{\alpha\mu}g_{\beta\nu} - g_{\alpha\nu}g_{\beta\mu}) \tag{17}$$

describe the contributions of the traceless Ricci tensor

$$C_{\alpha\beta} = \tilde{R}_{(\alpha\beta)} - \frac{\tilde{R}}{4}g_{\alpha\beta}$$

and curvature scalar. The three last terms in (14) do not have analogues in the Riemannian space, these are identically zero when torsion is trivial. These terms are completely determined by the irreducible part of the tensor

$$D_{\mu\nu} = \frac{1}{2}\varepsilon^{\alpha\beta\gamma}{}_{\mu}\tilde{R}_{\alpha\beta\gamma\nu} \tag{18}$$

where $\varepsilon_{\alpha\beta\mu\nu}$ is the completely antisymmetric tensor density of Levi-Civita. By definition, this describes the violation in U_4 of the Ricci identity

$$\tilde{R}^{\alpha}_{\mu\nu\beta} + \tilde{R}^{\alpha}_{\nu\beta\mu} + \tilde{R}^{\alpha}_{\beta\mu\nu} \neq 0.$$

Thus, three last irreducible curvature parts are as follows

$$T_{\alpha\beta\mu\nu} = \frac{1}{2}(g_{\alpha\mu}A_{\beta\nu} - g_{\alpha\nu}A_{\beta\mu} + g_{\beta\mu}A_{\alpha\nu} - g_{\beta\nu}A_{\alpha\mu}) \tag{19}$$

$$B_{\alpha\beta\mu\nu} = -\frac{1}{2}(\varepsilon_{\alpha\beta\mu}{}^{\lambda}B_{\lambda\nu} - \varepsilon_{\alpha\beta\nu}{}^{\lambda}B_{\lambda\mu}) \tag{20}$$

$$D_{\alpha\beta\mu\nu} = \frac{1}{2}\varepsilon^{\alpha\beta\gamma}{}_{\mu}\tilde{R}_{\alpha\beta\gamma\nu} \tag{21}$$

where the antisymmetric Ricci tensor part is equal to

$$A_{\mu\nu} = \tilde{R}_{[\mu\nu]} = -\frac{1}{2}\varepsilon^{\mu\nu\alpha\beta}D_{\alpha\beta} \tag{22}$$

and

$$B_{\mu\nu} = D_{(\mu\nu)} - \frac{1}{4}Dg_{\mu\nu}, D = g^{\mu\nu}D_{\mu\nu} \tag{23}$$

With the help of decomposition (4) the Riemann-Cartan curvature can be rewritten in the form

$$\tilde{R}^{\alpha}{}_{\beta\mu\nu} = R^{\alpha}{}_{\beta\mu\nu} + \nabla_{\mu}K^{\alpha}{}_{\beta\nu} - \nabla_{\nu}K^{\alpha}{}_{\beta\mu} + K^{\alpha}{}_{\sigma\mu}K^{\sigma}{}_{\beta\nu} - K^{\alpha}{}_{\sigma\nu}K^{\sigma}{}_{\beta\mu}. \tag{24}$$

2.2 Dynamics of the gravity theory with torsion

As one knows, the dynamics of the gravitational field (metric) in the general relativity is described by the Hilbert-Einstein action with the Lagrangian linear in curvature $L = -\frac{1}{2\kappa}R$. this is practically unique determined by the conditions of general covariance and second-order field equations. Contrary to general relativity, in the gravity with torsion there is considerable freedom in constructing the dynamical scheme, since one can define much more invariants from torsion and curvature tensor. There are two most attractable classes of models, namely: the Einstein-Cartan theory and the quadratic theories.

2.2.a The Einstein-Cartan theory

The Einstein-Cartan theory often also called the Einstein-Cartan-Sciama-Kibble (ECSK) model, is the minimal extension of GR (Trautman 1973; Hehl et al. 1976). The ECSK Lagrangian is of the Hilbert-Einstein form

$$L_{ECSK} = -\frac{1}{2\kappa}\tilde{R}. \tag{25}$$

The field equations are defined from independent variation of the total action with respect to the metric and torsion, and have the form

$$\tilde{R}_{\mu\nu} - \frac{1}{2}g_{\mu\nu}\tilde{R} = \kappa T_{\mu\nu}, \tag{26}$$

$$Q^{\lambda}{}_{\mu\nu} + 2\delta^{\lambda}_{[\mu}Q_{\nu]} = \kappa S^{\lambda}{}_{\mu\nu}, \tag{27}$$

Here the material sources are (provided the minimal coupling principle, see sect. 2.3) the metrical energy-momentum tensor $T_{\mu\nu}$ and the tensor of spin $S^{\lambda}{}_{\mu\nu}$.

In ECSK the interaction of torsion with spin is degenerate (contact) so that the torsion is simply proportional to spin. Here for the spinless matter (and in vacuum) ECSK equations reduce to that of GR. One can use the mentioned degeneracy of interaction as follows. The torsion can be computed from (27) in terms of spin and then with the help of (26) and (27) the first equation (26) is easily rewritten in the form of the standard Riemannian Einstein equations

$$R_{\mu\nu} - \frac{1}{2}g_{\mu\nu}R = \kappa T^{eff}_{\mu\nu}, \tag{28}$$

where the effective energy-momentum tensor contains quadratic spin contributions due to the contact spin-spin interactions of the matter source.

2.2.b Quadratic models

The Lagrangian of quadratic models includes all the possible invariants of order not greater than two which can be constructed from torsion and Riemann-Cartan curvature. Taking into account the symmetry properties, one can write the most general Lagrangian as the sum of 14 quadratic contractions, and add the Hilbert-Einstein and cosmological term (Ivanenko, Pronin & Sardanashvily 1985; Ponomariev, Barvinsky & Obukhov 1985):

$$\begin{aligned} L_2 \;=\; & -\frac{1}{2\kappa}[\tilde{R} - 2\Lambda + b_1\tilde{R}_{\alpha\beta\mu\nu}\tilde{R}^{\alpha\beta\mu\nu} + b_2\tilde{R}_{\alpha\beta\mu\nu}\tilde{R}^{\mu\nu\alpha\beta} + b_3\tilde{R}_{\alpha\beta\mu\nu}\tilde{R}^{\alpha\mu\beta\nu} + \\ & b_4\tilde{R}_{\alpha\beta}\tilde{R}^{\alpha\beta} + b_5\tilde{R}_{\alpha\beta}\tilde{R}^{\beta\alpha} + b_6\tilde{R}^2 + b_7\tilde{R}_{\alpha\beta}D^{\alpha\beta} + b_8\tilde{R}_{\alpha\beta}D^{\beta\alpha} + b_9\tilde{R}D + \\ & a_1Q^{\alpha}{}_{\mu\nu}Q_{\alpha}{}^{\mu\nu} + a_2Q_{\alpha\mu\nu}Q^{\alpha\mu\nu} + a_3Q_{\mu}Q^{\mu} + a_4\varepsilon^{\mu\nu\rho\sigma}Q^{\alpha}{}_{\mu\nu}Q_{\alpha\rho\sigma} + a_5Q^{\mu}\bar{Q}_{\mu}]. \end{aligned} \tag{29}$$

Five last terms are clearly not invariant under the parity transformations. If we demand the parity symmetry as is usually done in the gravity theory with torsion than put $b_7 = b_8 = b_9 = a_4 = a_5 = 0$.

2.3 Interaction of spin with torsion: minimal, semi-minimal and non-minimal

The problem of introduction interaction of gravitational field with matter is still disputable in the literature (Bukhdal 1990). as a rule the minimal coupling principle is used which essentially means that the ordinary derivatives in action functional and field equations are replaced by the covariant derivatives, and space-time metric $\eta_{\mu\nu}$ is replaced by $g_{\mu\nu}$. This procedure appears to be most natural in the context of the general gauge field theory which includes both internal symmetry interactions (electrodynamics, chromodynamics etc.) and gravitation. In fact, the minimal coupling recipe is closely related to the equivalence principle in Einsteinian gravity.

Its application, though must be careful in more general theory with torsion. For example, apply the minimal coupling principle to the scalar field theory (i) on the action functional level, and (ii) directly in the Klein-Gordon equation. The substitution $\partial_\mu \to \nabla_\mu = \partial_\mu + I^b_a \tilde{\Gamma}^a_{b\mu}$ and $\eta_{\mu\nu} \to g_{\mu\nu}$ (where I^b_a are the generators of Lorentz group in suitable representation) in the former case yields only the change the volume $d^4x \to \sqrt{-g}d^4x$ of the kinematic part of the action $\int d^4x(\partial_\mu\varphi)(\partial^\mu\varphi)$. While in the latter case the flat d'Alember operator $\partial_\mu\partial^\mu$ is replaced by $(\nabla_\mu\nabla^\mu + 2Q^\mu\nabla_\mu$. The last term makes the scalar field operator non-self-adjoint, and such an equation cannot be derived from a variational principle. Analogues inconsistencies are discovered also for the Dirac spinor field theory. Here one can prove that the interaction, resulting from the minimal coupling recipe, depends on the initial action functional. For example, if one starts from

$$S_{Dirac} = \int d^4x\sqrt{-g}\bar{\psi}(i\gamma^\mu\tilde{\nabla}_\mu - m)\psi, \tag{30}$$

the variational principle yields the covariant equation in the Riemann-Cartan space-time in the form

$$[i\gamma^\mu(\nabla_\mu + \frac{1}{2}\gamma_5\check{Q}_\mu + Q_\mu) - m]\psi = 0, \tag{31}$$

where, let us remind, ∇_μ is the torsionless spinor Riemannian derivative. But another choice

$$S_{Dirac} = \int d^4x\sqrt{-g}\{\frac{i}{2}(\tilde{\nabla}_\mu\bar{\psi}\gamma^\mu\psi - \bar{\psi}\gamma^\mu\tilde{\nabla}_\mu\psi) - m\bar{\psi}\psi\}, \tag{32}$$

yields the equation

$$[i\gamma^\mu(\nabla_\mu + \frac{1}{2}\gamma_5\check{Q}_\mu) - m]\psi = 0, \tag{33}$$

The first case displays an unnatural interaction of torsion trace vector with current of spinor field.

In our opinion the complete and correct formulation of the minimal coupling principle must include an additional demand that it is applied to the action functional which is explicitly written in the symmetrized self-adjoint type form. Supplementary arguments in favor of this formulation could be found in the studies of the consistency problem (Aragone & Deser, 1980; Buchdal, 1958,1962) and acausal anomalies (Obukhov, 1983).

The direct application of the minimal coupling principle to the gauge fields, such as electromagnetic and Yang-Mills, leads to new problems. In particular, the substitution $\partial_\mu \to \tilde{\nabla}_\mu$ in electromagnetic (Maxwell) theory gives for the field strength tensor

$$F_{\mu\nu} = \partial_\mu A_\nu - \partial_\nu A_\mu + 2Q^\lambda{}_{\mu\nu} A_\lambda, \tag{34}$$

which evidently is in conflict with the gauge invariance with respect to the transformation

$$A_\mu \to A'_\mu = A_\mu + \partial_\mu \alpha \tag{35}$$

and may cause the charge non-conservation (Aldersley, 1978) by the torsion produced via photon spin.

In connection with this many authors said the general belief that torsion does not interact with electromagnetic and Yang-Mills gauge fields (Trautman, 1973; Hehl et al., 1976; Kasya, 1975). On the other hand moving from classical to quantum theory one encounters the vacuum polarization effects, and these in principle could produce the interaction of photons with torsion (De Sabbata & Gasperini, 1980, 1981a-d). In particular, one can show that the vacuum connections induce in the right hand side the Maxwell equations an additional current of the form $F_{\mu\nu} = 2\partial_{[\mu} A_{\nu]}$. However, such an approach leads to a "restricted" interaction of photons with torsion (De Sabbata & Gasperini, 1980), which does not violate the gauge invariance of one imposes $\tilde{\nabla}_{[\alpha} F_{\mu\nu]} = 0$ from the beginning. This scheme is called the semi-minimal interaction.

In our opinion, the works of Hojman et al. (1978, 1979, 1980) shoe an example of consistent application of the minimal coupling principle to vector field in U_4. Certainly, the standard gauge invariant is violated, but instead it suggested to use the modified gauge transformations. For the electromagnetic case these read

$$A_\mu \to A'_\mu = A_\mu + e^{\phi(x)} \partial_\mu \alpha, \tag{36}$$

where $\phi(x)$ satisfies

$$Q^\lambda_{\mu\nu} + \delta^\lambda_\mu \partial_\nu \phi - \delta^\lambda_\nu \partial_\mu \phi = 0 \tag{37}$$

The Maxwell strength tensor remains invariant under such transformations. Analogous modification is formulated for Yang-Mills theory in U_4, however one should note that such a modification must be accompanied by a change of the interaction of fundamental matter with the gauge fields. For example, in the non-Abelian theory of matter multiplet ϕ^i interacting with the Yang-Mills field A^a_μ the covariant derivative and gauge field strength tensor must read

$$(D_\mu \phi)^i = (\delta^i_j \partial_\mu + g_0 e^{\phi(x)} A^k_\mu f^i_{kj}) \phi_j, \tag{38}$$

$$F^i_{\mu\nu} = \partial_\mu A^i_\nu - \partial_\nu A^i_\mu + g_0 e^{-\phi(x)} f^i_{jk}) A^j_\mu A^k_\nu. \qquad (39)$$

It is worth to note that the equation (37) can be interpreted as the restriction on the form of torsion. Solving (37) , one obtains

$$Q^\lambda_{\mu\nu} = \delta^\lambda_\nu \partial_\mu \phi - \delta^\lambda_\mu \partial_\nu \phi. \qquad (40)$$

and with such a torsion the last dangerous term in (34) is gauge invariant. this observation demonstrates a possibly complex nature of torsion and certain arbitrariness in the division of symmetries on "internal" and "external" ones.

Recently in work by Pronin (1988) it was proposed to introduced along with the gauge transformation (39) the transformation of torsion in the form (40). Then theory is explicitly gauge invariant, but the relation between torsion and internal symmetries arises.

As a final remark, we would like to draw attention to the observation that even in absence of tree-level interaction of torsion with electromagnetic field the one-loop calculations may reveal non-trivial radiative corrections of the form $\partial_{[\mu} Q_{\nu]} \partial^{[\mu} A^{\nu]}$ (Obukhov & Pronin, 1988) which can be eliminated only "on the mass-shell". The inclusion of such a terms into the initial Lagrangian (for preserving the renormalizability of the theory, see (Buchbinder & Odintsov, 1990)) violates both the principle of the minimal coupling and the Dirac's thesis about the beauty of a physical theory.

Completing this short discussion of interaction of spin with torsion we would like to mention the results of Gasperini (1984) who demonstrated the inducing of such interaction via the Lie-isotopic extension of the gauge group.

Non-minimal interaction of torsion with matter can arise in the same manner as in general relativity. For example, assuming the conformal invariance in the theory of massless scalar fields one should add the term of the form $\xi\varphi^2\tilde{R}$. this leads to the study of models with variable gravity constant in presence of non-trivial torsion. The latter is produced by scalar matter even in the ECSK theory.

2.4 Correspondence between the gravity with torsion and general relativity

Physical consequences of the gravity theory with torsion will be considered in the sections 3-5. However, since general relativity is already satisfactory proved in laboratory and planetary observations, it is important to check the agreement of the gravity with torsion with the results of Einsteinian general relativity in macroscopic domains for spinless matter (i.e when the average spin is zero). This correspondence problem was analyzed for the first time by Sandberg (1975) and Nester (1977).

For non-vacuum solutions one can show that torsion is essential only inside the gravitating bodies, and outside it rapidly decreases to zero. At the same time the metric feels the presence of torsion via non-linearity of the gravitational field equations. In the ECSK theory it is possible to establish the principally microscopic nature of the torsion field as a certain collective (or mean) field constructed from pairs of fermions (Gvozdev & Pronin, 1985).

The algebraic character of the spin-torsion coupling in the ECSK theory gives the direct correspondence with the general relativity equations with effective energy-momentum tensor. This fact underlines geometrical interpretation of several non-linear field theories. Let us formulate some most essential results in the form of theorems, though omitting proofs.

Theorem 1. The dynamics of the Dirac fermion massless field in ECSK is completely equivalent to the theory of non-linear Ivanenko-Heisenberg type spinor field in general relativity (Rodichev, 1961; Datta, 1971; Krechet & Ponomariev, 1976).

Analogous observation can be made for the interacting vector (electromagnetic) and gravitational fields (Ponomariev & Smetanin, 1978; Smetanin, 1982).

Theorem 2. Theory of vector (electromagnetic) field in ECSK theory is equivalent to the non-linear model of selfinteracting vector field in general relativity which is described by the Lagrangian

$$L = \hat{F}_{\alpha\beta}\hat{F}^{\alpha\beta} + 2b\hat{F}_{\alpha\beta}\hat{F}^{\beta\gamma}A^{\alpha}A_{\gamma}, \tag{41}$$

where

$$\hat{F}_{\alpha\beta} = F_{\alpha\beta} + \frac{2bA_{[\alpha}F^{\gamma}_{\beta]}A_{\gamma}}{1+bA_{\mu}A^{\mu}}, \quad F^{i}_{\mu\nu} = \partial_{\mu}A^{i}_{\nu} - \partial_{\nu}A^{i}_{\mu}, b = \frac{G}{c^4}. \tag{42}$$

The above mentioned results were obtained under the assumption of minimal coupling of matter and gravity. However, for example, the scalar field does not interact minimally with torsion. Let us consider the Jordan-Brans-Dicke type theory in U_4. The action function in general reads

$$S = \int d^4x\sqrt{-g}\{\frac{1}{2}\partial_{\mu}\phi\partial^{\mu}\phi - \xi\phi^2\tilde{R} - \frac{m^2}{2}\phi^2 - \frac{\lambda}{4}\phi^4\}, \tag{43}$$

The following result holds

Theorem 3. Theory of Jordan-Brans-Dicke in the Riemann-Cartan spacetime is equivalent to the RiemannianJordan-Brans-Dicke theory with "finitely-renormalized" coupling constants and field functions described by the action (Ivanenko, Pronin & Sardanashvily, 1985):

$$S' = \int d^4x\sqrt{-g}\{\frac{1}{2}\partial_{\mu}\hat{\phi}\partial^{\mu}\hat{\phi} - \xi\hat{\phi}^2\tilde{R} - \frac{m^2}{2}\hat{\phi}^2 - \frac{\hat{\lambda}}{4}\hat{\phi}^4\}, \tag{44}$$

where

$$\xi_1 = \frac{2\xi}{2-27\xi}, \hat{m}^2 = m^2\xi_1, \hat{\lambda} = 2\lambda\frac{\xi_1}{\xi}, \hat{\phi} = \phi\sqrt{\frac{2\xi}{\xi_1}}. \tag{45}$$

Important contribution to the understanding of the nature of torsion is made by a series of observations which could be formulated as "inverse quantum theorems" (Gvozdev & Pronin, 1985). Quantization of spinor field on the Riemannian background spacetime via the mean field method establishes the following result.

Theorem 4. Interaction of non-linear spinor field (with the four-fermion coupling) with the gravitational field in general relativity is equivalent (in the tree approximation in the mean field and 1-loop approximation in spinor field) to the Dirac theory in ECSK with effective cosmological term proportional to the square of torsion.

Contrary to ECSK theory it is more difficult to formulate general results about correspondence of quadratic models with general relativity. We will briefly review some well known facts.

Theorem 5. The quadratic model (29) with constants $\Lambda_3 \neq 0, \Lambda_1 = \Lambda_2 = \Lambda_4 = \Lambda_5 = \Lambda_6 = \mu_i = 0$ where $i = 1, 2, 3$, and

$$\Lambda_1 = 4(b_1 + b_2) + 2b_3 + b_4 + b_5,$$
$$\Lambda_2 = 4b_1 + b_3,$$
$$\Lambda_3 = 4(b_1 + b_2 + 2b_4 + itb_5 + 12b_6,$$
$$\Lambda_4 = 4b_1 - b_2) + b_4 - b_5,$$
$$\Lambda_6 = 4b_1 + b_3 + 2b_4,$$
$$\mu_1 = -1 + a_1 - a_2,$$
$$\mu_2 = 2 - \frac{1}{4}(2a_1 + a_2 = 3a_3),$$
$$\mu_3 = -(1 + \frac{a_1}{4} + \frac{a_2}{4}).$$

and conformal-invariant ($T^{\mu}_{\mu} = 0$) matter is equivalent to ECSK (Rauch, 1982; Neville, 1982).

Concretization of material source establishes further correspondence with general relativity (see above theorem 1-3). For the case of non-conformal matter ($T^{\mu}_{\mu} \neq 0$) there is also a complete correspondence with general relativity theory with effective energy-momentum tensor (Minkevich,1986). That is why suggested to call such a model "a minimally quadratic model".

The above theorems give the direct relation between the systems of the field equations in general relativity and in theory with torsion. However it is clear that another important issue is the study of correspondence between the relevant functional spaces of solutions in these theories, initialized by Frolov (1963, 1977). We will now discuss some of the relations of this type. To begin with, let us formulate non-vacuum results, generating with theorem 5.

Theorem 6. For spinless conformal invariant matter ($S^{\lambda}_{\mu\nu} = 0, T^{\mu}_{\mu} = 0$) quadratic models with parameters $\Lambda_1 = \Lambda_2 = \Lambda_4 = \Lambda_5 = 0, \Lambda_3 \neq 0, \mu_2 = 2, \mu_1 \neq \frac{2}{3}\Lambda\Lambda_3 \neq \mu_3$ are equivalent to general relativity theory with effective gravitational constant $G_{eff} = G(1 + \frac{2}{3}\Lambda\Lambda_3)^{-1}$ (Obukhov et al., 1989).

Torsionless solutions in quadratic models are contained to a considerable extent in the class of Einstein spaces of general relativity which satisfy the equations

$$R_{\mu\nu} = \frac{\Lambda}{4} g_{\mu\nu}, \Lambda = const. \qquad (46)$$

Theorem 7. For spinless ($S^{\lambda}_{\mu\nu} = 0$) matter the field equations of quadratic models with parameters $\Lambda_1 = \Lambda_3 = 0$ and any $T_{\mu\nu}$ or with parameters $\Lambda_1 = 0$ and conformal property ($T^{\mu}_{\nu} = 0$) are equivalent in absence of torsion to general relativity with effective gravitational constant $G_{eff} = G(1 + \frac{2}{3}\Lambda\Lambda_3)^{-1}$ (Obukhov, Ponomariev & Zhytnikov, 1989).

Results, covered by theorems, 1-7 refer to the non-vacuum case of correspondence between general relativity and gravity with torsion. The vacuum case $S^{\alpha}_{\mu\nu} = T_{\mu\nu} = 0$ is certainly of great importance also.

The vacuum ECSK equation is completely equivalent to general relativity.

In general vacuum quadratic models describe more complicated geometries than Einstein spaces. But for certain physically important cases the correspondence with general relativity is complete. Firstly, it is worth mentioning the generalized Birkhoff theorem results (Ramaswamy &Yasskin, 1979; Neville, 1980; Rauch & Nieh, 1981; Rauch, Shaw & Nieh, 1982;, Rauch, 1982; Baekler & Yasskin, 1984).

Theorem 8. (generalized Birkhoff's). The only $O(3)$-spherically symmetric vacuum solution of quadratic models is the torsionless gravitational field with the Schwarzschild metric valid in following cases (Obukhov, Ponomariev & Zhytnikov, 1989):

1. $\Lambda i = 0, \mu_i$ are arbitrary;
2. $\mu_1 = \mu_3 = -1, \mu_2 = 2, \Lambda_1 = \Lambda_2 = \Lambda_4 = \Lambda_5 = \Lambda_6 = 0, \Lambda_3 \neq 0$;
3. $\mu_2 = 2, \mu_3 = -1, \Lambda_2 = \Lambda_5$;
4. $\mu_2 = 2, \mu_3 = -1, \Lambda_2 = 0$;
5. $\mu_2\mu_3 \neq 0, \Lambda_1 = \Lambda_3 = \Lambda_6 = 0, \Lambda_2 = \Lambda_5$;
6. $\mu_2 = 2, \mu_3 = -1, \Lambda_i$ are arbitrary with additional asymptotic flatness condition;
7. $\mu_1\mu_2\mu_3 \neq 0, \Lambda_3 \neq 0, \Lambda_1 = \Lambda_2 = \Lambda_4 = \Lambda_5 = \Lambda_6 = 0$ with zero curvature scalar condition.

Notice that in 1-2 cases the Birkhoff's theorem also holds under a weaker assumption of $SO(3)$-symmetry.

Torsionless configurations are another important vacuum solutions of quadratic models (Frolov, 1977; Debney, Fairchald & Siklos, 1978. The following results can be established (Obukhov, Ponomariev & Zhytnikov, 1989).

Theorem 9. In general quadratic models the only vacuum torsionless solutions of the field equations are the Einstein spaces (46), except for the three cases which are characterized by the parameters $\varphi = 0, \varphi = \frac{2\Lambda}{3}$ or $\varphi = \frac{-4\Lambda}{3}$, where we denoted

$$\varphi = \frac{(1 + \frac{2}{3}\Lambda\Lambda_3)}{\Lambda_1}.$$

It is satisfactory to see that all the physically interesting quadratic models do not belong to these exceptional cases, hence their vacuum torsionless solutions reduce to Einstein spaces.

3 PHYSICAL EFFECTS IN EINSTEIN-CARTAN THEORY

We consider here only specific torsion effects and influence of spin of material sources on physics phenomenon.

The most characterizing property of Einstein-Cartan theory is the algebraic type of spin-torsion interaction. The torsion is equal zero outside of material source and may be found only through influence on the metric. In correspondence with the expressions (26, 27) the field equations in Einstein-Cartan are equivalent to Einstein equations with effective energy-momentum tensor

$$T^{eff}_{\mu\nu} = t_{\mu\nu}+\kappa(-4S^{[\alpha}_{\mu\beta}S^{\beta]}_{\nu\alpha}-2S_{\alpha\beta\mu}S^{\alpha\beta}_{\nu}+S_{\mu\alpha\beta}S^{\alpha\beta}_{\nu}+\frac{1}{2}g_{\mu\nu}(4S^{\gamma}_{[\alpha|\beta|}S^{\alpha\beta}_{\gamma]}+S_{\alpha\beta\gamma}S^{\alpha\beta\gamma})) \quad (47)$$

where $t_{\mu\nu}$ is the metrical energy-momentum tensor.

3.1 Effects of motionless particles

Let us consider the source of gravitational field in Einstein-Cartan theory in the form of the ideal spinning dust. Then (Hehl et al.,1976)

$$S^{\lambda}_{\mu\nu} = u^{\lambda}S_{\mu\nu}, \quad (48)$$

$$T^{eff}_{\mu\nu} = \rho u_{\mu}u_{\nu} - 2(g^{\alpha\beta}+u^{\alpha}u^{\beta})\nabla_{\alpha}[u_{(\mu}S_{\nu)}] - \frac{1}{2}\kappa(g_{\mu\nu}+2u_{\mu}u_{\nu})S_{\alpha\beta}S^{\alpha\beta}, \quad (49)$$

where u^{λ} is a 4-velocity and ρ is density of matter.

We may use the post-Newtonian approximation for the concrete case os source that consists of n point particles which have masses $m^{(a)}$ and spins $S^{(a)}$, $a = 1, \ldots.., n$ and $S_{\mu} = \frac{1}{2}\varepsilon_{\mu\nu\alpha\beta}u^{\nu}S\alpha\beta$.

The standard calculations (Castagnino et al., 1985, 1987, 1988) give us the next post-Newtonian metric components

$$g_{00} = 1+2U+2U^2-2G\sum_a \frac{m^{(a)}U_{(a)}}{|\mathbf{x}-\mathbf{x}_a|} - 3G\sum_a \frac{m^{(a)}v^2_{(a)}}{|\mathbf{x}-\mathbf{x}_a|}, \quad (50)$$

$$g_{0i} = \frac{G}{2}\sum_a \frac{m^{(a)}}{|\mathbf{x}-\mathbf{x}_a|}(7v_{(a)i}+(\mathbf{v}_{(a)}\mathbf{n}_{(a)})n_{(a)i}) - 2G\sum_a \frac{[\mathbf{n}_{(a)}\times\mathbf{S}^{(a)}]_i}{|\mathbf{x}-\mathbf{x}_a|^2}, \quad (51)$$

$$g_{ij} = -(1+2U)\delta_{ij} \quad (52)$$

where as usual

$$U = -G\sum_a \frac{m^{(a)}}{|\mathbf{x}-\mathbf{x}_a|}, \qquad U_{(a)} = -G\sum_{b\neq a}\frac{m^{(b)}}{|\mathbf{x}_a-\mathbf{x}_b|},$$

$$\mathbf{n}_{(a)} = \frac{\mathbf{x}-\mathbf{x}_a}{|\mathbf{x}-\mathbf{x}_a|}$$

and $\mathbf{v}_{(a)}$ is a 3-velocity of a-th particle.

It is easy to see that the quality discrepancy of (50-52) from PPN-metric for spinless particles consists in the existence of the last term in (51) that imitates an effect of rotation.

Using PPN-analysis it is possible to consider the interaction of two particles with spins $\mathbf{S}_1$ and $\mathbf{S}_2$ and masses $m_2 \gg m_1$. The acceleration of particle with mass m_1 caused by this interaction will be

$$\mathbf{a} = -Gm_2\frac{\mathbf{r}}{r^3} + \mathbf{a}_s,$$

where

$$\mathbf{a}_s(S_1, S_2) = \frac{4Gm_2}{m_1c^2r5}\left\{\mathbf{S}_1(\mathbf{S}_2\mathbf{r}) + \mathbf{S}_2(\mathbf{S}_1\mathbf{r}) - \frac{5(\mathbf{S}_1\mathbf{r})(\mathbf{S}_2\mathbf{r})}{r_2}\mathbf{r} + \mathbf{r}(\mathbf{S}_1\mathbf{S}_2)\right\}. \tag{53}$$

This result is an explicit illustration of natural linkage of spin and rotation in gravity with torsion. The main quality result is the influence of spin on physical phenomena and this influence looks like an imitation of rotation near the isolated sources in Einstein-Cartan theory. Particularly, if the source is a homogeneous ball with radii R, mass M and constant spin density that is described by the vector $\mathbf{S}$ (without any restriction we may suppose that $\mathbf{S} = (0, 0, S)$). Than the metric will be (Arkuszewski et al., 1974)

$$\begin{aligned} ds^2 &= c^2\left(1 - \frac{2MG}{c^2r}\right)dt^2 - \left(1 + \frac{2MG}{c^2r}\right)dr^2 - r^2(d\theta^2 + \sin^2\theta d\varphi^2) + \\ &\qquad \frac{16\pi G}{3c^2}R^3S\frac{\sin^2\theta}{r}dtd\varphi, \end{aligned} \tag{54}$$

where r, θ, φ are spherical coordinates. This metric is the metric of Lense-Thirring type for the rotational source with the angular momentum

$$\mathbf{L} = (0, 0, \frac{4\pi R^3}{3}S). \tag{55}$$

So, the verification of Einstein-Cartan theory may be made through investigation of motionless test bodies and electromagnetic radiation near polarized sources. To exclude possible additional electromagnetic effects due to spin-electromagnetic interaction it is need to put source in the box from superconducting material. The predicted effect is very small, for example, the total spin of iron ball with mass $M = 100kg$ and one polarized electron along z-axis per each atom is equal $0.5gcm^2sec^{-1}$. This is equivalent to very slow rotation of the same ball with angular velocity $\approx 1.8rad/year$.

3.2 Effects of moving bodies

We saw that the spin of central body leads to appearance of non-diagonal metric components. Let us consider as the introduced PPN-terms (50-52) change the geodesic equation that is the equation of motion for spinless particles

$$\frac{d^2x^\mu}{ds^2} + \Gamma^\mu_{\alpha\beta}\frac{dx^\alpha}{ds}\frac{dx^\beta}{ds} = 0. \tag{56}$$

In spite of that the torsion influence is in order $(\frac{v}{c})^4$ (Castagnino et al., 1985), we consider that the acceleration even the spinless particles in gravitational field depends on contact spin-spin interaction and is described by the next expression

$$\begin{aligned} \vec{a} &= \frac{d^2\vec{x}}{dt^2} = -\vec{\nabla}(\varphi + 2\varphi^2 + \psi) - \frac{d\vec{\xi}}{dt} + [\vec{v} \times [\vec{v} \times \vec{\xi}] + \\ & 3\vec{v}\frac{\partial \varphi}{\partial t} + 4\vec{v}(\vec{v}\vec{\nabla}\varphi) - v^2\vec{\nabla}\varphi, \end{aligned} \tag{57}$$

where φ denotes the terms of first order in $(\frac{v}{c})$ and ψ denotes the terms that have the four order in $(\frac{v}{c})$, and $\xi = g_{0i}$.

Now let us estimate this effect in proposal that the source of gravitational field is spin fluid with density of polarized particles $n_s \sim \frac{M}{\hbar^2}$. If we suppose that the test particles velocity and velocity of source particles are compatible in value than the relation of acceleration caused by spin influence to acceleration caused by spinless matter will be $\frac{a(s)}{a(m)} \approx 10^{-8}$ (Castagnino et, al., 1985, 1987, 1988).

On the other hand if we put the hyroscope near the spherically symmetrical source consists of particles with spin we may found the Schiff's effect due to non-diagonal components of metric. Now we write down the result without detail: the hyroscope with own angular momentum **J** apearance the precession that is described by the equation

$$\frac{d\mathbf{J}}{dt} = [\mathbf{J} \times \mathbf{K}], \tag{58}$$

in the field of central body with radii R and constant spin S, where the vector

$$\mathbf{K} = \frac{4\pi G R^3}{3c^2 r^3}\left(\mathbf{S} - \frac{3\mathbf{r}(\mathbf{S}\cdot\mathbf{r})}{r^2}\right) - \frac{3}{2}\frac{\mathbf{L}}{r^3}, \tag{59}$$

and $\mathbf{L} = m[\mathbf{r} \times \mathbf{v}]$ is the orbital momentum of hyroscope with the velocity $\mathbf{v}$.

This is the direct Schiff's effects. It is possible to additional experiment to find the influence of torsion on the metric. For example, the test particle with mass m and the velocity $\mathbf{v}$ feels the action of the Lorenzt type force

$$\mathbf{F}_L = m[\mathbf{v} \times \mathbf{B}], \tag{60}$$

where the vector field of "magnetic" type

$$\mathbf{B} = \frac{8\pi G}{3c^2}\left(\frac{R}{r}\right)^3\left(\mathbf{S} - \frac{3\mathbf{r}(\mathbf{S}\cdot\mathbf{r})}{r^2}\right) \tag{61}$$

is created by the spin density of source. In particularly, the Hall effects will take place under the force $\mathbf{F}_L$ for the electrons in the pies of iron.

But the effect is very small that has been showed on the example of iron ball in 3.1.

3.3 Interaction of electromagnetic waves with torsion

The problem of interaction of electromagnetic field with torsion in the Einstein-Cartan theory comes to the problem of waves motion in gravitational field of Lense-Thirring type that is described by the metric (54). The polarization vector of this wave is to rotate. The angle of rotation is calculated by the Skrotzkii's method (Skrotzkii, 1957) and will be

$$\frac{d\psi}{ds} = \frac{4\pi G}{3c^2}\left(\frac{R}{r}\right)^3 \left((\mathbf{Sn}) - \frac{3\mathbf{r}(\mathbf{Sr})(\mathbf{nr})}{r^2}\right). \tag{62}$$

here $\mathbf{n}$ is the tangential vector and s is the parameter along the light beam.

The existence of rotational effects for the polarization vector may be found in the experiments of the next type. Let us put the source of the light beam on the surface of body that acquires in the moment the spin directed along the light beam. In this case near the surface of central body

$$\Delta\psi = \frac{4G}{c^3 r^2}\Delta z S, \tag{63}$$

where $\frac{G}{c^3} \sim 10^{-39}, \Delta z \sim 1$ and $\frac{S}{r^2} \sim nR$ and n is the density of gravitational source. For neutron stars $R \approx 10^6$ and $n \approx 10^{14}$ and consequently $\Delta\psi \approx 10^{-19}$. This is far from the modern experiments.

Now let us consider the physical effects in the quadratic models.

4 PHYSICAL EFFECTS IN QUADRATIC MODELS

In the quadratic models the torsion is the dynamical field and may propagate out material sources. To consider experiments in the frame of Solar system it is need to do general analysis of post-Newtonian contributions in the theory. On the base of this it will be possible to predict the new effects. The distinctive property of the quadratic models is the existence of torsion waves that contribute a new effect in the gravitational waves experiments and will be described in the last part of this section.

4.1 General PPN-analysis and "fifth force" effects

It is well known that in the metric theories of gravity the PPN-metric is defined by the set of ten post-Newtonian parameters $\alpha_i, \beta, \gamma, \xi, \zeta_n, i = 1,2,3, n = 1,..,4$ and by the ten potentials, for example

$$U = -G \int d^3x' \frac{\rho(\mathbf{x}',t)}{|\mathbf{x}-\mathbf{x}'|}, \tag{64}$$

and $\Phi_k, \Phi_W, A, B, V_i, W_i$ and $k = 1,...,4$. The exact form of the last potentials is well known.

In gravitational theory with torsion the PPN-structure of metric is more complicated and this fact is caused by the torsion (Gladchenko et al., 1990). The concrete form of the post-Newtonian contributions reflects the influence of torsion degree of freedom on the gravitational interaction. We write down below the PPN-form of the metric (00)-component in the case $\Lambda_1 = \Lambda_3 = 0$ (Gladchenko et al., 1990)

$$\begin{aligned} g_{00} \;=\; & 1 + 2U + 2\beta U^2 + 2\xi\Phi_W - (2\gamma + 2 + \alpha_3 + \zeta_1 - 2\xi)\Phi_1 \\ & -2(3\gamma - 2\beta + \zeta_2 + \xi)\Phi_2 - 2(1+\zeta_3)\Phi_3 + 2(3\gamma + 3\zeta_4 - 2\xi)\Phi_4 \\ & +(\zeta_1 - 2\xi)A + 2\sum_{n=1}^{3} \tau_n E_n - \sum_{A=1}^{7} \sigma_A D_A - \delta_1 X_5 - \delta_2 X_6 \\ & +2c_1 G^2 \rho + 2c_2 K[O(v^4)], \end{aligned} \tag{65}$$

where the first two lines describe the usual post-Newtonian terms and the last line contains the so-called contact contributions. The third line contains the twelve new PPN-terms that are absent in the Riemannian version of PPN-algorithm. These new terms are

$$\begin{aligned} E_n &= (G)^n (\partial^n_{i_1,\dots i_n} U)^2, n = 1,2,3, \\ D_A &= (G)^{2+A} \textstyle\int d^3x' \frac{1}{|\mathbf{x}-\mathbf{x}'|} (\partial^{A-1}_{i_1,\dots i_{A-1}} \rho(\mathbf{x}',t))^2, A = 1,2,3, \\ D_{A-3} &= (G)^{1+A} \textstyle\int d^3x' \frac{1}{|\mathbf{x}-\mathbf{x}'|} (\partial^A_{i_1,\dots i_A} \rho(\mathbf{x}',t))(\partial^A_{i_1,\dots i_A} U), A = 1,2,3, \\ D_7 &= G^2 \textstyle\int d^3x' \frac{\partial_0\partial_0\rho(\mathbf{x}',t)}{|\mathbf{x}-\mathbf{x}'|}, \\ X_5 &= \frac{G^3}{4\pi} \textstyle\int d^3x' \frac{\partial_0\partial_0\rho(\mathbf{x}',t)e^{-m^*|\mathbf{x}-\mathbf{x}'|}}{|\mathbf{x}-\mathbf{x}'|}, \\ X_6 &= \frac{G}{4\pi} \textstyle\int d^3x' d^3x'' \frac{\partial_0\partial_0\rho(\mathbf{x}',t)e^{-m^*|\mathbf{x}-\mathbf{x}'|}}{|\mathbf{x}-\mathbf{x}'||\mathbf{x}'-\mathbf{x}''|}, \end{aligned} \tag{66}$$

The other g_{0i} and g_{ij} metric components coincide with the usual PPN-terms and we don't write them here.

The complete PPN-analysis of quadratic models includes the PPN-structure of torsion. But the expressions for the torsion components are very cumbersome and we don't discuss them here because the torsion influences only on spin of particle.

The analysis of standard relativistic gravitational effects in Solar system we may find the previous estimations on the constants b_i and a_i in the Lagrangian (29). These estimations have been found in papers by Obukhov and Pronin (1991).

The effect of fifth force type can be found in the quadratic models even for the spinless sources. The central spherical symmetric body creates the gravitational potential of the form (Obukhov et. al, 1989)

$$\phi(r) = -\frac{GM}{r}\left(1 + \frac{4}{3}g_2 e^{-m_2 r} + \frac{1}{3}g_0 e^{-m_0 r}\right). \qquad (67)$$

So, the changing of Newtonian potential takes place for the spinless source. But the specific of gravity with torsion consists in the consideration of spin as the source of gravitational field. it is useful to consider the spin nature of interaction potential.

The most famous source of torsion field is the fermions. The spin of the last is equal to pseudovector of axial current of Dirac field. The spin interacts with pseudo trace of torsion (10). Let us consider the contribution of torsion-torsion exchange in the dynamics of fermions. We put $g_{\mu\nu} \approx \eta_{\mu\nu}$. Then we find from the Lagrangian (29) that the Lagrangian of total theory in this approximation will be

$$\begin{aligned} L_{eff} &= -\frac{1}{2\kappa}\Big\{\frac{2}{9}\Lambda_5[-\partial_\mu \check{Q}_\nu \partial^\mu \check{Q}^\nu + +(\partial^\mu \check{Q}^\mu)^2] - \\ &\quad \frac{\Lambda_4}{3}(\partial^\mu \check{Q}^\mu)^2 - \frac{2\mu_1}{3}(\check{Q}^\mu)^2 + \frac{\Lambda_3}{27}(\check{Q}^\mu)^2\Big\} - \check{Q}^\mu \check{S}_\mu, \end{aligned} \qquad (68)$$

where $\check{S}_\mu = \frac{1}{2}\varepsilon_{\mu\nu\alpha\beta}S^{\nu\alpha\beta}$ is pseudotrace of fermion spin.

The exchange of torsion quanta leads to specific spin-spin torsion interaction. It is not so difficult to find in non-relativistic approximation

$$V_{\sigma_1\sigma_2}(r) = -\frac{3G}{4\mu_1}\left\{-m_1^2\frac{(\sigma_1\sigma_2)}{r}e^{-m_1 r} + (\sigma_1\nabla)(\sigma_2\nabla)\left[\frac{e^{-m_1 r} - e^{-\tilde{m}_0 r}}{r}\right]\right\}. \qquad (69)$$

where σ_1 and σ_2 are the spin matrix of fermions and we denoted $m_1^2 = \frac{3\mu_1}{\Lambda_5}, \tilde{m}_0^2 = \frac{2\mu_\cdot}{\Lambda_4}$.

The spin-spin interaction is very small and leads to supertune splitting of energy levels in atoms. The analysis of standard data (Lautrup et al., 1972) and comparison with predictions caused by (69) gives the some restrictions on the constants b_i and a_i. Unfortunately these estimations are not exact and depends on the energy levels (see Yakushin, 1991).

4.2 Spin precession

In gravitational theory with torsion most characterizing property is the influence on spin motion. The test bodies in the Riemann-Cartan space-time move in correspondence with Mathisson-Papapetrou equations (Yasskin & Stoeger, 1979). The main observable effect is the spin precession. For example, the next equation takes place in the Minkowski-Cartan space-time (Pronin, 1976)

$$\frac{d\mathbf{S}}{dt} = [\check{\mathbf{Q}} \times \mathbf{S}], \tag{70}$$

where $\check{\mathbf{Q}} = (0, \check{\mathbf{Q}}_1, \check{\mathbf{Q}}_2, \check{\mathbf{Q}}_3)$ and spin vector $\mathbf{S} = (S_{23}, S_{31}, S_{12})$. The analogous effect may be found for arbitrary torsion (Adamowics, 1975).

Using the post-Newtonian metric (50 - 52) we may get the most general equation for the spin precession (Gladchenko et al., 1990)

$$\frac{d\mathbf{S}}{dt} = [\mathbf{\Omega} \times \mathbf{S}], \tag{71}$$

where the angulkar velocity is

$$\mathbf{\Omega} = \frac{1}{2}\{[\mathbf{v} \times \mathbf{a}] - \mathbf{rot}\xi - (1 + 2\gamma + 2q_1 - q_3)[\mathbf{v} \times \nabla U] - (q_5 + q_6)\mathbf{rot}\mathbf{v}\}, \tag{72}$$

and $\mathbf{v}$ and $\mathbf{a}$ are velocity and acceleration and q_i are additional PPN-parameters that define the PPN-torsion structure.

The main difficulty in realization is the principle impossibility to extract the spin precession effect from the analogous effects under influence of rotational metrics.

4.3 Dispersion of electromagnetic waves

Let torsion interacts with electromagnetic directly and this interaction is ruled by the equations

$$\nabla_{[\alpha}F_{\mu\nu]} = 2Q^{\lambda}{}_{[\alpha\mu}F_{\nu]\lambda}, \nabla_{\alpha}F^{\beta\alpha} + T^{\beta}{}_{[\lambda\alpha}F^{\lambda\alpha} = 4\pi j^{\beta}, \tag{73}$$

where

$$T^{\beta}_{\lambda\alpha} = Q^{\beta}_{\lambda\alpha} + 2\delta^{\beta}_{[\lambda}Q_{\alpha]}.$$

Then after decomposition of vector-potential in the form

$$A^{\mu} = Re\{(a^{\mu} + b^{\mu} + c^{\mu} + \ldots)e^{i\theta}\},$$

where $b^{\mu} \sim \frac{\bar{\lambda}}{\lambda_0}$, $c^{\mu} \sim (\frac{\bar{\lambda}}{\lambda_0})^2$, λ is a wave length, $\lambda_0 = min(R, L)$, R is curvature and L is distance on that the spreading of light is considered, $\bar{\lambda} = \frac{\lambda}{2\pi}$. and introduction of polarization vector

$$f^{\mu} = \frac{a^{\mu}}{\sqrt{|a^{\mu}a^{*}_{\mu}|}}$$

and wave vector $k_{\mu} = \partial_{\mu}\theta$ it is possible to transform (73) to the expression (De Ritis et al., 1988)

$$\nabla_{\alpha}(a^2k^{\alpha}) = K^{\alpha\mu\nu}(a^{*}_{\alpha}k_{[\mu}a_{\nu]} + a_{\alpha}k_{[\mu}a^{*}_{\nu]}),$$

$$k^\alpha \nabla_\alpha f^\beta = \frac{1}{2} f^\beta \{(2Q^{\lambda\alpha\mu} + T^{\mu\alpha\lambda})(f^*_\lambda k_{[\mu} f_{\alpha]} + f_\mu k_{[\lambda} f_{\alpha]}) - \frac{1}{2}(Q^{\lambda\alpha\rho} + T^{\rho\alpha\lambda})(k_\alpha f_\rho - k_\rho f_\alpha)\}.$$

These two equations in general relativity take the form $\nabla_\alpha(a^2 k^\alpha) = 0$ and $k^\alpha \nabla_\alpha f^\beta = 0$. These are equivalent to the energy conservation law and parallel transport of polarization vector along light beam. Both of these "laws" will be destroyed in the gravity with torsion.

To find the possible effects of breakdown of the fotons number conservation law we suppose that torsion wave interacts with torsion in "semi-minimal" way. The torsion is also the plane wave. This situation is described by the Lagrangian

$$L = -\frac{c^4}{16\pi G}\left\{R - \frac{3}{2}\partial_\alpha \phi \partial^\alpha \phi\right\} - \frac{1}{16\pi}(F_{\mu\nu})^2 + \frac{g_0}{16\pi}\varepsilon^{\mu\nu\alpha\beta} A_\mu F_{\nu\alpha} \partial_\beta \phi,$$

where we put $\check{Q}_\alpha = \partial_\alpha \varphi$

Then as it was demonstrated by Wolf (1986) for the electromagnetic wave with components $F_{12} = B_z(x,t), F_{20} = E_y(x,t)$ interacting with torsion wave that is ruled by the equation

$$\frac{1}{c^2}\frac{\partial_2 \phi}{\partial t^2} - \frac{\partial_2 \phi}{\partial x^2} = \frac{8G g_0}{c^5} E_y B_0,$$

in presence of constant magnetic field $F_{13} = B_0$ the fotons will be splitted on the waves with frequencies $\omega_1 = \omega_0 + \Delta\omega$, and $\omega_2 = \omega_0 - \Delta\omega$, where ω_0 is the original frequency and

$$\Delta\omega = \frac{g_0 B_0 \sqrt{G}}{2c}$$

Supposing $B_0 = 10^{13}, g_0 \sim \frac{1}{137}$ we get $\Delta\omega \approx 10{-}3$. If $\omega_0 \approx= 10^8$ we may see that $\frac{\Delta\omega}{\omega_0} \approx 10^{-11}$. This value is in the possibilities of modern experimental technical.

4.4 Gravitational wave experiments

Specific of gravity with torsion consists in the consideration in gravitational wave experiments special polarized effects. In general relativity the possible polarization of weal gravitational waves investigated in details but in gravity with torsion the situation is considerably complicated because we even the spinless source radiates the torsion waves. Moreover even we use a detector which does not interact with torsion we will find nontrivial discrepancies from general relativity.

It is possible to show that the power of gravitational radiation is described by the expression (Obukhov et al., 1989)

$$\begin{aligned} \frac{dE}{dt} &= -\frac{G}{5} < \left[(1+g_2)\left(\dddot{D}^2_{ik} - \frac{1}{3}\dddot{D}^2\right) + \right. \\ & \left. \left(\frac{g_0+g_1}{18}\right)(\dddot{D}^2_{ik} + 3\dddot{D}^2) + \frac{30 g_1^2}{\Lambda_5}\dot{\Sigma}_i^2\right] > \end{aligned} \qquad (74)$$

where $g_1 = \frac{\mu_3}{\mu_3 - \mu_1}$, D^{ik} is quadruple mass momentum and Σ_i is the total 3-vector of spin.

This formulae gives us the possibility to do some physical conclusion on the observability of gravitational-wave effects in gravity with torsion. For example, the gravitational pulsar is the exactly spherically symmetrical compact object. The expression (74) for this configuration gives

$$\frac{dE}{dt} = -\frac{G}{27}(g_0 + g_1) < \dddot{D}^2 >$$

In Einstein general relativity this source does not radiate the gravitational waves. It is crucial discrepancy in predictions of these two theories.

5 QUANTUM EFFECTS OF SPIN-TORSION INTERACTION

The source of torsion field in gravity is the spin of particle. The spin of particle has essentially quantum nature. This result leads to some problems in the quantum measurability of contact spin-spin interaction.

5.1 Quantum nature of torsion in ESCK theory and problem of measurability

To clean the physical picture we restrict our consideration by the case of flat manifold $g_{\mu\nu}$ with totally antisymmetrical torsion field. Then the Hamiltonian of spin torsion interaction will be

$$H_{int} = \alpha Q\lambda\mu\nu S^{\lambda\mu\nu} \tag{75}$$

To measure the torsion field it is necessary to take particle with spin $S \sim \hbar$ and investigate its movement. Let the particle is localized in volume V. The interaction with torsion field will be resulted in some change of total energy δE, that is in order of

$$\delta E = H_{int}V$$

It follows from the Heisenberg's relation $\Delta t \Delta E \sim \hbar$ that

$$\Delta(\delta E)\delta t \to V\Delta H_{int}\Delta t \geq \hbar \tag{76}$$

Taking into consideration the equation (75) and using (76) we get

$$V\Delta t(S\Delta Q + Q\Delta S) \geq \hbar \tag{77}$$

Inserting (77) into (75) it ia easy to calculate that

$$\Delta Q_{\alpha\beta\nu} \sim \frac{\lambda_{\alpha\beta\nu}}{\Delta t} + Q_{\alpha\beta\nu},$$

where λ is some constant of order one.

The first term depends on a period of observation Δt and its can be made as small as it is need due to enlarging of Δt. The second term does not depend on time and it is proportional to torsion field $Q_{\lambda\mu\nu}$.

So, the uncertainty in measurement of torsion can not be less the strength of torsion field. This fact is right only in the Einstein-Cartan theory frame for particles with elementary spin. This result reflect the contact nature of spin-spin gravitational interaction and quantum origin of torsion field (Gvozdev &Pronin, 1984, 1985;Soffel, Müller & Greiner, 1982). The classical torsion field consideration is allowed only in quadratic models and it is some analogy of spin waves. The problem of measurability of torsion field is more complicated then it can be viewed at the first stage. To measure torsion field directly it is need to use the particles with spin of order $S \simeq N\hbar$. Then we can get from () $\Delta Q \leq \frac{Q}{N}$. To do this we are to prepare device in the form of test body with oriented spins. But in this case the problem of spin particle correlation and self-interaction of spins in test body is arisen. We think that it is not realistic way to measure torsion effects on quantum level. More realistic way is the using of interferometer experiments with polarized neutron beams.

5.2 Neutron interferometry

Neutron interferometry is powerful method to verification of gravitational spin-torsion interaction (De Sabbata et al., 1991). to study the most general situation we introduce in expression for covariant derivative the new constant of spin torsion interaction. To separate the spin-torsion effect, we restrict consideration to Minkowski-Cartan space. Then after presentation of Dirac field in bispinor form

$$\psi = \begin{bmatrix} \phi \\ \chi \end{bmatrix}, \tag{78}$$

we get the Pauli equation

$$i\hbar \frac{\partial \phi}{\partial t} = \left\{ -\frac{\hbar^2}{2m} \nabla^2 + g_0(\sigma \bar{\mathbf{Q}}) \right\} \phi + O(\hbar^3 g_0^2), \tag{79}$$

Now let us introduce the spin $\vec{S} = \frac{\hbar}{2} \phi^+ \vec{\sigma} \phi$ of test particle. Here

$$\phi = \begin{bmatrix} \cos\frac{\theta}{2} \exp\left[i\left(\frac{\xi+\eta}{2}\right)\right] \\ i \sin\frac{\theta}{2} \exp\left[i\left(\frac{\xi-\eta}{2}\right)\right] \end{bmatrix}, \tag{80}$$

The equation of spin motion that is consequence of Pauli equation will be

$$\frac{d\mathbf{S}}{dt} = \mathbf{T} + \frac{2g_0}{\hbar} [\check{\mathbf{Q}} \times \mathbf{S}], \tag{81}$$

where the term

$$\mathbf{T} = \frac{1}{m\rho}\left[\mathbf{S}\times\sum_i \frac{\partial}{\partial x^i}\left(\rho\frac{\partial \mathbf{S}}{\partial x^i}\right)\right], \tag{82}$$

describe the so called quantum self-influence of spin. The external torsion create the precession due to second term.

Let us consider the interferometer experiments. We assume that the neutron beams (I and II) are polarized in the antiparallel direction to the z axis. Then the spinor normalized function are

$$\psi_1 = |\downarrow_z> = \begin{bmatrix}0\\1\end{bmatrix}, \quad \psi_2 = e^{i\theta}|\downarrow_z> = e^{i\theta}\begin{bmatrix}0\\1\end{bmatrix}, \tag{83}$$

where θ is the phase shift.

If the external torsion field equals zero, the degree of polarization of the beam after interferometry is

$$\mathbf{P} = \frac{\psi^+_{fin}\sigma\psi_{fin}}{\psi^+_{fin}\psi_{fin}} = (0,0,-1), \tag{84}$$

But in the situation when one (or both) neutron beam(s) interact with torsion the states should be changed as

$$\psi_1' = \sqrt{\epsilon}|\uparrow_z> + \sqrt{1-\epsilon}|\downarrow_z>, \quad \psi_1' = e^{i\theta}(\sqrt{\epsilon}|\uparrow_z> + \sqrt{1-\epsilon}|\downarrow_z>), \tag{85}$$

Then the interferation of these beams leads to

$$\mathbf{P}' = (2\sqrt{\epsilon(1-\epsilon)}, 0, -1+2\epsilon), \tag{86}$$

So we can observe the effect of the polarized rotation plane due to quantum interferometry, which is caused by the previous interaction with torsion.

Let us give estimation of the effect: we consider $\varepsilon = \frac{1}{2}$ then we get the connection between density of particles in source and constant of spin torsion interaction.

$$n \approx \frac{10^{17}}{2g_0}$$

If $g_0 \ll 1$ that is right very far from source then effect is negligible small. But we may to change the region at that the torsion is not equal zero. For example, if the characteristic length of experiment will be on order $10^{-13}cm$ the effect will be observable but we have to overcame the very big difficulties in the kipping of fermions in so small region.

References

[1] Aragone C. & Deser S. (1980), Nuovo Cimento, **B57**, 33.

[2] Arkuzsewski W., Kopxzynski W., & Ponomariev V.N. (1974), Ann. Inst. H.Poincaré, **A21**, 89.

[3] Baekler P. & Yasskin P. (1984), Gen. Rel. Grav., **16**, 1135.

[4] Brodskii A.M., Ivanenko D.D., & Sokolik H.A. (1962), ZhETF, **41**, 1307 (in Russian).

[5] Buchbinder I.L. & Odintzov S. (1990), Class. Quant. Grav., **9**, 789.

[6] Buchdal Y.A. (1958), Nuovo Cimento, **10**, 96.

[7] Buchdal Y.A. (1962), Nuovo Cimento, **25**, 488.

[8] Buchdal Y.A. (1989), J. Math. Phys., **30**, 700.

[9] Cartan É. (1922), C.R. Acad. Sci. (Paris), **174**, 593.

[10] Cartan É. (1923), Ann. Éc. Norm., **40**, 325.

[11] Cartan É. (1924), Ann. Éc. Norm., **41**, 1.

[12] Cartan É. (1925), Ann. Éc. Norm., **42**, 17.

[13] Castagnino M., Levinas M.L., and Umerez N. (1985), Gen. Rel. Grav. bf 17, 683; (1987), ibid. **19**, 545; (1988), ibid. **20**, 715.

[14] De Ritis R., Scudellaro P., & Stornaiolo C. (1988), Phys. Lett. **A126**, 389.

[15] De Sabbata V. & Gasperini M. (1980), Lett. Nuovo Cimento, **28**, 181; ibid. 228.

[16] De Sabbata V. & Gasperini M. (1981), Lett. Nuovo Cimento, **30**, 363; ibid. 503; ibid.1933

[17] De Sabbata V., Pronin P.I., & Sivaram C. (1991), Int. J. Theor. Phys., **30**, 1671.

[18] Debney G., Faerchaild E.E., & Siklos S.T.C. (1978), Gen. Rel. Grav., **9**, 879.

[19] Einstein A. (1955), The meaning of relativity, (Princeton Univ.: Princeton).

[20] Frolov B.N. (1977), Izvestia VUZov, Fizika, **3**, 154.

[21] Gasperini M. (1984), Hadronic J., **7**, 650.

[22] Gladchenko M.S., Ponomariev V.N., & Zhytnikov V.V. (1990), Phys. Lett., **B241**, 67.

[23] Gvozdev A.A. & Pronin P.I. (1985a), Izvestia VUZov, Fizika, **1**, 7 (in Russian).

[24] Gvozdev A.A. & Pronin P.I. (1985b), Vestnik Mosc. Univ. ser. Fis. Astr., **1**, 29 (in Russian).

[25] Datta B.K. (1971), Nuovo Cimento, **B6**, 1.

[26] Hehl F.W., v.d.Heude P., Kerlick G.D., & Nester J.M. (1976), Rev. Mod. Phys., **48**, 393.

[27] Hojman S., Rosenbaum M., Ryan M.P., & Shepley L.S. (1978), Phys. Rev., **D17**, 3141.

[28] Hojman S., Rosenbaum M., Ryan M.P., (1979), Phys. Rev., **D19**, 430.

[29] Hojman S., Mukhy C., & Sayed W.A. (1980), Phys. Rev., **D22**, 1915.

[30] Ivanenko D.D., Pronin P.I., & Sardanasvily G.A. (1985), The gauge gravitational theory, Moscow Publ House: Moscow (in Russian).

[31] Kasuya M. (1975), Nuovo Cimento, **B28**, 127.

[32] Kibble T.W.B. (1961), J. Math. Phys., **2**, 472.

[33] Krechet V.G. & Ponomariev V.N. (1976), Phys. Lett., **A56**, 14.

[34] Minkevish A.V. (1986), Izvestia AN BSSR, **5**, 100 (in Russian).

[35] Nester J.M. (1977), Phys. Rev., **D16**, 2395.

[36] Neville D.E. (1980), Phys. Rev., **D21**, 2770.

[37] Neville D.E. (1982), Phys. Rev., **D26**, 2638.

[38] Obukhov Yu.N. (1983), J.Phys., **A16**, 3895.

[39] Obukhov Yu.N. & Pronin P.I. (1988), Acta Phys. Pol., **B19**, 341.

[40] Obukhov Yu.N. & Pronin P.I. (1991), in Vol. "Class. Field Theory and Grav.Theory", **2**, 112 (in Russian).

[41] Obukhov Yu.N., Ponomariev V.N., & Zhytnikov V.V. (1989), Gen. Rel. Grav., **21**, 1107.

[42] Ponomariev V.N., Barvinsky A.O., & Obukhov Yu.N. (1985), Geometrodynamical methods and gauge approach to gravitational theory, Atomizdat:Moscow (in Russian)

[43] Ponomariev V.N. & Smetanin E.V. (1978), Vestnik Mosk. Univ., ser. Fiz., Astr., **5**, 29 (in Russian)

[44] Pronin P.I. (1976), Proc. of 3-d Soviet Grav. Conf., 121 (in Russian).

[45] Pronin P.I. (1988), Hadronic J., **11**, 291.

[46] Ramaswamy S. & Yasskin P. (1979), Phys. Rev., **D19**, 2264.

[47] Rauch R.T. (1982), Phys. Rev., **D26**, 932.

[48] Rauch R.T., & Nieh H.T. (1981), Phys. Rev., **D24**, 2029.

[49] Rauch R.T., Shaw J.C., & Nieh H.T. (1982), Gen. Rel. Grav., **14**, 331.

[50] Rodichev V.I. (1961), ZhETF, **40**, 1469 (in Russian).

[51] Sandberg V.D. (1975), Phys. Rev.,**D12**, 3013.

[52] Schrödinger E. (1946), Proc. roy. Irish Acad., **49**, 275

[53] Sciama D.W., (1958), Proc. Cambr. Phil. Soc., **54**, 72.

[54] Skrotzki G.B. (1957), Dokl. Acad. Nauk USSR, **114**, 73.

[55] Smetanin E.V. (1982), Izvestia VUZov, Fizika, **1**, 30 (in Russian).

[56] Stueckelberg E.C.G. (1948), Phys. Rev., **73**, 808.

[57] Trautman A. (1973), Symposia Mathematica, **12**, 139.

[58] Wolf C. (1986), Nuovo Cimento, **B91**, 231.

[59] Utiyama R. (1958), Phys. Rev., **101**, 1597.

[60] Yasskin P.B. & Stoeger W.R. (1979), Phys. Rev., **D21**, 2081

ANOMALOUS SPIN I: EXPERIMENTS WITH A POLARIZED-MASS TORSION PENDULUM

ROGERS C. RITTER, GEORGE T. GILLIES, AND LINDA I. WINKLER
Department of Physics
University of Virginia
Charlottesville, Virginia 22901
e-mail: rcr@virginia.edu

ABSTRACT

Two torsion pendulum experiments having unique "macro-electron" masses with ~10^{23} polarized electrons provide limits on some hypothetical anomalous spin couplings. Minimal magnetic shielding is needed for these self-compensating masses, which exhibit no experimentally measurable magnetic moment, so that normal electromagnetic effects do not interfere in the searches for non-magnetic spin coupling. The coherent interaction of the large number of polarized spins in such a mass allows these macroscopic experiments to have intrinsic sensitivity similar to that published data indicates for atomic traps. The present studies are designed especially to take advantage of symmetries that can be set up with small polarized masses. Experiments having two intrinsic forms of spin interactions will be discussed, *i.e.* the spin-spin (dipole-dipole) and mass-spin (monopole-dipole) interactions between masses. Considered as P and T violating interactions between particles, the second of these experiments is also motivated by the well-known "strong CP problem." One result of the monopole-dipole experiment was the establishment of the first limits to axionic coupling in the "Turner Window" in astrophysical data. Experiments of the two forms will be described, in both of which a torsion pendulum is operated in the dynamic (time-of-swing) mode. Design optimization methods, experimental and analytical techniques, and results of these and other experiments are discussed.

Contents

I. Introduction

A broad motivation for the present work is the question of inclusion of spin in general relativity, which was much discussed in this School. It is sometimes summarized in the observation that the Poincaré group (inhomogeneous Lorentz group) has two invariants: mass and spin (helicity for massless systems), the latter not having yet been observed in the context of general relativity experiments.

Here we are interested primarily in the experimental approach to this question, and how it has started to become of practical interest in the past two decades, mostly in the context of searching for hypothetical anomalous particle interactions. Moody and Wilczek were concretely involved in this new surge of interest, and discussed it in the 1983 Pacific Coast Gravitation Meeting. They pointed out that macroscopic experiments, for example those using torsion pendulums, could take advantage of the coherent interaction of the many particles in polarized masses to search for axion-like interactions with considerable sensitivity, if these interactions have "long" ranges, say a few cm or more. In fact, the group of Newman had started a spin experiment based on this concept.

Earlier (1964), Leitner and Okubo [1] had produced a heuristic gravitational potential modified by a $\sigma \cdot \mathbf{r}$ term to yield a simple symmetry violation. It was later extended by several authors to include other symmetry violating terms, ending in the more generalized CPT form,

$$U_{CPT}(r) = U_0(r) \{1 + A_1 \sigma \cdot \mathbf{r} + A_2 \sigma \cdot (v/c) + A_3 \mathbf{r} \cdot [(v/c) \times \sigma]\}, \qquad (1)$$

where $U_0(r)$ is the usual gravitational potential. Here the first added term violates P and T, the second P and C, and the third C and T. Klein and Thorsett [2] analyzed the Right- and Left- circularly polarized light difference from a pulsar to set a stringent limit on A_1.

A more explicit context for the two types of experiment to be discussed here is the work of Moody and Wilczek [3] on fundamental fermions weakly coupled by very light bosons with long interaction range (of mass $<10^{-5}$ eV). Two years before the advent of the "fifth force" [4], their paper pointed out that the G(r) experiment of Spero et al. [5], a measurement of the inverse square gravitational force dependence, set a limit of $10^{-41.5}$ on an anomalous Yukawa coupling at a scale of 3 cm, close to a test of possible strengths of axion-mediated forces. While this was important in establishing the sensitivity and potential experimental viability of macroscopic measurements, it was performed with unpolarized particles. Other suggestions for very light, weakly coupled bosons include familons [6], majorons [7], arions [8] and spin-1 antigravitons [9].

The Moody-Wilczek calculations deal with the possible one-boson couplings of fermions shown in Figure 1, which can allow for three different types of forces. The monopole-monopole case is that of G(r), with two scalar vertices, and having no polarization properties. The dipole-dipole case, with two pseudoscalar vertices, has tensor form, and forms the basis for our first experiment.

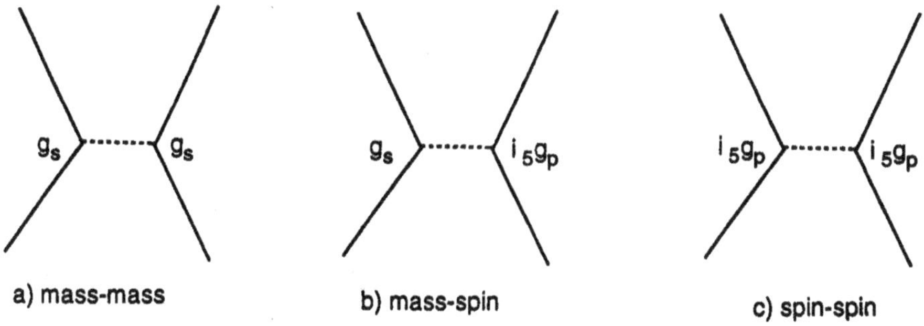

Figure 1. Three types of interaction from single-boson exchange interaction between fermions. These have G(r), $\sigma \cdot \mathbf{r}$, and $\sigma \cdot \sigma$ character, respectively.

II. The Two Spin Experiments

Before taking up these experiments, we will discuss briefly the nature of the sensitivity of torsion pendulum experiments, expanding on the suggestion in [1] about the coherence of interaction in massive polarized bodies. Phillips [10] had much earlier used a dipole magnet on a torsion fiber, with a superconducting shield, to search for a preferred direction in space, essentially an experiment dealing with the second term of equation 1. The dipole magnet had aligned intrinsic spins, and the superconducting magnetic shield (operating on currents of Cooper pairs) does not shield out nonmetric magnetic effects. The suggestion of Ni [11] to use magnetically compensated materials makes it practical to construct a torsion pendulum with minimal shielding, greatly extending the practical aspect of this instrument. It is possible to have huge numbers of spin-aligned electrons, now approaching 10^{24}, in small masses suitable for mounting on a torsion fiber. As a bench mark on its sensitivity, we compare this with one sensitive NMR experiment [12] in which 5000 ions in a trap provided the detector spins interacting with the Earth as a source. Essentially, with a given source the intrinsic sensitivity of the atomic experiment would have to be 10^{20} times greater than the torsion pendulum to compensate for the smaller number of interacting particles. In the section discussing experimental results we will expand on this comparison.

Spin masses on a torsion pendulum will permit experiments with interactions displayed as the second two of the diagrams of Figure 1, shown in simplest arrangements in Figure 2 below.

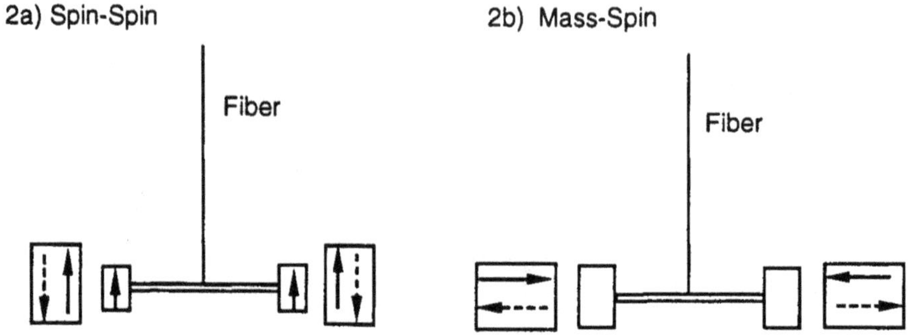

Figure 2. Arrangement of masses for the two University of Virginia spin experiments. In the spin-mass experiment unpolarized masses (blank figures) are used on the pendulum and the axes of the external spin source masses are oriented toward these detector masses. The interaction potential of this experiment has the form $\sigma \cdot r$ and reversal of the source mass spin directions changes the sign of such an interaction.

III. "Macroscopic Electron" Spin Masses

Following the suggestion of Ni, and with his assistance, we constructed the first masses of Dy_6Fe_{23} for use with torsion pendulums. This phase of DyFe is found to have magnetic compensation between opposing intrinsic and orbital spin at near room temperature (in our experiment, exactly at -24 °C). The model for this behavior, consisting of two opposing Brillouin functions, is found to accurately depict experiments with single crystals of the material (Herbst and Croat [13] and Bara & Pedziwiatr [14]). These two functions have different temperature scales, and the zero - crossing of the slope of their difference yields the compensation point. Near crossing, the value of the slope provides an estimate of the degree of polarization of the material. In our case this was 75% of the single crystal value. Along with the Landé g-factor, we then estimated the electron spin polarization of the single-crystal, and hence of our masses, yielding (0.4 ± 0.1) electron per configuration as the polarization of the masses.

The masses were constructed as follows. 50 mesh crushed Dy_6Fe_{23} purchased from Research Chemicals (Phoenix, AZ) was pressed into aluminum cylinders at about 7×10^7 Pa and sealed. Preferring to operate at room temperature rather than -24°C, we used thin 0.9999 pure Fe cylinders around these, followed by aluminum spacing cylinders and then finally μ-metal cylinders to shield against the small uncompensated moment. After the pendulum was constructed and its sensitivity calibrated, test fields from the surrounding Helmholtz coils were used to determine that the residual laboratory fields would develop torques $< 10^{-4}$ of the experimental torque sensitivity.

IV. The Spin-Spin Experiment

At the time of the Moody-Wilczek work only the earlier atomic experiment of Ramsey [15] had direct bearing on this kind of force. This

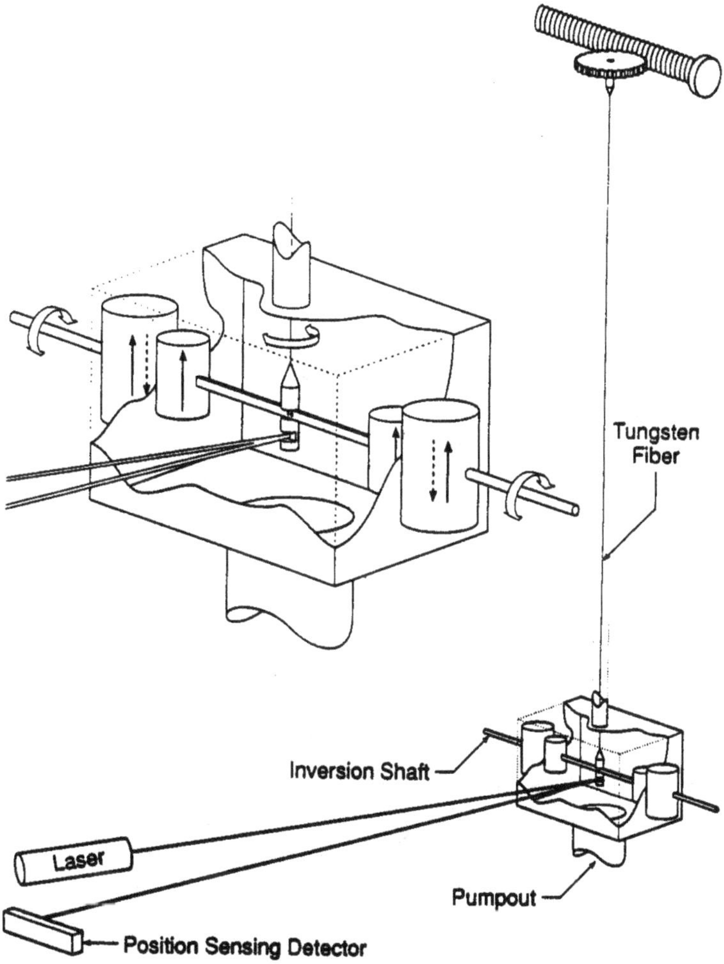

Figure 3. Arrangement of the four spin masses, fiber, position sensing and vacuum pumping of the spin-spin experiment, in a form which preferentially selects the $\sigma \cdot \sigma$ term of the tensor form.

experiment sets a limit on any nonmagnetic interaction between two protons in the hydrogen molecule at $<4 \times 10^{-4}$ that of the proton magnetic moments. Newman [16] was in the process of a first search for a macroscopic dipole-dipole form of the anomalous spin coupling with a shielded low temperature torsion pendulum. Independently, Ni [11] had proposed the use of magnetically self-compensating materials for such experiments. During his visit to our laboratory in 1985 he helped initiate the $\sigma \cdot \sigma$ experiment described here [17].

This first University of Virginia spin experiment, in pursuit of an anomalous spin-spin interaction, was operated in the dynamic mode, with an oscillation period of 576 seconds (a frequency of 1736 μHz). The torsion pendulum consisted of two larger source masses, M, which would hypothetically attract (or repel on inversion) the smaller, "detector" masses, m, supported on the fiber. The mass properties and their equilibrium positioning are listed in Table 1. The pendulum fiber was 25 μm in diameter, of tungsten 0.5 m long. A vibration-isolated turbopump maintained a vacuum of 6×10^{-6} Torr, yielding a pendulum decay time of 4.9×10^{5} seconds. (The damping was partly due to internal fiber friction.) A magnetic damper at the top of the fiber selectively reduced motion in the simple pendulum mode by a factor of 10^{-4}, while negligibly affecting the torsional mode.

Great care was taken to keep out any permanent and permeable moments in materials for the rest of the apparatus except the masses. Except for the mass shields, no known permeable material was present. This dictated the use mostly of aluminum, and sensitive testing of the brass stock for the few such items used. Previously discussed magnetic mass shielding reduced the free-field magnetic interaction between source and detector masses to $< 10^{-14}$ of the magnetic interaction due to the bare polarized electrons in them. However, ambient fields lead to forces between the shields on the masses. Consequently, large Helmholtz coils were put around the experiment, and a stable, precision flux-gate magnetometer continuously monitored the three components of the field at the region of the masses. Except during rare "magnetic storms" the fields were kept to less than 10 nT at the masses, with these coils. With test currents in the Helmholtz coils the torque on the fiber from magnetic interaction between masses was calibrated. (The interaction potential between shielded masses is quadratic in the Helmholz currents.) From this the ferromagnetic interaction between masses due to maintaining the 10 nT limits was determined at the value quoted.

Table 1 - Properties and Locations of the Spin Masses

Source Masses - Dy-Fe Mass	45.2 g
Detector Masses - Dy-Fe Mass	9.65 g
Source Masses - Number of Aligned Electrons	1.36×10^{23}
Detector Masses - Numbered of Aligned Electrons	2.91×10^{22}
Distance - Fiber to Detector Mass Axis (l)	3.94 cm
Distance - Source Mass Axis to Detector Mass Axis (L)	3.36 cm

The raw data in this experiment were the pendulum positions, sampled every 2.5 seconds, accurately triggered by a cesium beam frequency standard. A single run consisted of 4096 points, and was followed immediately by a like run but with the source mass spins inverted. A total of 77 such run pairs were recorded, and the signal of interest from these was the frequency difference from each source mass inversion. Analysis of each data point used the area under the fundamental peak in the power spectrum, calculated by an FFT. For sufficient precision, four known major biases or errors were considered and accounted for in the FFT [18], so that a frequency measurement error less than 1 μHz was obtained. This error was completely dominated by the ordinary experimental fluctuations (such as seismic noise and incompletely compensated temperature variations) combining to 0.021 μHz. The final mean differential frequency for the parallel-antiparallel difference, after all corrections, was $\Delta f = (0.005 \pm 0.021)$ μHz. This is (3 ± 12) ppm of the pendulum frequency of 1736 μHz. This level, related to the gravitational interaction of the masses is $\Delta G/G = (1.9 \pm 7.9) \times 10^{-3}$, where G is the Newtonian gravitational constant.

The signal was then evaluated in terms of an asymptotic form of the potential

$$V = \alpha \mu_e^2 \sigma \cdot \sigma / r^3 , \qquad (2)$$

where μ_e is the magnetic moment of the electron, r is the distance between interacting electrons and α is the fractional anomalous interaction between electrons relative to their intrinsic magnetic interaction. The finite-size masses were integrated over using a Monte Carlo technique [19]. By these methods [17] the above data yield

$$\alpha = (1.6 \pm 6.9) \times 10^{-12}. \qquad (3)$$

After presenting the University of Virginia spin-mass experiment we will discuss both results along with other experiments of similar type or intention. At that time the dimensionless coefficient α will be placed in terms of the usual pseudoscalar coupling coefficient g_P^2.

V. The Spin-Mass Experiment

An experiment of the second type, a spin-mass experiment for a potential with one scalar and one pseudoscalar vertex, can gain the same basic knowledge as the first, but also provides additional information. Before discussing that, we note that the spin-mass experiment should provide a more sensitive measure of the pseudoscalar coupling coefficient, since it contains only one pseudoscalar vertex. Krause and Fischbach [20] have discussed a mechanism which constrains the strength of macroscopic-ranged pseudoscalar forces, stemming from parity non-conservation. For one-boson scalar-scalar exchange they find:

$$V_{ss} = - g_s^2/4\pi r, \tag{4}$$

where g_s is the dimensionless scalar coupling constant. For the scalar-pseudoscalar case the potential V_{sp} is greater by

$$V_{sp}/V_s \sim (g_p/g_s)\,[1/(2mr)] \sim (g_p/g_s) \times 10^{-13}, \tag{5}$$

where the suppression factor [1/(2mr)] is evaluated at one meter. [Here we use a different notation from Ref. 20; V_{sp} for us means the mixed vertex case, scalar-pseudoscalar, hence we take a ratio of different quantities and get the single-vertex suppression factor.] That factor has its origin in parity conservation. The scalar coupling has absorption or emission in an s-state, while the pseudoscalar coupling must be in a p state with suppression represented by [1/(2mr)]. In the scalar-pseudoscalar case, suppression occurs only at one vertex, but in the spin-spin interaction it will occur at a level $\sim 10^{-26}$.

The spin-mass experiment [21] is set up to be of the "efficient" form $\sigma \cdot \mathbf{r}$, as illustrated in Figure 1b above. Only two major changes in the pendulum were needed to convert this experiment from the previous. The small polarized detector masses in the pendulum were replaced by unpolarized copper masses containing 2.4×10^{25} nucleons, and the axes of the external source masses were rotated to intersect the equilibrium position of the center of the detector masses. A new pendulum fiber was emplaced, 22 μm in diameter and of a very stable material ("nicotine") similar to the metal used in mechanical clocks. From these changes the pendulum oscillation period increased to 714 s. The experimental sensitivity of this new arrangement increased by a factor of 4000 due partly to this change, but more importantly to the fact that the unpolarized copper masses had several thousand more interacting nuclei than the polarized electrons in the masses they replaced.

Four run series, each of about two months duration, were carried out. These were taken with all possible relative arrangements of the four source mass spins. Two of these were taken for background purposes, with the two sides of the pendulum having opposite relative mass directions so that any $\sigma \cdot \mathbf{r}$ forces would cancel to first order. The other two were "real" experiments in which the $\sigma \cdot \mathbf{r}$ forces of the two sides would add, and differed only in the exchange of position of the two source masses. In summing the results of the two different "real" series, this interchange would cancel any effects of geometric differences of the two sides which might lead to a bias in the raw inversion signal [21]. Analysis of these data led to a mean inversion signal of $\Delta f = (-0.160 \pm 0.003)$ μHz (with our sign convention).

This large raw inversion signal could have significant artifactual content. In fact, further study considered the possibilities of axial offset ε of the center of mass of each of the spin masses from their centers of gravity. Given the known gravitational sensitivity of the system, it was possible to predict a value of ε from the experimental result of Δf = -0.160 μHz to be -0.68 or -0.72 mm, by two algebraic methods. To test this inference, a very light pan balance was constructed, and the offsets calculated from these balance measurements had a mean ε of (-0.70 ± 0.05) mm [22]. This confirmation was sufficient to conclude that the axial offsets were this value, and therefore should be compensated in our data.

Following this compensation, we made a standard set of estimates of other errors in the experiment, shown in Table 2. Along with the above considerations, this leads to the corrected experimental frequency shift from mass inversion of Δf = (-0.010 ± 0.047) μHz.

Table 2. Error Budget for the $\sigma \cdot \mathbf{r}$ Experiment

Source	Estimated Uncertainty, μHz
Fluctuations (thermal, seismic, gravity gradient variation)	0.022
Temperature variation	0.007
Frequency evaluation error (FFT)	0.005
Magnetic Variation	<0.0005
Pendululm mass decentering	0.020
Nonlinearity correction incompleteness	0.015
Uncorrected radial mass offset	0.002
Uncorrected pendulum angular offset	0.003
Axial mass offset correction error	0.031
Monte Carlo geometrical correction error	0.008
Total (rms)	0.047

A final corrected limiting value of $|\Delta f| < 0.060$ (at the one sigma level) takes into account all known errors. This limit is interpreted with the Moody-Wilczek potential for two interacting fermions,

$$V = (g_s g_p/\hbar c)(\hbar^2/8\pi m_p)\, \sigma \cdot \mathbf{r}\, (1/\lambda r + 1/r^2) e^{-r/\lambda}, \qquad (6)$$

where m_p is the polarized particle (electron) mass and λ is the range of the interaction.

The second derivative of V, integrated over the six coordinates of the two finite masses, is the restoring coefficient of the macroscopic potential. With the definition of a particle interaction strength factor,

$$D = g_p g_s/(8\pi m_p \hbar c), \qquad (7)$$

and the assumption that the interacting masses are at a point, the frequency shift for the asymptotic value of the interaction, $\lambda >> r$, can be written in the form,

$$\Delta f = (\hbar^2 D l L N_p N_m)/[(2\pi^2 f I(L-l)^4], \qquad (8)$$

where we have inserted the trigonometry associated with the small angular oscillation of the torsion pendulum, and evaluate it with a first order expansion. Here l and L are the mass location distances given in Table 1, N_p and N_m are the number of detector and source particles given in Table 1, f is the pendulum frequency (1400 μHz), and I is the pendulum moment of inertia, $2m_d l^2$. Here m_d is the mass of one detector mass.

To account for the finite mass sizes, the number for Δf stated above and used in equation 7 was first modified by a correction factor derived for the finite mass sizes by the technique of Monte Carlo integration, as was done for the first experiment, but with the new geometry. Inserted in equation 7, the final experimental limit on Δf yields the asymptotic value of $g_s g_p/4\pi\hbar c = 3 \times 10^{-27}$.

VI. Discussion of Results, Conclusion

The number of laboratory measurements of anomalous spin-spin interactions is small, and only a few of these were done with a torsion pendulum. A 1991 review by Adelberger *et al.* [23] includes six measurements, four for $\sigma \cdot \sigma$ and two for $\sigma \cdot \mathbf{r}$. A later work by Wineland *et al.* [12] refers to four more results of the spin-spin type and four more spin-mass experiments. In addition, Ref. 12, an NMR measurement, discusses boson exchanges between masses other than electrons. In Tables 3 and 4 we list these results and other later ones. We also note that the units of sensitivity for spin-mass experiments in Ref. 12, kg^{-1}, cause relative and absolute shifts as compared with $g^2/4\pi\hbar c$ by inclusion of the mass of the polarized particle in the denominator, in kilograms.

Given the order of magnitude interest in comparing these results with theory, we do not present the usual plots of coupling constant limits as a function of range. Instead, we list the important elements numerically. The Tables list the reported or quoted asymptotic value of each result, along with the approximate minimum range of interaction at which the experiment would set a limit no greater than 10 times its asymptotic value.

Table 3. Sensitivities of various spin-spin experiments.

Experimental Method	Ref	Asymp. $g_p^2/4\pi\hbar c$	min λ, m
Molecular spectra	Ramsey 79 [15]	5×10^{-5}	5×10^{-8}
NMR	Ansel'm 83 [12]	10^{-10}	
S.C. Tors. Pend.	Newman 85 [16]	3×10^{-10}	0.02
Induction	V & G 88 [12]	5×10^{-16}	0.1
Induction	Bobrakov 89 [12]	10^{-16}	
Induction	Hawkins 89 [23]	2×10^{-9}	
Tors. Pendulum	Ritter 90 [17]	10^{-13}	0.02
Tors. Pendulum	Pan, Ni 92 [24]	1.5×10^{-14}	~ 0.02
Induction	Chui, Ni 93 [25]	5×10^{-16}	~ 0.05
Induction	Ni 94 [26]	2×10^{-16}	~ 0.05

The induction experiments are clearly the most sensitive of this group. In these experiments a polarized source hypothetically exchanges a boson with a detector inside a magnetic shield (metric in character), which would be penetrated by an anomalous coupling. The anomalous interaction is periodic, and induces a periodic polarization in the detector material. It is to be noted that some of the early versions of this experiment used ferromagnetic detector masses in which the induced polarization would also be magnetic, and is read by sensitive magnetometers. A difficulty is that these very sensitive limits infer a non-response level of this material as much as 10^{-5} of the tested lower limit of permeable response of the material. Ni [25, 26] avoids this problem by using a paramagnetic salt as the detector material.

A special point about many of these macroscopic spin experiments follows from the large pseudoscalar suppression factor discussed above. Although the torsion balance spin-spin experiment has the masses arranged with σ orthogonal to the vector $\mathbf{r}$ between source and detector, this is strictly true only for point masses. With finite masses there is at least some component of $\mathbf{r}$ parallel to σ for all of the particles except those exactly at the mass centers of a perfectly aligned system. In addition to the spin-aligned electrons, the polarized masses usually contain slightly polarized or essentially unpolarized nuclei. The cosθ component for the electron-nucleus pairs in interacting masses, even for θ near 90 degrees, is still much larger than the suppression factor, so the $\sigma \cdot \sigma$ experiments (including others of those published which use macroscopic masses) do in fact more likely provide some unspecified limit on $\sigma \cdot \mathbf{r}$. In the case of induction measurements, the orientation of inducing spin sources determines the "efficient" geometry either for a $\sigma \cdot \sigma$ or $\sigma \cdot \mathbf{r}$ experiment, but this "macroscopic impurity concept" is still applicable for finite size sources and detectors.

The $\sigma \cdot \mathbf{r}$ results have special interest because they can relate directly to the axion. We first discuss these experiments more generally. An early experiment by Wineland and Ramsey set a limit of 10^{-4} Hz on the shift of the deuteron Larmor frequency in response to the Earth's gravitational field. Experiments with the Earth as a source have the most sensitive asymptotic results in Table 4. But they loses sensitivity for ranges much below dimensions of the earth, Hsieh et al.; in 1989, performed the first mechanical tests in this category, with a pan balance. We note that, except for Youdin, the atomic experiments listed all used the Earth as the source mass. Hence their sensitivity is only significant for force ranges $\sim 10^6$ meters or greater. A consequence of the large source, $\sim 10^{51}$ nucleons (and considerably more electrons) to interact with the detector is the increased asymptotic sensitivity.

Returning to the question of basic sensitivity of atomic and pendulum spin experiments, we compare the Wineland [12] and Ritter [21] results (the only ones for which the number of detector particles are known to us). The experimental asymptotic sensitivity ratio, atomic/pendulum, is $3 \times 10^{-27}/4 \times 10^{-35} \sim 10^8$. The product of the number of interacting particles is: 1) $5 \times 10^3 \times 10^{51}$ for the NMR experiment, and 2) $3 \times 10^{22} \times 2 \times 10^{25}$ for the pendulum experiment. This ratio of 5×10^{54} to 6×10^{47} particles is surprisingly close to the experiment limits ratio. Such a discussion does not include technical questions such as detector noise, etc; which bear on future prospects for such experiments. Nor does it include the special "tricks" that might be available to increasing future sensitivity by each method. In this regard, we mention the recent ion trap measurement of Youdin, in which special ion trap arrangements were made so that local sources could be used, rather than the Earth, thereby permitting the first shorter range results for this type of experiment. Those authors also point out, based on theoretical arguments, that coupling to a neutron is much more sensitive than to an electron. [The polarized particle, with the pseudoscalar vertex, is listed in the first column of Table 4.]

Table 4. Sensitivities of various $\boldsymbol{\sigma} \cdot \mathbf{r}$ experiments.

Polarized Particle	Ref	Asymp. $g_p g_s/4\pi\hbar c$	$\lambda(0.1)$, m
Proton	Velyukov 68 [12]	3×10^{-25}	$\sim 10^6$
Proton	Young 69 [12]	3×10^{-31}	$\sim 10^6$
Deuteron	Wineland 72 [12]	3×10^{-34}	$\sim 10^6$
Electron	Hsieh [23]	3×10^{-25}	0.5
Electron (via Be^{9})	Wineland 91 [12]	4×10^{-35}	$\sim 10^6$
Neutron	Venema (92) [27]	3×10^{-36}	$\sim 10^6$
Electron	Ritter 93 [21]	3×10^{-27}	0.02
Electron	Jen, Ni (94) [28]		~0.02
Electron (induction)	Ni 96 [29]	8×10^{-30}	~0.05
Neutron	Youdin 96 [30]	5×10^{-29}	0.2

The 1996 experiment of Ni used no polarized particles. Rather, an unpolarized copper source is rotated around a mass of magnetically shielded paramagnetic salt detection material (TbF_3). A hypothetical $\sigma \cdot \mathbf{r}$ interaction would penetrate the shield and create a rotating polarization in the salt. A sensitive magnetometer (loop on a dc SQUID) measures the projection of the rotating induced magnetic field on the loop.

Comparison of Tables 3 and 4 supports Fischbach's finding that an added pseudoscalar vertex decreases the sensitivity by roughly a factor of 10^{-13}. Furthermore, the G(r) or fifth force results compiled in Ref. 23 (composition-dependent monopole-monopole interactions or G(r) measurements) show approximately another such factor of sensitivity greater than these $\sigma \cdot \mathbf{r}$ results.

We can consider further the limit of $|A_1| < 4 \times 10^{-12}$ that was set on the first added term of equation 1 by comparing Right- and Left-Hand Circular Polarization of radiation from the millisecond pulsar PSR 1937 + 21. Since the gravitational dimensionless internucleon coupling is $\sim 10^{-38}$, this value for A_1 would seem to infer a limit of about 10^{-50} for the dimensionless axionic coupling. It is not clear to us what restrictions, if any, might be present in the interpretation of the pulsar results for electromagnetic radiation. Such an extremely strong limit is below most principle-of-equivalence violating measurements of any kind. However, some of the composition-dependent results of Ref. 23 and those of more recent publications, are, in fact, approaching that level. But experimental measurement of anomalous spin-dependence at such levels, at least for an interaction with ~cm or shorter range, awaits some radical improvement.

References

[1]. J. Leitner and S. Okubo, *Phys. Rev.* **136** (1964) B1542.
[2]. J.R. Klein and S.E. Thorsett, *Phys. Lett.* A **145** (1990) 79.
[3]. J.E. Moody and Frank Wilczek, *Phys. Rev.* D **30** (1984) 130.
[4]. E. Fischbach *et al. Phys. Rev. Lett.* **56** (1986) 3.
[5]. R. Spero *et al*, *Phys. Rev. Lett.* **44** (1980) 94.
[6]. F. Wilczek, *Phys. Rev.* **49** (1982) 1549.
[7]. G. Gelmini and M. Roncadelli, *Phys. Lett.* **99B** (1981) 411.
[8]. A. Ansel'm, *Pis'ma Zh. Eksp. Teor. Fiz.* **35** (1982) 266.
[9]. J. Scherk, *Phys. Lett.* **88B** (1979) 265.
[10]. P.R. Phillips and D. Woolum, *Il Nuovo Cimento* **LXIV** (1969) 28.
[11]. Wei-Tou Ni, private communication, 1984.
[12]. D.J. Wineland *et al*, *Phys. Rev. Lett.* **67** (1991) 1735.
[13]. J.F. Herbst and J.J. Croat, *J. Appl. Phys.* **55** (1984) 3024.
[14]. J.J. Bara and A.T. Pedziwiatr, *J. Magn. Magn. Mater.* **17** (1982) 172.
[15]. Norman F. Ramsey, *Physica* **96A** (1979) 185.
[16]. R.D. Newman, in *Proceedings of the Third Marcel Grossman Meeting on the Recent Developments of General Relativity*, Shanghai, People's Republic of China, ed. Hu Ning (North-Holland, Amsterdam, 1983), p. 1497.

[17]. Rogers C. Ritter *et al, Phys. Rev.* **D 42** (1990) 977.
[18]. C.E. Goldblum, R.C. Ritter, and G.T. Gillies, *Rev. Sci. Instrum.* **59** (1998) 778.
[19]. L.I. Winkler and C.E. Goldblum, *Rev. Sci. Instrum.* **63** (1992) 3556.
[20]. D. E. Krause, E. Fischbach, and C. Talmadge, in *Perspectives in Neutrinos, Atomic Physics, and Gravitation*, ed. J. Tran Thanh Van, T. Damour, E. Hinds and J. Wilkerson, (Editions Frontieres, Gif-sur-Yvette Cedex - France, 1993), p 455.
[21]. R.C. Ritter, L.I. Winkler, and G.T. Gillies, *Phys. Rev. Lett.* **70** (1993) 701.
[22]. Rogers C. Ritter, in *Perspectives in Neutrinos, Atomic Physics, and Gravitation*, ed. J. Tran Thanh Van, T. Damour, E. Hinds and J. Wilkerson, (Editions Frontieres, Gif-sur-Yvette Cedex - France, 1993), p 427.
[23]. E.G. Adelberger *et al, Ann. Rev. Nucl. Part. Sci.* **41** (1991) 269.
[24]. Sheau-Shi Pan, Wei-Tou Ni, and Shen-Che Chen, *Mod. Phys. Lett.* **A 7** (1992) 1287.
[25]. T.C.P. Chui and Wei-Tou Ni, *Phys. Rev. Lett.* **71** (1993) 3247.
[26]. Wei-Tou Ni, T.C.P. Chui, Sheau-Shi Pan, and Bo-Yuan Cheng, *Physica* **194B** (1994) 153.
[27]. B.J. Venema *et al, Phys. Rev. Lett.* **68** (1992) 135.
[28]. Tsung-Haw Jen et al, in *Proceedings of the SixthMarcel Grossman Meeting on General Relativity*, (World Scientific, Singapore, 1996).
[29]. Wei-Tou Ni, *Chinese Jour. Phys.*, **34** (1996) 962; Wei-Tou Ni et al., in *Proceedings of the SeventhMarcel Grossman Meeting on General Relativity*, ed. Robert T. Jantzen and G. Mac Keiser (World Scientific, Singapore, 1997) p 1625.
[30]. A.N. Youdin *et al, Phys. Rev. Lett.* **77** (1996) 2170.

ANOMALOUS SPIN II: Search for Galactic Dark Matter Interacting with a Spin Pendulum

Rogers C. Ritter, George T. Gillies, and Linda I. Winkler
University of Virginia

Abstract

A torsion pendulum, having masses with huge numbers of spin-aligned electrons, and used to investigate the existence of weak spin-dependent forces at cm distance, has been modified to seek an anomalous spin interaction with a hypothetical galactic dark matter halo. In the redesign, the pendulum is operated in the static mode, and rotation of the Earth causes the pendulum to sweep out the sky, with a torsion pattern predicted from a $\boldsymbol{\sigma} \cdot \mathbf{r}$ form of interaction. The masses are arranged in the pendulum to form a macroscopic "anomalous spin quadrupole." The predicted pattern will have a 12-hour period in sidereal time, and any potential signal is derived from the time series of the pendulum position compared with the pattern. This study is therefore more specific, and sensitive, than several previous general searches for spatial anisotropy. At galactic source distances, the pendulum is operated at the extreme limits of its sensitivity, and a small signal-to-noise ratio would be expected for any conventional run period. Consequently, it has been run for 790 sidereal days in a first series, and a possible signal at 1/60 of the gravitational acceleration towards the galactic center is observed 2.6 times the standard deviation of the mean from zero. In the static mode, and with such a low signal-to-noise ratio, a pendulum is more vulnerable to some artifacts, and an extensive series of analytical tools have been applied in an effort to determine the origin of this apparent signal, including a "sidereal filter." Most recently, the quadrupole has been inverted in the plane perpendicular to the pendulum fiber, which would reverse the sign of any true signal.

Contents

I. Introduction, Background

The flat distribution of stellar velocities in our galaxy suggests the existence a halo containing five to ten times more matter than has been observed with electromagnetic sensors. Recent Hubble searches for brown dwarfs, etc. have failed to supply any significant amount of new observable baryonic matter, leading to the supposition that a dark halo of exotic matter is a likely solution. Axions have been a prominent candidate for this matter, particularly in view of the elegance with which their existence would solve a major particle physics problem.

Exchange particles such as axions, associated with anomalous spin interactions at short distances that were the subject of the previous paper [1], while in themselves candidates for the missing matter, do not function as such in our experiment. Instead, they could only serve the function of coupling hypothetical exotic elements of the halo to our detector. But the conventional axion can not couple over such long distances. This follows from the well-known range limitations of such axionic forces, approximately 0.2 mm to 20 cm, as inferred by stellar evolution and other astrophysical data which leads to the Turner Axion Window [2]. Instead, we are motivated purely experimentally, by the presence of a sensitive spin-dependent detector which hypothetically could interact by means of some totally new anomalous spin interaction. More general theoretical considerations related to the dark matter question have been briefly discussed by De Sabbata et al. [3], and these suggest the need for more experimental searches for evidence concerning dark matter.

We have changed our instrument to make a particular type of search for a preferable direction towards our galactic center. Such an experiment can also be considered a specialized test of Local Lorentz Invariance.

II. An "Anomalous Spin Quadrupole"

Two experiments of a more general, but less sensitive, nature have preceded ours and have not found such a preferred direction. These, by Phillips [4] and by Ni [5], did not have a search for a dark matter halo as their purpose. Experimentally, they used spin dipole masses on torsion pendulums as the detectors. The spin mass in the experiment of Phillips was an ordinary permanent magnet, shielded from local magnetic interactions by means of superconducing shields. Ni employed compensated masses of the type we have discussed [1], and achieved slightly greater sensitivity than Phillips. Both of these experiments accomplished a directional search by using the rotation of the Earth, so that sidereal time is converted to direction. The accrued time series data, over 13.5 and 10 days, respectively, were analyzed by spectral methods, in a search for a peak that would relate to a specific direction. Because of the limited search time, and the potential natural breadth of a peak resulting from interaction with a halo, the signal-to-noise ratio, for the purposes of a dark matter search, would be limited.

The construction and noise level of the University of Virginia torsion pendulum is such that its intrinsic signal-to-noise ratio is very nearly the same as that of Ref. [4] and [5]. To execute a new experiment with a hope of observing the halo required two major differences in our operation. The first, obvious, one was to operate the pendulum for something like two orders of magnitude longer, thus averaging random interferences by a factor of approximately 10.

Perhaps more important, we chose to arrange our masses in a spin quadrupole arrangement, rather than the dipole of the previous torsion pendulum arrangements. In the absence of an absolute selection rule, such an array has a decidedly worse distance dependence, i.e. its sensitivity would fall off with distance to the interacting halo element by one order of magnitude faster, than would a dipole. Nevertheless, this concept allows a totally new degree of freedom in a spin experiment, and the quadrupole was fabricated from two spin masses as shown in Figure 1. This construction can be seen to be the spin counterpart of the electrostatic quadrupole made from a pair of antiparallel dipoles, often called the "square quadrupole." It is, in fact, an anomalous spin quadrupole, since the only physical spin analog of $\mathbf{p} \cdot \mathbf{r}$ for each electric dipole is the anomalous $\sigma \cdot \mathbf{r}$ in our spin dipole. It is an unfamiliar arrangement, given our common association of magnetism with spin, and the absence of magnetic monopoles.

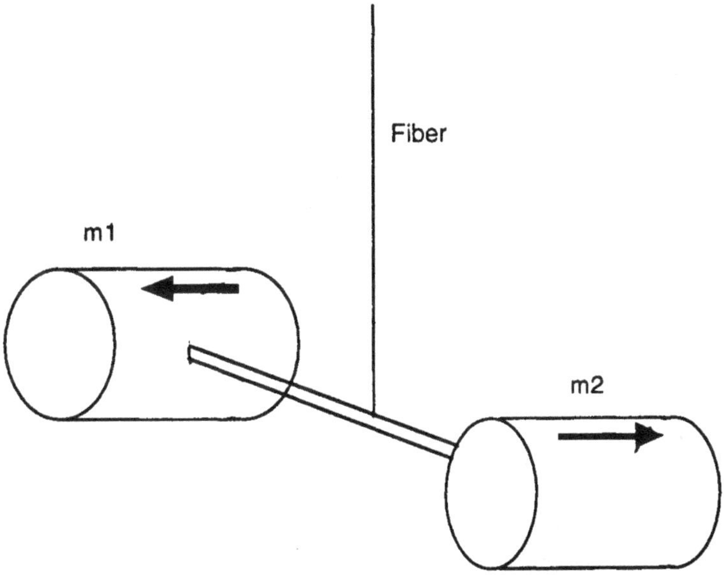

Figure 1. Quadupole arrangement of two spin masses on a fiber. Such a macroscopic object can only exist as a quadrupole for an anomalous interaction of the masses with distant matter, in the absence of the existence of "spin monopoles."

From the spacings and spins of the individual masses, the quadrupole spin tensor of the mass system can be calculated. In a coordinate system with the mass centers along the x-axis, the spin directions along +y and -y directions, and z along the fiber axis it takes the form:

$$Q(\sigma) = \begin{pmatrix} 0 & 6\sigma L & 0 \\ 6\sigma L & 0 & 0 \\ 0 & 0 & 0 \end{pmatrix} \tag{1}$$

The value of the moment $6\sigma L$ is approximately $3.1 \times 10^{23} \hbar$-cm.

III. Sidereal Aspects

The trigonometry of the torsion pendulum, with spin axes in the East-West direction, gives the pattern shown in Figure 2 for the torque of the quadrupole interacting with a point at the center of our galaxy.

Also displayed in Figure 2 are numerical integration over the halo of a form quoted by Stubbs [6], and one 24-hour experimental run. Since we have chosen a completely experimental point of approach, the theoretical fit has only the form $\sigma \cdot \mathbf{r}$, but no conventional distance dependence or scaled coupling factor. Therefore the scale of this interaction pattern is arbitrary, which requires special analysis of the pendulum data, as will be seen.

The means of locating direction in this analysis will be the comparison of a timed pendulum signal with this sidereal pattern prediction. It was more convenient to operate the experiment in UTC, given the timing of the computer clock, the sampling trigger clock (a cesium-beam standard) and ancillary equipment (e.g. temperature measurement), so all data were taken with 1000 s intervals in UTC time. For the analysis, these were converted to sidereal time, and all subsequent aspects of the analysis were maintained in sideral time, referenced to noon on Vernal Equinox. At this writing, more than three full days difference in the two systems has evolved since the start, but synchronization is maintained in analysis to about 0.1 s or better.

Advantages can accrue from the specificity of this search in sidereal time. A major benefit comes from the ability to separate, to some extent, interference from events occurring in UTC time, having periodicities not related to sidereal time. This is particularly useful in the analysis with a "sidereal filter" to be described below. Another benefit is with the gravitational interaction of astronomical bodies. In Table 1 we list several of these potentially interfering interactions.

Table 1. Accelerations on a detector mass on Earth from several objects.

Earth	$1.66\ cm/s^2$
Sun	0.59
Galaxy	1.85×10^{-8}
Galaxy Dark Matter (Stubbs est. [6])	5×10^{-9}

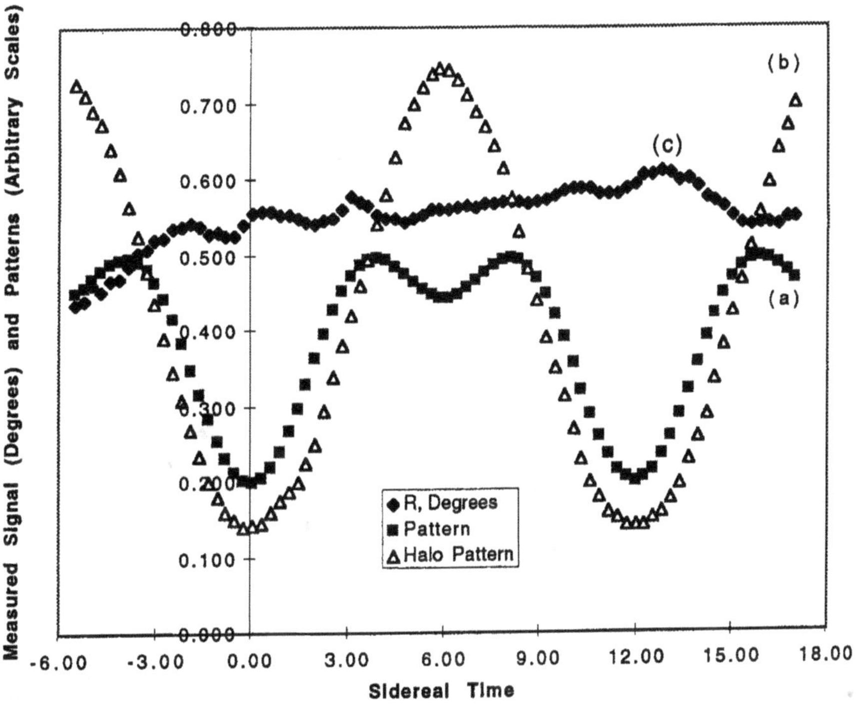

Figure 2. Pattern for a hypothetical interaction of the pendulum: a) with a point in the center of the galaxy, and b) with a halo centered around that point. Also shown: (c) ~24-hour pendulum position signal R .

For purposes of analysis, the primary difference in the two halos is the reduction of higher frequency components by the integration over the halo, as might be expected. This can be observed directly in the two amplitude spectra of Figure 3. For some time we analyzed data with both of these patterns but found, within our errors, no significant difference in the results. Based on the uncertainty of the actual form of the halo, we chose to use the point source pattern for the analysis of the reported data.

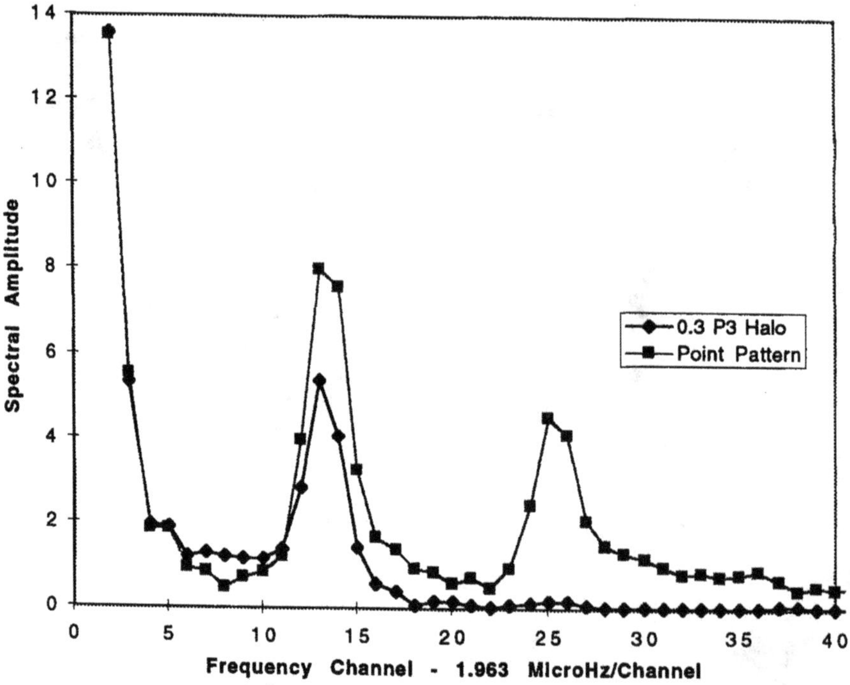

Figure 3. Amplitude Spectrum of the pattern from point spectrum and from a numerically integrated halo model [6]. The peaks at channels 13 and 25 correspond to periods of 24 hours and 12 hours, respectively. There is no significant power above the noise in the higher channels. The high dc level arises from our arbitrary zero of the patterns.

IV. The Experiment

The spin quadrupole is mounted on a fiber system modified from the previous University of Virginia torsion pendulum experiments [1]. There are no local source masses as there were previously, and the pendulum is operated in the static mode. Sampling occurred at 1000 s intervals, as before, but the system is run at atmospheric pressure to provide partial damping in the torsional mode. The pendulum is still significantly underdamped (Q = 63, decay time 6503 s). This allows the development of noise-driven oscillations at the 650 second period of the pendulum, which obscures any real or artifactual torques on the pendulum. Fourier components in the predicted signal are unimportant for times shorter than an hour, so both a preconditioning low-pass analog filter and a digital data filter are used to remove these oscillations. The analog filter has a cutoff frequency of 0.001 Hz and a roll-off of 48 db per octave. This also has the effect of preventing aliasing that could otherwise arise from higher frequency noise.

In the first 50 or so days of pendulum operation, many ancillary diagnostic runs were made; a number of problems were found and corrected, and much of the pendulum was characterized. Following this, approximately 100 days of running were used to learn about the new static mode of pendulum operation, to establish a final method of data acquisition and handling, and to obtain preliminary estimates of signal-to-noise ratio. Preliminary results from 29 days of operation were reported at the Moriond workshop on Particle Astrophysics, Atomic Physics, and Gravitation [7]. After this, arbitrarily, March 19, 1994 was chosen as the starting day for a long "official" series.

V. Analysis, Conclusion

At this writing, the experiment has been running for 1260 sidereal days. Because of unavoidable interruptions, including one large fire in the physics building, not all of these days have given useful data. In spite of best efforts, and purchase of uninterrupted power backup equipment, there have been a number of power spikes which "leak" through this protection, and cause loss of a full day's data, or worse. For analysis, eighty seven 1000-second samples (one day) are lumped into five successive one-day units, in all cases synchronized to sidereal time. Interruption of one or more days in a five-day series can be accounted for in synchronization, but we have chosen to limit such data, by starting the 5-day series over if more than one day is missed. Consequently, the overall "duty cycle" of successful experimental data-taking has been about 50%.

The 5-day raw pendulum position signal is first corrected for fiber drift, for temperature variation and for its derivative. Corrections are based on methods developed from a large amount of diagnostic testing. This corrected signal is labelled R in Figure 2. Even with such corrections, it is apparent that the signal-to-noise ratio is small, as we have expected.

The experimental quantity of interest, which we will call S, is the fit of R to the predicted pattern. Because the latter is an unscaled quantity, the slope of such a fit cannot be used directly to provide a numerical signal. Instead we use the spread of the data (its standard deviation) in degrees, multiplied by a statistical factor that measures the fraction of the spread attributable to the fit, which we call the signal S for each five-day run. The factor can be either of two standard statistical parameters: the Pearson Correlation Coefficient of a fit line, or the standard correlation between the two sets, R and the pattern. Except for a special statistical procedure to account for the number of parameters in the former, these are equivalent. Moreover, the slope of the fitted line has about a 75% correlation with either of these two correlation parameters.

After insertion of the position-detector calibration, the quantity S has units of angle of the torsion pendulum displacement. This is converted to torque with the fiber constant, from which the acceleration of each of the two masses is determined. In further statements we will use the unit of acceleration for the signal quantity S.

Running plots were kept of the sequential values of S and of various averages and statistical quantities. After about 2.5 years a steady trend toward an average signal >2 standard deviations away from zero became apparent [8]. That is, the accumulating average of 5-day signals levelled to a roughly constant value, but its standard deviation of the mean (standard error) was falling as the expected $1/N^{1/2}$. The signal-to-noise ratio for each five-day run averaged about 4% as deduced from Pearson's Correlation Coefficients. A number of artifactual errors from ambient causes were suspected as sources of the fluctuations, which were considerably greater than the usual fluctuation noise of the pendulum [1], and these would be expected to occur with some dependence on Universal Time (UTC).

To cope in some degree with what was anticipated to be the bulk of such "artifactual noise," we devised a sidereal filter. This is based on three factors: 1) after one-half year's passing, Universal Time has a 12-hour shift relative to sidereal time, 2) many unwanted effects from interferences in the vicinity of the experiment are diurnal in Universal Time, and 3) a real quadrupole signal would have a 12-hour pattern. Operation of the filter depends on comparing sets of data which are 183 sidereal days apart. Two five-day run series of R values (corrected as stated above), separated in time by 183 sidereal days, are used to compute each such-filtered signal--actually two signals. First the two five-day signal trains were added, giving a data series we call S^+. Then they were subtracted, giving a series S^-. Since a true galactic spin-quadrupole signal must have a 12 sidereal hour period, S^+ can contain it, while S^- can not. Conversely, any artifactual signal which is odd in a diurnal period, i.e. in a UTC day, would vanish in S^+ but survive in S^-. Uncorrected temperature effects would be expected to be largely odd in a 24 hour period, as might a number of others. Subsequently we have analyzed all data in this way, with two final "signal" sets S^+ and S^-. This operation requires a half-year lag after the start, before the first such signal set is acquired.

This first run series was continued for 53 five-day pairs, approximately 800 sidereal days, to yield a final signal for this series. At the end of that period, the final mean signals were determined and reported [9]. A few subsequent, very minor corrections to those figures leave the values at $\langle S^+ \rangle = (-3.1 \pm 1.2) \times 10^{-10}$ cm/s^2, and $\langle S^- \rangle = (2.2 \pm 1.4) \times 10^{-10}$ cm/s^2. Thus $\langle S^+ \rangle$ exhibits a 2.6 σ value, potentially of some significance. The gravitational acceleration of a mass toward the galactic center is 1.85×10^{-8} cm/s^2. From $\langle S+ \rangle$ we can write the result of this experiment as the acceleration on each spin mass as:

$$a = a_g[1 - (0.017 \pm 0.007)]. \qquad (2)$$

Subsequently, we have sought ways to understand and reduce the noise of this 53-run experiment, while commencing a second series, described below. A first consideration is the effectiveness of the sidereal filter. Power spectra were taken of the averages of the 53 run pairs, $\langle S+ \rangle$ and $\langle S- \rangle$, shown in Figure 4, along with those of the point and halo patterns. In the differenced series a peak at 1 day (near channel 6) is significant, but it is small in the summed series. This indicates the effectiveness of the sidereal filter. All four plots have a significant peak at the 1/2 day location, which would contain data for any real signal. Its significant appearance in the differenced spectrum, which cannot contain a true signal, shows a limit to the sidereal filtering, and also that we cannot expect to benefit strongly by using the FFT as a band-pass filter on that peak to enhance the signal-to-noise value of the $\langle S^+ \rangle$ spectrum. The strong peak at 1/4 day in the point pattern spectrum, but with only noise at that location in the signal spectra, supports our finding that integrating over the halo of the point sidereal pattern makes little difference in our signals.

The modest degree of effectiveness in the sidereal filter method is apparent from the extended FFT spectral comparisons of S^+ and S^- (not shown here), where the noise level ratio S^-/S^+ on the extended spectra is only about two. After subtracting that background, the ratio S^+/S^- in the 12-hour peak is about 3.5, although the adjoining shoulder at 2/3 day complicates this a bit. Overall, this suggests that the sidereal filter noise reduction is in the range 2 to 3.5, and is different for artifactual and random types of noise.

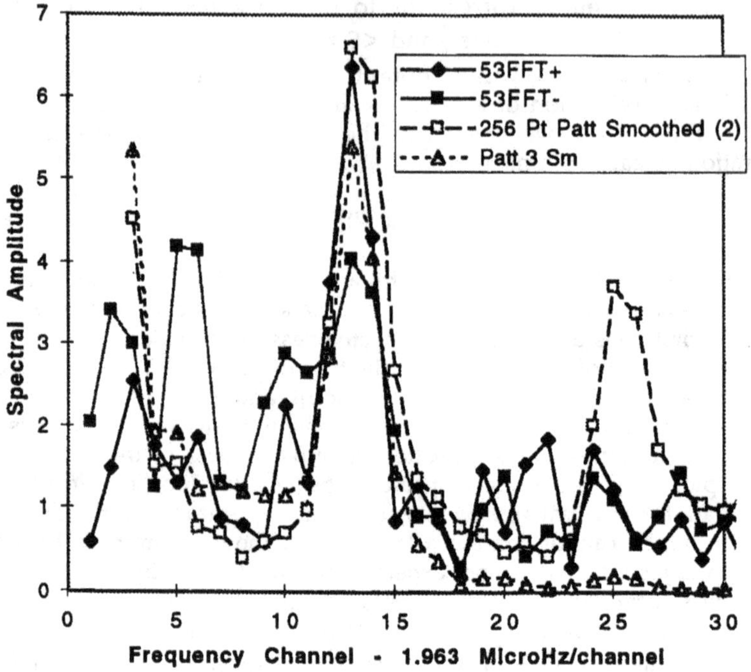

Figure 4. Spectral amplitudes of <S+> and <S->, and the point and halo patterns. The patterns were smoothed to remove FFT aliasing. Shown are the lower frequency part of 256 channel spectra in which resolution is limited by 1000 s signal sampling time. Peaks approximately at channels 5.5, 12.5 and 25.5 correspond to events with one-day, one-half day, and one-fourth day periods. At frequencies off-scale the spectra are almost all noise. Possible strength at roughly 2/3 day in <S+> and <S-> is mysterious.

Among other concerns for artifactual effects or faulty analysis were a possible bias in data handling, mass imbalance, biased signal construction, lack of normal distribution of data points, and the low signal-to-noise ratio of the individual data sets.

Several of the analysis concerns were tested by generation of 60 random number data sets of the approximate length of our runs, and scaled with a standard deviation several times larger than that of the series S. These would lack a time scale, but otherwise would represent our system as noise without a signal. These data, which we call T, were treated in sum and difference pairs, and analyzed just as the pendulum data. The results were $\langle T^+ \rangle = (0.93 \pm 3.2) \times 10^{-10}$, and $\langle T^- \rangle = (-0.028 \pm 3.6) \times 10^{-10}$. We can conclude that our signal construction procedure does not generate a bias in random data.

The following considerations apply to a mass imbalance question. The masses had been constructed and put into machined, 0.9999 purity iron cups, and then into fabricated μ-metal shields, with thin, machined, aluminum intervening layers. The finished masses matched in weight to <0.5%. However, on rotation of the Earth any mass imbalance is an exact odd function in any 24-hour sidereal period, and therefore its effects will completely cancel in each S^+, while doubling in S^-. The 24-hour peak at channel 5.5 for S^- in Figure 4 most likely illustrates this fact.

VI. A New Experiment

A simple change converts the experiment to reverse the sign of the interaction $\sigma \cdot \mathbf{r}$. Such a change would go much further than any amount of analysis to establish the potential reality of the observed signal. To accomplish this change, the plane of the quadrupole was inverted about 350 sidereal days ago.

Since the 183 day lag of the filter, there have been about 160 days of running, from which 22 sets of S^+ and S^- have been completed and analyzed. The fluctuation level of this set of data, i.e. the standard deviation of its mean, is approximately 50% higher than that at 22 points in the original run series. This can delay considerably the development of a meaningful trend. Such an increase is beyond the second moment of the statistics, and must have a physical source. We are studying this issue, while continuing the experiment.

References

[1]. R.C. Ritter, G.T. Gillies, and L.I. Winkler, "Anomalous Spin I: A Polarized-Mass Torsion Pendulum," previous paper this volume.

[2]. M.S. Turner, *Phys. Rep.* **197** (1990) 67 .

[3]. A. De Sabbata, V.N. Melnikov, and P.I. Pronin, *Progr. Theor. Phys.* **88** (1992) 623.

[4]. P. R. Phillips, *Phys. Rev. Lett.* **59** (1987) 1784.

[5]. Shih-Liang Wang, Wei-Tou Ni, and Sheau-Shi-Pan, *Mod. Phys. Lett.* **A 8** (1993) 3715.

[6]. C.S. Stubbs, *Phys. Rev. Lett.* **70** (1993) 119.

[7]. R.C. Ritter, L.I. Winkler, and G.T. Gillies, in *Particle Astrophysics , Atomic*

Physics, and Gravitation, ed. J. Tran Thanh Van, G. Fontaine, and E. Hinds, (Editions Frontieres, Gif-sur-Yvette Cedex - France, 1994), p 441.
[8]. R.C. Ritter, L.I. Winkler, and G.T. Gillies, in *Dark Matter in Cosmology, Quantum Measurements, and Experimental Gravitation*, ed. R. Ansari, Y. Giraud-Heraud, and J. Tran Thanh Van, (Editions Frontieres, Gif-sur-Yvette Cedex - France, 1996), p 417.
[9]. R.C. Ritter, L.I. Winkler, and G.T. Gillies, in proc. of XXXInd Rencontres de Moriond, Les Arc, France, Jan. 18-25, 1997, in press.

Acknowledgments

We wish to express our appreciation to Prof. Venzo de Sabbata and the "Ettore Majorana Centre for Scientific Culture" for the invitation to this stimulating School, and for their gracious hospitality during the Course.

Project SEE (Satellite Energy Exchange): Proposed Space-Based Method for More Accurate Gravitational Measurements

Alvin J. Sanders, Dept. of Physics and Astronomy,
University of Tennessee, Knoxville, Tennessee 37996-1200 USA
ASanders@utk.edu

George T. Gillies, Dept. of Mechanical, Aerospace, & Nuclear Engineering,
University of Virginia, Charlottesville, Virginia 22901 USA
GTG@Virginia.edu

Presented at the International School of Cosmology and Gravitation, Erice, Sicily, May 13-20, 1997

Abstract

Project SEE (Satellite Energy Exchange), an international effort to organize a new space mission for fundamental measurements in gravitation, is described.

Project SEE entails launching a dedicated satellite and making detailed observations of free-floating test bodies within its experimental chamber. This has been described in detail elsewhere (Sanders & Deeds 1992a and 1992b; Sanders & Gillies 1996).

A SEE mission has eight measurement goals:

1. Test the Equivalence Principle (EP) at intermediate distances (few meters) by a search for violations of the inverse-square law (ISL) at these distances.

2. Test the EP at long distances (radius of the Earth) by a search for ISL violations at these distances. This would be indicated by anomalous precession of perigee.

3. Test the EP at intermediate distances (few meters) by a search for composition dependence (CD) in the attraction between alternative test bodies.

4. Test the EP at long distances (radius of the Earth) by a search for CD in the attraction of the Earth on the test bodies. This would be indicated by apparent violations of Kepler's third law.

5. Measure the Newtonian gravitational constant G to one part in 10^6.

6. Measure (or test for) the secular time variation of G.

7. Test for the PPN (Parameterized Post-Newtonian) anisotropy parameter α_1 and other post-Einsteinian effects.

8. Test for anomalous $\boldsymbol{\sigma \cdot r}$ coupling of the quantum-mechanical spins in a test body to the mass of the Earth. This is described in another paper by the authors in this volume.

Thus, a SEE mission would obtain accurate information self-consistently on a number of distinct gravitational effects. Test #2 and #3 would chiefly provide confirmation of earlier, more precise experiments. All the other tests would significantly improve our knowledge of the nature of gravity (Sanders & Deeds, 1992, and Sanders & Gillies, 1996).

Our interpretation of the EP tests (#1-4 above) will be conventional--i.e., we posit a possible new force which can be parameterized as if due to the exchange of a corresponding Yukawa-type mediator.

$$U(r) = (GMm/r) \times [1 + \alpha \exp(-r/\Lambda)]$$

An EP violation which is not consistent with the hypothesis of a new Yukawa-type particle could have extremely profound consequences.

The long-standing goal of a more precise determination of G has taken new urgency now with the sharply conflicting results from recent high-precision determinations (Gillies 1997 and elsewhere in this volume).

Despite the 60-year history of interest in the possibility of a time-varying G and continuing theoretical impetus in this direction (Melnikov 1994), experimental results for the time variation of G have been rather inconsistent and confusing (Gillies 1997). SEE would provide the first controlled measurement for G-dot.

Damour and Esposito-Farèse have pointed out that non-Einsteinian effects may be revealed by perturbation analysis of satellites in Earth orbits (1994a and 1994b) with resonance-enhanced methods, due to the high precision of ground tracking now available. A SEE mission can test for the PPN (Parameterized Post-Newtonian) anisotropy parameter α_1 and other PPN parameters However, this would require a higher orbit than currently planned (2200 km *vs.* 1500 km) (Sanders & Gillies, 1996).

As originally proposed (Sanders & Deeds, 1992a and 1992b), SEE did not include a test for spin-dependent gravity. Elsewhere in this volume, we now introduce a possible method for testing for anomalous spin coupling ($\sigma \bullet \mathbf{r}$) on a SEE mission.

Description of the SEE method

Here we briefly describe the rather paradoxical relative motion of the two test bodies--the large "shepherd" and the small "particle"--during a SEE encounter: Two co-orbiting satellites--one trailing the other--may exchange substantial gravitational energy if their orbits are nearly identical, so that they remain very close together for several orbits around the Earth. If the body in the lower (and therefore faster) orbit approaches the other from behind, the trailing body is picking up energy from the leading body and, after the passage of some time, may acquire sufficient energy to rise above the leading body (stated more precisely, the semi-major axis of the orbit of the trailing body may grow to exceed that of the leading body). At this point the trailing body will begin to *fall back*, while still continuing to pick up energy from the leading body. The relative trajectory of the two bodies can be very smooth if their pre-encounter orbits are both circular (or otherwise very similar); otherwise the trajectory has a cycloidal character. The behavior of bodies in SEE encounters has been confirmed by extensively modeling, both at the University of Tennessee prior to the original SEE publication (Sanders & Deeds 1992) and subsequently by a group led by Prof. Melnikov in Moscow (Alekseev *et al.* 1994, Bronnikov *etal.* 1993, and Melnikov *etal.* 1993), and it has also been confirmed in nature by the co-orbiting satellites of Saturn (Dermott & Murray 1981). Please see note added in proof.

We note that, although the gravitational force is of course always attractive, a SEE encounter gives the paradoxical appearance of mutual *repulsion* by the two bodies. This can be understood in terms of the virial theorem.

It is convenient to scale the experimental masses and distances so that the duration of a SEE encounter is typically about one day, the encounter length is typically 5-10 meters, the change in the altitude of the small test body (the particle) during an encounter is typically a small multiple of 10 cm, and the instantaneous relative speeds are typically 100 to 300 microns/sec. This requires the shepherd mass to be ~100 to 250 kg. The change of the altitude of the small body during an encounter is proportional to the mass of the shepherd and approximately inversely proportional to the distance of closest approach.

Note that a SEE-encounter trajectory is rather long and narrow--in fact, it is virtually one-dimensional. The extremely narrow shape of a SEE trajectory has an important consequence: Nearly all the information is contained in the *separation* of the test bodies as a function of time. This fact greatly simplifies data analysis, in that it results in very relaxed pointing requirements for the SEE satellite. This fortuitous circumstance is rather unusual among high-precision space experiments.

Satellite design The SEE satellite is essentially in the shape of a long cylinder scaled to enclose the long, narrow SEE encounters. The experimental chamber itself is a closed cylinder approximately 1 meter diameter and 5 to 10 meters long. Temperature control is provided by a number of closely-spaced coaxial open cylinders which surround the experimental chamber and overhang it by staggered amounts. The outermost cylinder is about 2 meters diameter and 10 to 15 meters long. The total mass is 2 to 4 metric tons.

The mass distribution of the SEE satellite is configured in order to produce zero internal gravitational force within the experimental chamber in principle. That is, the satellite is "gravitationally invisible" to anything inside the chamber (Sanders & Deeds, 1992).

Great care is taken to minimize the photon pressure on the test bodies, especially the particle. The capsule of course shields the test bodies from direct sunlight. Protection from internal blackbody radiation is achieved by making the temperature inside the experimental chamber highly uniform, constant, and equal to that of the shepherd. Temperature control is almost completely passive (Sanders & Gillies, 1996; Schunk & Sanders, 1997).

The systems for precision measurement of temperature and of test-body position are compatible with the requirements for temperature control and gravity-nulling.

Jerk-free torque for orientation of the SEE satellite is supplied by interaction of currents in external coils with the Earth's magnetic field. Jerk-free station-keeping thrust is supplied by "natural solar sailing"--i.e., the reflection of solar photons off the outer surfaces of the satellite (Sanders & Deeds, 1992).

Experimental Measurements of a SEE Mission

This section describes the measurements that would be made on a SEE mission.

The SEE concept is rooted in the tradition of orbital-perturbation analysis. Thus, we seek to make very precise measurements of small effects, by allowing time to magnify them naturally. As with all such analyses--from the discovery of Uranus to the explanation of the perihelion precession of Mercury--our analysis methods will disentangle the sought-after effects from each other and from various background effects (such as the influence of the Moon and of the Earth's harmonics). Although in some cases the background effects may be large, they will generally be calculable and--since SEE provides for *controlled* experiments--we will often have the added luxury of being able to choose the phases of the effects under investigation, relative to each other and relative to the unwanted background effects. Finally, although we begin with specific hypotheses, if neither these nor other existing hypotheses provide satisfactory fits to the data, and if exhaustive searches for further systematic errors prove fruitless, then this circumstance will invite theorists to posit new hypotheses.

Experimental Measurements I: SEE Encounters (for *G*, EP, and spin-dependent tests)

The bulk of the scientific investigation on a SEE mission would entail precise analyses of the relative motion of the two test bodies--the large "shepherd" and the small "particle"--during a number of SEE (Satellite Energy Exchange) encounters, as described above. To wit, the first five experiments and the eighth experiment described above are based on detailed analysis of ultra-precise *on-board* measurement of test-body trajectories during a number of SEE (Satellite Energy Exchange) encounters within the SEE satellite. In this section we present the basic principle for measuring each individual effect, although of course no one effect can be isolated and measured in the absence of competing effects. The proposed measurements are outlined in some detail in the original SEE proposal (Sanders & Deeds, 1992), page 491 and pages 501-502.

The value of *G* is obtained from the accelerations of the particle during a SEE encounter. Thus, account is taken of the peculiar and counter-intuitive dynamics which results from the fact that both bodies are in orbit around the *Earth* and are chiefly under the influence of its gravity rather than each other's (and which may be understood in terms of the virial theorem). More particularly, the measured value for *G* is inversely proportional to the square root of the time required to complete any prescribed portion of a SEE encounter.

The intermediate-range (~meters) EP test based on an inverse-square-law (ISL) test will straightforwardly compare the measured values of *MG* (*M* is the shepherd mass) obtained at various locations along the trajectory of each SEE encounter, and then search for apparent variation of *MG* as a function of the separation of the test bodies. We note that errors in the shepherd mass *M* drop out.

The long-range (~radius of the Earth) EP test based on an ISL test takes advantage of the fact that a relative precession of the perigees of the test bodies would be caused by the perturbing force of a putative Yukawa-type particle. This precession is due mainly to the cubic term in the force, in exact analogy to the anomalous precession of Mercury predicted by General Relativity.

The intermediate-range (~meters) EP test based on composition dependence (CD) is done essentially by comparing the values of *G* obtained with particles of different composition. Complete SEE encounters may be used. We note here that the shepherd mass is again unable to contribute any error. Moreover, by accurately replicating the trajectories within the capsule, we virtually eliminate any contribution of the mass-distribution errors in the capsule walls to the CD-test error. This is because whatever Newtonian perturbations may exist along a given trajectory should be the same, *ceteris paribus*, for any particle on the same trajectory.

The long-range (~radius of the Earth) EP test based on composition dependence (CD) is obtained by searching for an apparent violation of Kepler's third law for simultaneously-orbiting test bodies within the capsule. That is, the relationship between orbital radii and orbital periods will differ very slightly between two test bodies if their composition difference causes them to fall at different rates in the Earth's field. This test is closely analogous to laser radar ranging (LLR) measurements of lunar parallactic inequality. Even the techniques for obtaining extreme accuracy are similar in principle: In the LLR case, the *relative* distance of the Earth and the Moon from the Sun is inferred essentially by comparing the Earth-Moon distances at new moon and full moon, and this can be measured to within 1.3 cm--many orders of magnitude more accurately than the various astronomical distances involved can be known (see, for example, Nordtvedt, 1996a and 1996b). Likewise, SEE will measure the *differential values* of the orbital radii and periods of the test bodies, using on-board precision-measurement systems, and the result will be many orders of magnitude more accurate than the absolute values of these quantities based on ground tracking.

The search for spin-dependence in gravity is based on detailed observation of the differential perturbation of two particles, one with its quantum-mechanical spin highly polarized and the other spinless. This is described further in this volume in another paper by the authors.

A strength of SEE is that a rather large number of different materials may be used for the particles.

They are small enough that several dozen different particles can easily be stowed. Hence, SEE can do the composition-dependent EP tests with a large number of different materials. Moreover, if *two* particles are used in simultaneous SEE encounters with the shepherd, then myriad pairs of materials are available.

Experimental Measurements II: Tests for *G*-dot and Post-Einsteinian Effects

Searches for secular time variation of *G* and for various post-Einsteinian effects would be carried out on a SEE mission by precise measurements of the perturbations of the Earth orbit of the shepherd, as determined by ground tracking, rather than by analysis of on-board trajectories of the test bodies during SEE encounters.

For *G*-dot, what is to be measured is the shepherd's orbital period. An anomalous secular increase would indicate that the product $M_E G$ is decreasing. Although we cannot separate *G*-dot from M_E-dot, a result of either a time-varying or constant $M_E G$ would be of enormous interest. With centimeter-level tracking, a relative change in $M_E G$ of less that a few parts in 10^{13}/yr could be detected within a year. This exceeds the sensitivity needed to distinguish among the predictions of a number of different theories, which typically predict that *G* is changing at a few parts in 10^{11}/yr.

We are indebted to Prof. Ritter for pointing out during this summer school that great care will be necessary to account for the time-varying potential of the Earth as regards its impact on the planned *G*-dot experiment. Consequently, studies are now in progress to assess the feasibility of accounting for this variability with sufficient accuracy to permit a cosmologically significant measurement of *G*-dot.

Perturbations in various orbital parameters may indicate violations of general relativity, as outlined in recent papers by Thibault Damour and colleagues (Damour, 1994a and 1994b). Choosing specific initial values of the orbital elements would result in resonance conditions, which may be necessary to observe such post-Einsteinian effects in LEO. Further, Sanders and Gillies (1996) have pointed out that a sun-synchronous orbit can combine two or more resonances, thus enhancing the observability of these effects.

Moreover, it may also be possible to detect a small anisotropy of space by slight fluctuations in the period of the shepherd's Earth orbit, since the orientation of the shepherd's orbital plane in solar-system coordinates is subject to an annual cycle. However, it will be of utmost importance to distinguish any such observed effect from a possible seasonally-varying systematic error.

It is very important that substantial capacity for self-calibration be incorporated to the greatest possible extent in the design of any experiment for making measurements of Newtonian gravity in space. In particular, *in-situ* analysis of the mass distribution of the capsule and large test body is essential, since it is unlikely that this distribution can be modeled *a priori* to sufficient accuracy. SEE has this capability intrinsically, as a consequence of the large variation in both the separation of the test bodies and in their positions within the capsule.

Acknowledgments

We wish to express our appreciation to Prof. Venzo de Sabbata and the "Ettore Majorana Centre for Scientific Culture" for the invitation to this stimulating school and for their gracious hospitality during the Course. We are pleased to acknowledge financial support from the Marshall Space Flight Center of NASA, Oak Ridge National Laboratory, Oak Ridge Center for Manufacturing Technologies, the Tennessee Higher Education Commission Science Alliance, and the Dean of Arts of Sciences, the Office of Research, and the Department of Physics and Astronomy of The University of Tennessee.

****

Note added in proof: Shortly after completion of the summer school, it was announced that a recently-discovered asteroid is in fact co-orbiting with the Earth, interacting with it according to the SEE-encounter principle (Wiegert *et al.* 1997).

References

A.D. ALEKSEEV, K.A. BRONNIKOV, N.I. KOLOSNITSIN, V.N. MELNIKOV, and A.G. RADYNOV; *Meas. Tech.* **37** (No. 1), 1-5 (1994) [translated from " *ИЗМЕРИТЕЛЬНАЯ ТЕХН* (No. 1), 3-5 (Jan., 1994)].

K.A. BRONNIKOV, N.I. KOLOSNITSYN, M.Yu. KONSTANTINOV, V.N. MEL'NIKOV, and A.G. RADYNOV; *Meas. Tech.* **36** (No. 9), 951-957 (1993) [translated from *ИЗМЕРИТЕЛЬНАЯ ТЕХН* (No. 9), 3-6 (Sept., 1993b).

T. DAMOUR and G. ESPOSITO-FARÈSE; *Phys. Rev. D* **49** (No. 4), 1693-1706 (1994a).

T. DAMOUR and G. ESPOSITO-FARÈSE; *Phys. Rev. D* **50** (No. 4), 2381-2389 (1994b).

S.F. DERMOTT and C.D. MURRAY; *Icarus* **48**, 12-22 (1981).

G.T. GILLIES; *Rep. Prog. Phys.* **60**, 151-225 (1997).

V.N. MELNIKOV, M.Yu. KONSTANTINOV, N.I. KOLOSNITSIN, K.A. BRONNIKOV, A.G. RADYNOV, A.D.ALEKSEEV, and P.N. ANYONYUK; "Report for Project SEE"; Moscow (1993).

V.N. MELNIKOV, "Fundamental constants and their stability: A review"; *Internat. J. Theor. Phys.* **33**, 1569-1579 (1994).

K. NORDTVEDT; *Class. Quantum Grav.* **13**, 1309-1316 (1996a).

K. NORDTVEDT;*Phys. Today* **49**, 26-31 (May, 1996b).

A.J. SANDERS and W.E. DEEDS; *Phys. Rev. D* **46** (No. 2), 489-504 (1992a).

A.J. SANDERS and W.E. DEEDS; *Bull. Am. Phys. Soc.* **37** (No. 7), 1675 (1992b).

A.J. SANDERS and G.T. GILLIES; *La Rivista del Nuovo Cimento* **19** (No. 2), 1-54 (1996).

R.G. SCHUNK and A.J. SANDERS, NASA Technical Memorandum, forthcoming.

P.A. WIEGERT, K.A. INNANEN, and S. MIKKOLA; *Nature* **387**, 685-686 (12 June 1997)

A Comparative Survey of Proposals for Space-Based Determination of the Gravitational Constant G

Alvin J. Sanders, Dept. of Physics and Astronomy,
University of Tennessee, Knoxville, Tennessee 37996-1200 USA
ASanders@utk.edu

George T. Gillies, Dept. of Mechanical, Aerospace, & Nuclear Engineering,
University of Virginia, Charlottesville, Virginia 22901 USA
GTG@Virginia.edu

Presented at the International School of Cosmology and Gravitation, Erice, Sicily, May 13-20, 1997

Abstract

Proposals for determining the gravitational constant G in space are compared and contrasted. We find that only three proposals have carefully treated the many physical processes in the space environment which might potentially vitiate the sought-for accuracy. The capability of the proposed missions to make other gravitational measurements and tests is described.

Since virtually the beginning of the space age, researchers have been attracted to the prospect of utilizing the relatively quiet environment of space for making an accurate determination of the Newtonian gravitational constant G, which has proved to be exceptionally elusive in terrestrial experiments. The need for an improved measurement of G is well known. It is the least precisely determined of all the fundamental constants of nature, having the presently accepted value [Cohen & Taylor (1986 and 1987)] of $G = (6.67259 \pm 0.00085) \times 10^{-11}$ m^3 kg^{-1} s^{-2}. The relative uncertainty in this value, 128 parts per million (ppm), is 10^2 to 10^5 times larger than that of most of the constants which arise in atomic and nuclear physics[1].

Moreover, the seven highest-precision measurements of G, each claiming an uncertainty of $\approx$100 ppm, all exclude each other within the limits of their quoted errors (Gillies, 1997). These measurements generally differ from the CODATA value by multiple standard deviations. The largest departure, some 0.6% (6000 ppm), is exhibited by the result of the PTB in Braunschweig, while the one-standard-deviation scatter among the remaining six values which claim $\approx$100 ppm is in fact about 500 ppm. Some might argue that the size and nature of these discrepancies may be such as to reopen the possibility that there is new physics to be explored in this regime of interaction strength. Whether or not this is so, these results would seem to indicate that we may now be reaching the limit of performance with the paradigm of terrestrially-based mechanical instrumentation as used to determine the absolute value of G.

Over the past 30 years, a number of different proposals have been made for the determination of the Newtonian gravitational constant, G, on board a satellite in Earth orbit. The isolation of the gravitational interaction between the test bodies and the elimination of many types of manmade

---

[1] Until recently the gas constant R and some 10 physicochemical constants derived from it were also conspicuous for their large errors. Now a recent reduction in the uncertainty in R, which will correspondingly reduce those of the derivative physicochemical constants in the next CODATA least-squares adjustment to be issued in 1997 [Cohen & Taylor (1995)], means that G and its derivative constants, *viz.*, the Planck mass, time, and length, will soon stand almost completely alone in having large uncertainties: Only two other constants will still have uncertainties over 1.8 ppm: the Stefan-Boltzmann constant and the anomalous magnetic moment of the muon, both of which will have uncertainties of about 7 ppm.

and natural disturbances are reasons often cited for proposing such experiments. While the environment of space indeed offers many advantages over that on the surface of the Earth, orbitally-based determinations of G nevertheless face a number of significant challenges, both conceptual and practical in nature. In a recent article of the same title of this presentation (Sanders & Gillies, 1996), we present an overview of the various space-based experiments proposed to date. We especially consider how background disturbances and other interactions might affect the ability of these experiments to achieve their desired accuracy. We find that most of the early proposals focus almost totally on *Gedankenexperimenten*, but that three of the most recent proposals extend the scope of these efforts by including well-developed physical bases, which have been described in detail in the literature.

The various published proposals for measuring G in space fall naturally into two groups: The first group, which we dub "Exploration of Principles under Idealized Conditions" (EPIC), concentrated on describing the basic physical principles of the proposed experiments. The authors typically chose to forgo assessments of the precision-measurement aspects of their experiments, and instead concentrated on the underlying physical mechanisms.

The hallmark of the second group of experiments is that they not only treat the conceptual essence of the proposed experiment but also attempt to identify and deal with the many other physical interactions which occur in an actual experiment. Accounting for the effects of such interactions is essential to achieving the sought-for accuracy.

Given this set of criteria, the second group is comprised, then, of only three proposals, namely the NEWTON proposal of the University of Pisa [Nobili *et al.* (1989, 1990, and 1993)], the Satellite Energy Exchange (SEE) proposal of the University of Tennessee [Sanders & Deeds (1992a and 1992b) and Sanders *et al.* (1993)], and the proposed G/ISL (G/Inverse-Square-Law) experiment of the STEP project [Blaser *et al.* (1993), Paik & Blaser (1993), and Paik (1994)].

None of these proposals has yet reached the level of specificity entailed in engineering feasibility studies (although planning for many aspects of the overall STEP mission is in a very advanced stage). Moreover, given the minute size of the gravitational force and the high-accuracy goal of the proposed experiments (~1 part in 10^6), it is possible that important effects still remain to be identified.

The common practice of approaching the kinematics of the test body interaction from the point of view of an idealized force-free region has caused considerable difficulty in conceptualizing the test-body interactions and in deriving and interpreting the equations of motion. This approach requires viewing the Earth's gravity gradient as a "problem." At best this has resulted in the expenditure of a great deal of effort, and it has in some instances led investigators to the problematic step of proposing artificial restraints to force the motion of the test bodies to imitate that which would result in an idealized force-free region. In contrast, the point of view that both of the test bodies are in orbit around the Earth and are only slightly perturbed by each other is much more natural, both mathematically and conceptually.

In any G-in-space experiment, it is important to minimize vibrations of the capsule and to maintain its thermal stability. The motivation for thermal stability is to minimize both the changes in the size and shape of the capsule and the net force on the test bodies due to blackbody radiation anisotropy. A reasonable and achievable stipulation is that thermal stability be sufficient to maintain rigidity (all intra-capsule distances) to within about one optical wavelength. Vibration avoidance also requires essentially jerk-free thrust and torque.

The satellite housing a G experiment must be drag free, which entails counteracting atmospheric drag itself and also compensating for micrometeorite strikes, solar-radiation pressure, and Earth-radiation pressure. Except at very low altitudes, solar-radiation pressure is the dominant effect, and at extremely high (geosynchronous) orbits, it blows an uncompensated satellite out of its orbital plane by ten meters or more.

Careful choice of orbit is a critical element in a strategy for achieving these ends. Sun-synchronous orbits are well suited to these purposes. Such an orbit was prescribed in the SEE and STEP proposals. The problems accompanying geosynchronous orbits raise serious questions about their practicability.

The issue of mass distribution errors in the large test mass, due to both density inhomogeneities and shape errors (such as non-sphericity), warrant close scrutiny, especially in experiments which entail small separations between the test bodies. The use of cylindrical geometry for the test masses, as in Cook-Marussi stack, may substantially alleviate errors in the latter regard.

It is very important that substantial capacity for self-calibration be incorporated to the greatest possible extent in the design of any experiment for making measurements of Newtonian gravity in space. In particular, in-situ analysis of the mass distribution of the capsule and large test body is essential, since it is unlikely that this distribution can be modeled *apriori* to sufficient accuracy. SEE has this capability intrinsically, as a consequence of the large variation in both the separation of the test bodies and in their positions within the capsule.

Finally, the instrument-design requirements for measuring G in space overlap substantially with those for accomplishing several related objectives which also entail stringent precision-measurement capability. These include measurements of drag and radiation pressure, trials of next-generation accelerometers, exploration of selected harmonics in the Earth's geopotential field through resonances, and tests for various effects bearing on fundamental aspects of gravitation theory, including possible violations of the equivalence principle, time variation of G, and non-Einsteinian values of the PPN parameters. NEWTON, SEE and STEP are all capable of pursuing a number of these objectives.

Acknowledgments

We wish to express our appreciation to Prof. Venzo de Sabbata and the "Ettore Majorana Centre for Scientific Culture" for the invitation to this stimulating school and for their gracious hospitality during the Course. We are pleased to acknowledge financial support from the Marshall Space Flight Center of NASA, Oak Ridge National Laboratory, Oak Ridge Center for Manufacturing Technologies, the Tennessee Higher Education Commission Science Alliance, and the Dean of Arts of Sciences, the Office of Research, and the Department of Physics and Astronomy of The University of Tennessee.

References

A.D. ALEKSEEV, K.A. BRONNIKOV, N.I. KOLOSNITSIN, M.Yu. KONSTANTINOV, V.N. MELNIKOV, and A.G. RADYNOV; paper presented to Eighth Russian Gravitational Conference (Moscow, May, 1993a)

A.D. ALEKSEEV, K.A. BRONNIKOV, N.I. KOLOSNITSYN, V.N. MEL'NIKOV, and A.G. RADYNOV; *Meas. Tech.* **36** (No. 10), 1070-1077 (1993b) [translated from *ИЗМЕРИТЕЛЬНАЯ ТЕХН.* (No. 10), 6-9 (Oct., 1993b)].

A.D. ALEKSEEV, K.A. BRONNIKOV, N.I. KOLOSNITSYN, V.N. MEL'NIKOV, and A.G. RADYNOV; *Meas. Tech.* **37** (No. 1), 1-5 (1994) [translated from *ИЗМЕРИТЕЛЬНАЯ ТЕХН.*

(No. 1), 3-5 (Jan., 1994)].

P.N. ANTONYUK, K.A. BRONNIKOV, and V.N. MEL'NIKOV; *Meas. Tech.* **36** (No. 8), 837-844 (1993) [translated from *ИЗМЕРИТЕЛЬНАЯ ТЕХН.* (No. 8), 3-6 (Aug., 1993)].

P.N. ANTONYUK, K.A. BRONNIKOV, and V.N. MEL'NIKOV; *Astron. Lett.* **20** (No. 1), 59-61 (1994) [translated from *ПИСЬМА В АСТРОН. Ж.* **20** (No. 1), 72-75 (1994)].

J.-P. BLASER, M. BYE, G. CAVALLO, T. DAMOUR, C.W.F. EVERITT, A. HEDIN, R.W. HELLINGS, Y. JAFRY, R. LAURENCE, M. LEE, A.M. NOBILI, H.J. PAIK, R. REINHARD, R. RUMMEL, M.C.W. SANFORD, C. SPEAK, L. SPENCER, P. SWANSON and P.W. WORDEN; STEP (Satellite Test of Equivalence Principle): Report on the Phase A Study, ESA/NASA report SCI (93) 4 (March, 1993).

E.R. COHEN and B.N. TAYLOR; *Physics Today* **48** (No. 8, part 2), BG9-BG16 (1995).

G.T. GILLIES; "The Newtonian gravitational constant: recent measurements and related studies;" *Rep. Prog. Phys.* **60**, 151-225 (1997).

M.L. LIDOV and M.A. VASHKOV'YAK; *Astron. Lett.* **20** (No. 2), 188-198 (1994) [translated from *ПИСЬМА В АСТРОН. Ж.* **20** (No. 3), 229-240 (1994)].

V.N. MELNIKOV, M.Yu. KONSTANTINOV, N.I. KOLOSNITSIN, K.A. BRONNIKOV, A.G. RADYNOV, A.D. ALEKSEEV, and P.N. ANTONYUK; "Report for SEE Project"; (1993).

A.M. NOBILI, A. MILANI, E. POLACCO, I.W. ROXBURGH, F. BARLIER, K. AKSNES, C.W.F. EVERITT, P. FARINELLA, L. ANSELMO, and Y. BOUDON; "NEWTON: A manmade planetary system in space to measure the constant of gravity *G*: Proposal for the new medium-size mission of the European Space Agency (ESA)" (1989).

A.M. NOBILI, A. MILANI, E. POLACCO, I.W. ROXBURGH, F. BARLIER, K. AKSNES, C.W.F. EVERITT, P. FARINELLA, L. ANSELMO, and Y. BOUDON; *ESA J.* **14**, 389-408 (1990).

A.M. NOBILI, A. MILANI, E. POLACCO, D. BRAMANTI, G. CATASTINI, I.W. ROXBURGH, F. BARLIER, K. AKSNES, C.W.F. EVERITT, P. FARINELLA, L. ANSELMO, and Y. BOUDON; "NEWTON: A manmade planetary system in space to measure the constant of gravity *G*: Proposal for the M3 medium-size mission of ESA" (May, 1993).

A.V. OSIPOVA; *Meas. Tech.* **36** (No. 12), 1305-1310 (1993) [translated from *ИЗМЕРИТЕЛЬНАЯ ТЕХН.* (No. 12), 3-6 (Dec., 1993)].

H.J. PAIK; *Class. and Quantum Grav.* **11**, A133-A144 (1994).

H.J. PAIK and J.-P. BLASER, proceedings of the STEP symposium, University of Pisa, April, 1993, in press.

A.J. SANDERS and W.E. DEEDS; *Phys. Rev. D* **46** (No. 2), 489-504 (1992a).

A.J. SANDERS and W.E. DEEDS; *Bull. Am. Phys. Soc.* **37** (No. 7), 1675 (1992b).

A.J. SANDERS, W.E. DEEDS and G.T. GILLIES; in *The Earth and the Universe: Festschrift in honour of Hans-Jürgen Treder*, edited by WILFRIED SCHRÖDER (International Association of Geomagnetism and Aeronomy, Bremen-Rönnebeck, Germany, 1993), pp. 360-365.

A.J. SANDERS and G.T. GILLIES; *Rivista del Nuovo Cimento* **19** (No. 2), 1-54 (1996).

Space-Based Measurements of Spin in Gravity

Alvin J. Sanders, Dept. of Physics and Astronomy,
University of Tennessee, Knoxville, Tennessee 37996-1200 USA
ASanders@utk.edu

George T. Gillies, Dept. of Mechanical, Aerospace, & Nuclear Engineering,
University of Virginia, Charlottesville, Virginia 22901 USA
GTG@virginia.edu

Presented at the International School of Cosmology and Gravitation, Erice, Sicily, May 13-20, 1997

Abstract

Proposals for space-based tests for spin-dependent gravitation are reviewed in the context of ground-based results, and a method for making such a test on a SEE (Satellite Energy Exchange) mission is suggested.

Within the past decade several investigators have proposed methods for searching for spin-dependent gravitational interactions in space. We briefly review these proposals here. A review of recent theory and the experimental situation for spin-dependent gravity appears elsewhere in this volume (Ritter *et al.*, I and II).

Experimental tests for spin-dependent gravity require that the magnetic effects associated with spin polarization be removed to a very high order, since electromagnetic forces are so much greater than those of gravity (ordinary mass-mass coupling, and especially any forces hypothesized for spin-mass or spin-spin coupling). Two basic techniques may be used to exclude magnetic effects:

1. Superconducting shielding around the body in which spins have been aligned by inducing a magnetic moment.

2. "Spin-compensation." This means composing the test body of transition elements in proportions such that the magnetic effects of the quantum-mechanical spins are approximately compensated by those of the orbital angular momenta.

The latter approach has been used in a series ground-based experiments at the Jesse Beams Laboratory at the University of Virginia, including an ongoing search for spin coupling to a hypothetical dark-matter halo centered on our galaxy (Ritter, Goldblum, Ni, Gillies, & Speake, 1990), (Ritter, Gillies, & Winkler, 1993, 1994, 1996, and 1997). These experiments used the compound Dy_6Fe_{23}. Residual external magnetism was virtually eliminated by shielding with very thin (< 1 mm) layers of pure iron and mu metal.

The proposed STEP (Satellite Test of the Equivalence Principle) satellite includes plans for an accelerometer designed to search for spin-dependent gravity (Blaser *et al.*, 1993). The STEP accelerometer uses the former approach--superconducting shielding--to isolate the detector mass from the magnetic effects of the spin-source body. The basic principle of this experiment is as follows: The polarization in a spin-source body is controlled by a current through a coil around the body, and the acceleration of a surrounding cylindrical test body is measured by SQUIDS. One then searches for correlations between this acceleration and the polarization of the spin-source body. This experiment is described elsewhere in detail. (Blaser et al, 1993).

We note that the STEP approach entails controlling spins by a current and then measuring the resulting acceleration of a mass. That is, the current is the independent variable, while the acceleration of the mass is the dependent variable. Ni has proposed a converse experiment to search for spin-mass coupling. The principle is essentially as follows: Move a source mass (by whatever means) and search for changes in the polarization of a nearby test mass using a pick-up coil (Ni, 1996).

Ni points out that this experiment has the advantage that it can be tested on the ground. We would suggest that this could be carried a step further: Since the source mass is not free-floating, it would seem that this scheme does not require weightless conditions, and therefore it may be possible to carry out the entire experiment quite satisfactorily on the ground.

Project SEE (Satellite Energy Exchange) of the University of Tennessee and Oak Ridge National Laboratory (Sanders & Gillies, 1996) is described elsewhere in this volume. As originally proposed (Sanders & Deeds, 1992a and 1992b), SEE did not include a test for spin-dependent gravity. We now introduce a possible method for testing for anomalous spin coupling ($\sigma \bullet \mathbf{r}$) on a SEE mission.

The method relies on observing an orbital perturbation due to the supposed anomalous spin coupling to the mass of the Earth. If the spin σ of an orbiting test body lies in the plane of its orbit, then the force resulting from any spin coupling to the mass of the Earth of the form $\sigma \bullet \mathbf{r}$ will always have the same sense with respect to the spin direction. To see the resulting effects on the orbit--and for convenience of discussion only--we consider a polar orbit, and we choose the spin direction to be parallel to the Earth's rotation axis. Then, assuming the coefficient of $\sigma \bullet \mathbf{r}$ is positive, the effect on the body's orbit will be a secular decrease in the orbital radius at the southward crossing of the equator (since this is the point 90 degrees downstream from the point where σ and $\mathbf{r}$ are parallel) and a corresponding secular increase in the orbital radius at the northward equatorial crossing. This will be manifested as either (a) a change in eccentricity, if the perigee is at the equator, or (b) a precession of perigee, if the perigee is located at one of the Earth's poles, or (c) a combination of these two effects, if the perigee is located elsewhere. We reiterate that the poles are used only as arbitrary references; any fixed direction in the orbital plane could have been used.

The resulting perturbation can be most readily observed experimentally by comparison with the orbit of a co-orbiting spinless body. Of course, collisions would be avoided by using slightly different Earth orbits. However, the orbits and the spin directions must be chosen carefully to assure that the mutual perturbation due to Newtonian gravity cannot masquerade as effects due to spin.

On a SEE mission the method of choice for suppressing magnetic effects is the use of compensation, since the SEE geometry and temperature preclude the use superconducting shielding. Candidate compensation compounds with different compensation temperatures are available (Herbst & Croat, 1984).

The pair of test bodies whose orbits are to be compared for a $\sigma \bullet \mathbf{r}$ experiment on a SEE mission could be either (1) two particles, one with spin and the other spinless, or (2) the (spinless) shepherd and a particle with spin . Because the use of a shepherd with spin would be likely to undermine confidence in the validity of any other non-Newtonian effects (those not involving spin) which might be discovered on a SEE mission, we consider only the former arrangement.

Specifically, the two particles could be placed in slightly eccentric orbits with identical semi-major axes, but with their perigees 180 degrees apart (we assume for discussion that the shepherd is in a perfectly circular orbit). Continuing the use of the Earth's poles as reference points for convenience in visualizing spatial relationships, if the perigee of one particle is at the North pole and that of the second is at the South pole, then the first particle will always lie generally to the south of the second particle. Thus, the Newtonian attraction of the second particle upon first will always have a northerly sense. This is easily seen when both particles are over either pole, since one particle is at its perigee and the other at its apogee. This is also true at all intermediate times, since the particle which is at perigee at one point pulls ahead of the other particle for the next half orbit. For example, at the equatorial crossing the two particles will have the same orbital radius, and the one which was more recently at perigee will be leading (by exactly twice the amount of the vertical separation at the time of perigee).

Because the sense of the Newtonian force between the two particles with opposing (180 degrees apart) perigees is always the same, it follows that the resulting orbital perturbation is a common precession of perigees (and *not* a change in eccentricity). Thus, it follows that, when testing for a $\sigma \cdot \mathbf{r}$ interaction, the preferred orientation of the spin is such that resulting orbital perturbation will be a secular perturbation of the eccentricity rather than a perigee precession, lest the Newtonian attraction be misinterpreted as a $\sigma \cdot \mathbf{r}$ effect. Successful implementation of this method requires that the perigees of the two particles be accurately known and that, if not exactly 180 degrees apart, corrections be applied. It is expected that the accuracy of the internal tracking in the SEE capsule will be sufficient to make the necessary corrections.

Acknowledgments

We wish to express our appreciation to Prof. Venzo de Sabbata and the "Ettore Majorana Centre for Scientific Culture" for the invitation to this stimulating school and for their gracious hospitality during the Course. We are pleased to acknowledge financial support from the Marshall Space Flight Center of NASA, Oak Ridge National Laboratory, Oak Ridge Center for Manufacturing Technologies, the Tennessee Higher Education Commission Science Alliance, and the Dean of Arts of Sciences, the Office of Research, and the Department of Physics and Astronomy of The University of Tennessee.

References

J.-P. BLASER, M. BYE, G. CAVALLO, T. DAMOUR, C.W.F. EVERITT, A. HEDIN, R.W. HELLINGS, Y. JAFRY, R. LAURENCE, M. LEE, A.M. NOBILI, H.J. PAIK, R. REINHARD, R. RUMMEL, M.C.W. SANFORD, C. SPEAKE, L. SPENCER, P. SWANSON and P.W. WORDEN; STEP (Satellite Test of Equivalence Principle): Report on the Phase A Study, ESA/NASA report SCI (93) 4 (March, 1993).

J.F. HERBST and J.J. CROAT; *J. Appl. Phys.* **55**, 3024 (1984).

WEI-TOU NI; *Class. Quantum Grav.* **13**, A135-A141 (1996).

R.C. RITTER, C.E. GOLDBLUM, WEI-TOU NI, G.T. GILLIES, and C.C. SPEAKE; *Phys. Rev. D.* **42** (No. 4) 977-991 (1990).

R.C. RITTER, G.T. GILLIES, and L.I. WINKLER; *Phys. Rev. Lett.* **70** (No. 6), 701-704 (1993).

R.C. RITTER, G.T. GILLIES, and L.I. WINKLER; *Proceedings of the XVth International School of Cosmology and Gravitation*; submitted for publication.

R.C. RITTER, G.T. GILLIES, and L.I. WINKLER; *Proceedings of the XVth International School of Cosmology and Gravitation*; submitted for publication.

R.C. RITTER, L.I. WINKLER, and G.T. GILLIES; in J. Trân Than Vân, G. Fontaine, and E. Hinds, eds., *Particle Astrophysics, Atomic Physics, and Gravitation: Proceedings of the XXIXth Rencontre de Moriond (XIVth Moriond Workshop)* (Editions Frontières, Gif-sur-Yvette, France, 1994), pp. 441-444.

R.C. RITTER, L.I. WINKLER, and G.T. GILLIES; in J. Trân Than Vân, G. Fontaine, and E. Hinds, eds., *Dark Matter in Cosmology - Quantum Measurements Experimental Gravitation: Proceedings of the XXXIst Rencontre de Moriond (XVIth Moriond Workshop)* (Editions Frontières, Gif-sur-Yvette, France, 1994), pp. 417-422.

R.C. RITTER, L.I. WINKLER, and G.T. GILLIES; Proceedings of the XXXIInd Rencontre de Moriond (XVIIth Moriond Workshop), submitted for publication.

A.J. SANDERS and W.E. DEEDS; *Phys. Rev. D* **46** (No. 2), 489-504 (1992a).

A.J. SANDERS and W.E. DEEDS; *Bull. Am. Phys. Soc.* **37** (No. 7), 1675 (1992b).

A.J. SANDERS and G.T. GILLIES; *Rivista del Nuovo Cimento* **19** (No. 2), 1-54 (1996).

MACH'S PRINCIPLE AND TORSION IN GENERAL RELATIVITY

Yu Xin (Alfred Yu)

Hong Kong Institute of General Relativity and Cosmology and

Department of Astronomy,

Beijing Normal University, Beijing, PRC

ABSTRACT

The gravitational sub-theory of the "General Relativity on Spinor-Tensor Manifold" is here presented in the context of Mach's Principle and is shown to predict all the known physical consequences of that principle in a covariant way.

1. Preamble

Any relativistic theory of spacetime that purports to embrace the microscopic domain (Quantum Theory, QT) and the macroscopic universe in a unified scheme must have Mach's Principle at its foundation if Einstein's philosophy is to be followed. For, in his successive attempts at a unified field theory, Einstein sought to derive the properties of matter from spacetime geometry in order to avoid the duality between metric and matter, which duality is anti-Machian. If the term "properties of matter" in the above is extended to include the quantum, electromagnetic as well as inertial manifestations, then we would have a truly Einsteinian approach to a modern Unified Field Theory (UFT) which is also Machian; according to Einstein[1]. "In order to develop this (Mach's) idea within the limits of the modern theory of action through a medium, the properties of space-time continuum which determine inertia must be regarded as field properties of space, analogous to the electromagnetic field". (For this point of view, see M. Sachs)[2]. However, although Mach's Principle (MP) and the Equivalence Principle (EP) served as the two principal guides in Einstein's search for a spacetime theory of gravitation, only the latter has attained the status of a verifiable physical principle. The MP has remained controvertial (Reinhardt[3]) not least because GR, though originally inspired by MP has been found to be ultimately anti-Machian. In one particular case it is actually shown that GR, whilst satisfying EP goes against MP at the same time, though MP in general is supposed to imply EP. This paper retraces some of these anti-Machian features of GR and seeks to discover the reasons behind the GR failure to meet the Machian ideas. It then goes on to show that solidly based on EP and MP, a UFT can be established whose sub-theory, the alternative General Relativistic Theory of gravitation embraces all of the physical consequences of MP as well as the verifications of the three classical tests and the gravitational radiation of PSR 1916+13.

2. Mach's Principle and Einstein's GR

That Mach's Principles implies the EP has been explicated by Reichenbach[4] thus: "The equality of inertia and gravity is the strict formulation of Mach's principle in the narrower sense. It implies that every phenomenon of inertia observable in an accelerated system can also be explained as a gravitational phenomenon . . . ." And yet, Einstein's GR though based on the EP has been found to violate the MP in at least the following important instances: (Woodward and Youngrau[5], Rindler[6] and Torretti[7])

(a) There exist nonsingular solutions beside Minkowski space of the unmodified Einstein equation with $T_{\mu\nu} = 0$. (e.g., Taub-NUT space and Ozsváth-Schücking space). Hence, the same matter configuration (viz., no matter) can give rise to inequivalent non-flat spacetimes. We attribute this to the non-existence of gravitational field momentum tensor (the analogue of the electromagnetic field momentum) on the RHS of Einstein's equation. For, even at spacetime points where there is no ponderable matter, there should still be gravitational field due to the masses in the rest of the universe and hence $T_{\mu\nu}$ (of gravitational field) $\neq 0$ which condition would then be demanded by Mach's Principle.

(b) According to R. Torretti[7], ". . . it was shown by C.H. Brans (1962) that the first Machian effect derived by Einstein from his field equations, viz., the inertia of a particle is increased by piling up masses in its neighborhood, can always be wiped out by a suitable choice of coordinate system and does not even show up in a co-ordinate-free description. For, in a Fermi chart along the world line of any particle, relative to which all the gravitational field components $\Gamma^{\alpha}_{\beta\gamma}$ vanish, the particle's behaviour appears to be unaffected by the gravitational influence of the rest of (distant matter in the universe) and its inertial tendency to persevere in or relapse into a geodetic path cannot be linked to that influence . . . the Principle of Equivalence implies that physical phenomena referred to such charts obey the laws of Special Relativity and show no effects of the surrounding mass distribution. In this sense, as Steven Weinberg has ably observed, "The equivalence principle and Mach's principle are in direct opposition". We attribute this unwitting contradiction between MP and EP in Einstein's GR to (i) the use of Levi-Civita connection (hence Riemannian spacetime) which was the only type of connection known Einstein in 1916 and (ii) the consequent identification of gravity with the metric $g_{\mu\nu}$. We shall see later that an extension of this connection could resolve this difficulty.

(c) The electromagnetic-like inductive effects of inertia as consequences of MP in the form of Lense-Thirring effects viz., (i) that a centrifugal force field as well as a Coriolis field are generated in a massive rotating hollow sphere and (ii) that a massive rotating body induces an acceleration on nearby test particles can indeed be obtained by the linearized field equations. The linearization, however, presupposes a flat background metric, in contradiction to MP. Indeed, these effects are derived from the components of the Levi-Civita connection (see e.g. Einstein[1]) which can also be transformed away as in the case of (b). The cause of the failure of Einstein's GR to comply with the inductive effects of inertia as a consequence of MP can be made good as suggested in (b) above.

(3) As a consequence of the absolute acceptance of Einstein's GR in its present form, its anti-Machian solutions have led to the doubts of MP as a viable physical principle (see, e.g. Reinhardt[3]). Against this doubt we assert that: (i) MP is truly

relativistic and is hence reasonable. For, according to Mach (Rindler) ". . . it does not matter if we think of the earth as turning round on its axis, or at rest while the fixed stars revolve around it . . . The law of inertia must be so conceived that exactly the same thing results from the second supposition as from the first." It is hard to imagine a counter statement to this. Indeed, as shown earlier, MP actually implies EP. (ii) MP is necessary as the foundation of any covariant theory of spacetime. For, we have the following logical sequence:

$$MP \Rightarrow Covariance \Rightarrow spacetime\ is\ curved.$$

MP certainly implies covariance of field equations; but a field equation written in tensor form in general curvilinear coordinates does not suffice since this can also be done in Minkowski spacetime. It requires no less than a curved spacetime to guarantee general covariance in a theory, and (iii) a generalized MP serves as a philosophical as well as physical foundation for any covariant UFT in the Weyl, Eddington-Einstein conception. The Generalized Mach's Principle (GMP) may be stated thus:

In the absence of all matter (including its inertial, electromagnetic and quantum manifestations) in the universe, spacetime has no meaning and hence, does not exist. On the contrary, spacetime is the manifestation of matter (including all of its inertial, electromagnetic and quantum manifestations). That is spacetime $\Leftrightarrow$ matter.

Hence GMP implies unity of the universe with its various manifestations and hence the existence of a covariant UFT based on spacetime geometry.

4. General Relativity on Spinor-Tensor Manifold (GRSTM) — Its Gravity Sub-Theory.

(a) One such UFT in accord with GMP was recently proposed by Yu[8](1996) in the name of "General Relativity on Spinor-Tensor Manifold" which contains a Nondualistic Einstein's equation, Maxwell's equation as well as Dirac's equation all derived from the single geometry of a Spinor-Tensor manifold. It is not the purpose here to repeat that theory, but is to show that, in the absence of the electromagnetism, its gravitational sub-theory predicts all the consequences of the restricted MP in a covariant way.

(b) The "pure" gravitational sub-theory of the GRSTM consists of the following basic field equations derived entirely from the Spinor-Tensor geometry: (Yu[8])

$$\overset{\circ}{R}_{\mu\nu} - \frac{1}{2} g_{\mu\nu} \overset{\circ}{R} = -6(h_\mu h_\nu - \frac{1}{2} g_{u\nu} h_\alpha h^\alpha) + \lambda g_{\mu\nu} \tag{4.1}$$

$$\nabla_\rho S_{\mu\nu}^{\cdot\cdot}\rho = 0 \tag{4.2}$$

where the axial vector h^λ is related to the torsion tensor $S_{\mu\nu}^{\cdot\cdot}\rho$ by

$$S_{\lambda\mu\nu} = t_{\lambda\mu\nu\rho} h^\rho \tag{4.3}$$

as required by the EP and λ is the cosmological constant. In (4.1), $\overset{\circ}{R}_{\mu\nu}$is the usual Ricci tensor in terms of the Levi-Civita connection.

In his "Fundamental Theory", Eddington[9]defines a "particle" as a "carrier of variates". Hence, the "V_{10}-particle" is defined by a momentum vector denoted by an "E-number" (an element of a variety of Clifford Algebra) of 10 components 4 of which denote the linear momentum vector of the particle and 6 denote the spin angular bi-vector. Further, a "V_{16}-particle" is denoted by a 16-component E-number with the extra 6 components denoted the electromagnetic field tensor. Following this philosophy, we shall call h^λ the "matter momentum vector" and since, according to EP, inertia and gravitation are mutually convertible we can generally "represent" h^λ by

$$h^\lambda = \frac{\nabla^\lambda \Omega}{N3} + t^{\lambda\mu\nu\rho} \nabla_\rho A_{\mu\nu} + \gamma t^{\lambda\mu\nu\rho} M_{\mu\nu\rho} + \gamma p^\lambda \tag{4.4}$$

where $\gamma = \frac{8\pi}{c}\sqrt{G}$; $G =$ gravitational constant and where

$$\Omega = \text{ scalar gravitational field} \tag{4.5}$$

$$A_{\mu\nu} = \text{ gravitational radiant bi-vector} \tag{4.6}$$

$$M_{\mu\nu\rho} = \text{ spin-angular momentum tri-vector} \tag{4.7}$$

and

$$p^\lambda = \text{ linear momentum 4 vector} \tag{4.8}$$

Thus, the first two terms on the RHS of (4.4) together form the "gravity part of h^λ" and the last two terms together the "inertial part of h^λ". Note that (4.4) corresponds, in the sense of Hestenes[10], the representation of a multivector A in 4 dimensions:

$$A = \langle A\rangle_0 + \langle A\rangle_1 + \langle A\rangle_2 + \langle A\rangle_3 + \langle A\rangle_4 \tag{4.9}$$

where $\langle A\rangle_r$ denotes an "r-blade" in 4 dimensions.

In (4.4)

$$\begin{aligned} \Omega &\leftrightarrow \langle A\rangle_0, \quad p^\lambda \leftrightarrow \langle A\rangle_1, \quad A_{\mu\nu} \leftrightarrow \langle A\rangle_2 \\ M_{\mu\nu\rho} &\leftrightarrow \langle A\rangle_3, \quad t^{\lambda\mu\nu\rho} \leftrightarrow \langle A\rangle_4 \end{aligned} \tag{4.10}$$

Thus, (4.4) represents a "complete momentum vector" of a matter field, which naturally includes both gravity and inertial field as required by the EP. Their respective proportions in any representation depend on the particular application.

Now, taking the divergence of (4.1), taking into account the contracted Biachi Identity and the condition $\nabla g_{\mu\nu} = 0$, one obtains

$$\nabla_\lambda h^\lambda = 0 \tag{4.11}$$

And substitution of (4.4) into (4.2) via (4.3) gives

$$\Box A_{\mu\nu} = \gamma \nabla_\rho M_{\mu\nu}^{\cdot\cdot}{}^{\rho} + \gamma\left(\frac{\partial P_\nu}{\partial x^\mu} - \frac{\partial P_\mu}{\partial x^\nu}\right) \tag{4.12}$$

Also, substitution of (4.4) into (4.11) gives

$$\Box\Omega = +\sqrt{3}\gamma\nabla_\lambda P^\lambda \tag{4.13}$$

Evidently, (4.11) expresses the law of conservation of the field momentum vector h^λ whilst (4.12) and (4.13) are the wave equations for the gravitational field $(\Omega, A_{\mu\nu})$.

Thus, the sources for the gravitational radiant field $A_{\mu\nu}$ are the divergence of spin angular momentum and the curl of linear momentum of ponderable matter. There is, however another way of looking at (4.12):

$$\nabla_\rho M_{\mu\nu}{}^{\cdot\cdot}\rho = \frac{1}{\gamma}\Box A_{\mu\nu} - (\frac{\partial P_\nu}{\partial x^\mu} - \frac{\partial P_\mu}{\partial x^\nu}) \tag{4.14}$$

which is the Law of Conservation of Angular Momentum: The total angular momentum of a material system is conserved if no gravitation is radiated as a result of its rotational motion and if its circulation vanishes.

Further, as has already been shown for a test particle moving slowly in a gravitational field.(Yu[8]). (4.12) also yields, on contraction with U^ν, the four velocity vector of the particle

$$U^\lambda B_{\lambda\nu} = \gamma\frac{DU_\nu}{ds} \tag{4.15}$$

where

$$B_{\lambda\mu} \equiv \frac{1}{m}(\Box A_{\lambda\mu} - \gamma\nabla_\rho M_{\lambda\nu}{}^{\cdot\cdot}\rho) \tag{4.16}$$

Hence, the particle moving in a gravitational field follows a geodesic path iff

$$U^\lambda B_{\lambda\nu} = 0 \tag{4.17}$$

A sufficient condition is

$$\Box A_{\lambda v} = 0 \quad \text{and} \quad \nabla_\rho M_{\lambda\nu}{}^{\cdot\cdot}\rho = 0 \tag{4.18}$$

That is, $A_{\lambda\mu}$ is in a plane wave condition and that its spin angular momentum vanishes or else conserved. In this case, only the Ω–field is felt.

(c) We first illustrate the above theory by considering a test particle moving in an

Ω-field of a spherical star in the case where

$$h^{\lambda} = \frac{1}{\sqrt{3}} \nabla^{\lambda} \Omega \tag{4.19}$$

so that (4.1) becomes

$$\overset{\circ}{R}_{\mu\nu} - \frac{1}{2} g_{\mu\nu} \overset{\circ}{R} = -2(\nabla_{\mu}\Omega \nabla_{\nu}\Omega - \frac{1}{2} g_{\mu\nu} \nabla_{\alpha}\Omega \nabla^{\alpha}\Omega) \tag{4.20}$$

and (4.13) becomes

$$\Box \Omega = 0 \tag{4.21}$$

In isotropic coordinates the solution for (4.20) is

$$ds^2 = e^{-2\Omega} dt^2 - e^{2\Omega} d\vec{r} \cdot d\vec{r} \tag{4.22}$$

and that for (4.21) is

$$\Omega = \frac{M}{r} + \int_{v} \frac{e}{r_1} dv; \tag{4.23}$$

e = density of distant star, M = mass of control star, r_1 = distance between particle and a distance star where $\int_v \frac{\rho}{r_1} dv$ is the potential due to the rest of the universe. In the case of the solar system, it is safe to put

$$\int_{v} \frac{\rho}{r_1} dv = 0 \tag{4.24}$$

The geodesic equation (Yu[11])

$$\begin{aligned} \frac{Du^{\lambda}}{ds} &= 0 \Rightarrow \\ \ddot{\vec{r}} + \frac{M}{r^3} \vec{r} &= \frac{M}{r^3} [\vec{L} \wedge \dot{\vec{r}} + (\vec{r} \cdot \dot{\vec{r}}) \dot{\vec{r}}] \end{aligned} \tag{4.25}$$

which, in the case of Sun-Mercury system, gives an estimate of the precession of the perihelion to the same degree of accuracy as Einstein's GR (Lee[12]). However, the full solution (4.23) does agree with Mach's Principle in which case we have

$$\ddot{\vec{r}} + \frac{M}{r^3}\vec{r} - \frac{M}{r^3}[\vec{L} \wedge \dot{\vec{r}} + (\vec{r}\cdot\dot{\vec{r}})\,\dot{\vec{r}}] = (1+\dot{\vec{r}}^2)\frac{\partial}{\partial \vec{r}}\int_v \frac{\rho}{r_1}dv - 2\,\dot{\vec{r}}\,\frac{\partial}{\partial s}\int_v \frac{\rho}{r_1}dv$$
$$= -(1+\dot{\vec{r}}^2)\int_v \frac{e\,\vec{r}}{r_1^3}dv + 2\,\dot{\vec{r}}\int_v \frac{\rho}{r_1^3}(\vec{r}\cdot\dot{\vec{r}})\,\vec{r}\,dv \tag{4.25a}$$

(d) In nearly flat spacetime viz.,

$$g_{\mu\nu} \simeq \eta_{\mu\nu} \text{ (Minkowski metric)} \tag{4.26}$$

with $M_{\mu\nu \cdot \cdot \lambda} = 0$, (4.12) gives (Yu[8])

$$curl\ \vec{e} = -\frac{1}{c}\frac{\partial \vec{b}}{\partial t} \tag{4.27}$$

$$curl\ \vec{b} = -\frac{1}{c}\frac{\partial \vec{e}}{\partial t} - \gamma\,\vec{p} \tag{4.28}$$

where

$$-\vec{e} \equiv A_{oi}, \quad \vec{b} = t_{ijk}A^{ik} \tag{4.29}$$

being the radiant vectors of the gravitational field. Thus, with the correspondence

$$(\vec{e}, \vec{b}) \leftrightarrow (\vec{E}, \vec{H}) \equiv \text{electromagnetic field} \tag{4.30}$$

$$(-\gamma\,\vec{p}) \leftrightarrow (\vec{J}) \equiv \text{ electrical current} \tag{4.31}$$

we have the identity

$$\frac{1}{c^2}\frac{\partial}{\partial t}(\vec{e} \wedge \vec{b}) = -\rho\,\vec{e} + \gamma\,\vec{p} \wedge \vec{b} + div\,\vec{T} \tag{4.32}$$

where

$$\vec{T} \equiv (\vec{e}\,\vec{e} + \vec{b}\,\vec{b}) - \frac{1}{2}\,\vec{I}\,(e^2 + b^2) \tag{4.33}$$

$$\vec{I} \equiv \text{unit tensor in Euclidean subspace} \tag{4.34}$$

$$\rho \equiv p_o$$

Thus, $\vec{e} \wedge \vec{b}$ can be identified as the radiant gravitational field momentum and the identity (4.32) resembles a kind of "equation of motion:" for the gravitational field transfer. It is likely to be important in the case of explosion of stars and the violent effects taking place in the active galactic nuclei.

5. Machian Consequences of the Theory

According to Einstein[1], the principle consequences MP may be stated as follows.

(1) The inertia of a body must increase when ponderable masses are piled up in its neighbourhood.

(2) A body must experience an accelerating force when neighbouring masses are accelerated, and, in fact, the force must be in the same direction as that acceleration.

(3) A rotating hollow body must generate inside of itself a "Coriolis field", which deflects moving bodies in the sense of the rotation, and a radial centrifugal force as well.

As a corollary of (3) we also have

(4) A massive rotating central body induces as acceleration on neighbouring particles.

(a) The first effect is clearly deducible from the present theory as explicated in (4.25a) which is the equation of motion of a test particle (of unit mass) in the presence of a central body of mass M and the distribution of matter ρ in the rest of the universe. For, according to Yu[8].

$$\frac{dt}{ds} = e^{2\Omega} \tag{5.1}$$

where Ω is given by (4.23) and thus for (4.25a),

$$\ddot{\vec{r}} \equiv \frac{d}{ds}\left(\frac{d\vec{r}}{ds}\right) = e^{2\Omega}\frac{d}{dt}\left(e^{2\Omega}\frac{d\vec{r}}{dt}\right) \tag{5.2}$$

which clearly demonstrates (1) in the conformally flat subspace as required by (4.23).

(b) In order to demonstrate the existence of the Coriolis-like force consider firstly the term

$$\vec{f} = \frac{M}{r^3}[\vec{L} \wedge \dot{\vec{r}} + (\vec{r} \cdot \dot{\vec{r}})\,\dot{\vec{r}}] \tag{5.3}$$

Let

$$\vec{r} = r\,\vec{e}\,; \quad \vec{e} = \frac{\vec{r}}{r} \tag{5.4}$$

and thus

$$\dot{\vec{r}} = \dot{r}\vec{e} + r\,\vec{w} \wedge \vec{e} \tag{5.5}$$

where $\vec{w}$ is the angular velocity of the radius vector $\vec{r}$.

Hence,

$$\vec{L} = \vec{r} \wedge \dot{\vec{r}} = r^2\,\vec{w}$$

and we have

$$\vec{f} = -\frac{M}{r}\,\vec{e} \wedge \vec{w} + \frac{M}{r^3}(\vec{r} \cdot \dot{\vec{r}})\,\dot{\vec{r}} \tag{5.6}$$

The first term is the Corilis-like force (similar to that derived by Lense and Thirring in the context of Einstein's Theory) and the second term (not present in the Lense-Thirring effect) is the force responsible for the deflection of the particle "in the direction of rotation". Thus the Machian effect (3) is here seen to be qualitatively correct if only M is present.

An expression almost identical to that of (5.3) due to the mass distribution in the rest of the universe can also be obtained from the RHS of (4.25a) (to be reported in a further paper) hence deducing the Machian effect (3) in full. This effect has not been obtained in Einstein's Theory in a covariant way.

(c) The Machian effect (2) is basically a Maxwell-like induction effect which can obviously be obtained by solving the gravito-Maxwell equations (4.27) and (4.28) in the usual way. Note that (4.25a) has been obtained from the scalar field Ω alone without the radiant field $A_{\mu\nu} \equiv (\vec{e}, \vec{b})$ and hence cannot in principle account for the inductive effect on its own.

6. Conclusions

In his eloguent statement of the "Generalized Mach's Principle", Mendal Sachs[2] has this to say: "When the principle that asserts this interpretation of inertial mass of matter - called the 'Mach Principle'— is incorporated with the theory of general relativity, it follows that the inertial manifestation of matter must be derivable from the field properties of space-time. Thus, the full exploitation of the Mach principle implies that inertia must be incorporated with all the other manifestation of interacting matter in this unified field approach. But the approach also implies that there are no intrinsic properties of elements of matter that would be definable in an atomistic theory. Thus, in addition to inertia,the electromagnetic properties of elementary matter, nuclear moments, etc., are all related to the dynamical coupling of this matter to the rest of a closed system. Such a view, which we may call the 'generalized Mach principle', then removes all recurrents of atomism from the theoretical description of matter, in terms of first principles" Evidently, whilst based on a different geometry (from that of M. Sachs), the "General Relativity on Spinor-Tensor Manifold" (Yu[8]) is in full accord with the above stated GMP. Moreover, its pure gravitation sub-theory demonstrates all the known physical consequences of the (restricted) MP in all their mathematical details and in a covariant way.

Note, however, that although (4.1) resembles Einstein's equation, it differs from Einstein's GR in the following significant ways:

(a) In GR, the gravitational field is identified with the metric tensor $g_{\mu\nu}$ which, at times, (when it is convenient) is also identified with the "inertial field". And yet, in application the "inertial field" is usually given by the matter Energy Momentum Tensor (EMT) $\rho u_\mu \, u_\nu$ which is quite distinct from $g_{\mu\nu}$ and which determines it via the Einstein equation. Further, in its application to "empty space" (which is not really empty, because there is gravitational field) the RHS of the Einstein equation is zero, because in GR, there is no EMT for the pure gravitational field. This fact, incidentally, makes it nonsense to speak of the transmission of gravitational radiation (Merceier[13] et al, Wald[14])

In the present theory, both the gravitational field and the inertial field are embodied in h^λ, the "momentum vector for field and matter" as represented in (4.4). And, since, according to the EP, field and matter are mutually convertible, the dissection between field momentum and matter momentum is determined in actual applications. Thus, even for "empty space" (with only gravitational field), the EMT for the gravitational field is not zero and for the statics gravitational field, it is given by the RHS of (4.20). The gravitational field is determined by (4.21), the generalized Laplace equation. The metric tensor $g_{\mu\nu}$ although not gravitational field itself, is nonetheless given explicitly in terms of it by (in isotropic coordinates).

$$g_{oo} = e^{-2\Omega}, \qquad g_{11} = g_{22} = g_{33} = e^{2\Omega} \tag{6.1}$$

(b) In Einstein's GR, the matter EMT is independently postulated whilst in the present theory, it is a consequence of the spacetime geometry via the axial vector h^λ of the torsion tensor as given by (4.3). One of the consequences of this is that in GR, the EMT of matter is not conserved a whilst in the present theory, the "momentum vector for field and matter" h^λ is conserved as shown by (4.11).

(c) In Einstein's GR, the gravitational radiation can only be obtained by linearizing the Einstein's equation hence bringing with it many unanswered questions (Merceier[13] et al, Wald[14]). In the present theory, gravitational radiation is covariantly described by (4.12) and, in the weak field limit, very elegantly by the gravito-Maxwell equations (4.27) and (4.28). The "quadruple formula" for gravitational radiation which predicts the observational results of PSR1916+13 can then be derived in a natural way (Yu[8]).

REFERENCES

[1] Einstein, A., "The Meaning of Relativity", Methuen, London, (1956)

[2] Sachs, M., "General Relativity and Matter", D. Reidel, Dordrecht, Holland (1982)

[3] Reinhardt, M., "Mach's Principle - a critical review", Ztschr. Naturf. Vol. 28a, pp.529-537 (1973)

[4] Reichenbach, H., "The Philosophy of Space and Time", Dover (1958)

[5] Woodward, J.F. and Yourgrau, W., "The Incompatibility of Mach's Principle and The Principle of Equivalence in Current Gravitation Theory", British J. for Phil. of Sci., Vol. 23, pp.111-116 (1972)

[6] Rindler, W., "Essential Relativity", Springer-Verlag, Berlin (1979)

[7] Torretti, R., "Relativity and Geometry", Pergannon, Oxford (1983)

[8] Yu, X., "General Relativity on Spinor-Tensor Manifold", Quantum Gravity (Int. School of Cosmology and Gravitation XIV Course) ed. P.G. Bergmann, V. de Sabbata and H.-J.Treder, World Scientific pp.382-411 (1996)

[9] Eddington, A.S., "Fundamental Theory", Cambridge V.P. (1948)

[10] Hestenes, D. and G. Sobczyk., "Clifford Algebra to Geometric Calculus", D. Reidel, Dordrecht, Holland (1984)

[11] Yu, X., Astrophysics and Space Science, Vol. 202 p. 237 (1993)

[12] Lee, W.T., Ph.D. Thesis, Department of Applied Mathematics, The Hong Kong Polytechnic University (1995)

[13] Mercier, A., H.-J. Treder and W. Yourgrau, "On General Ralativity" Akadomie-Verlag (19 79)

[14] Wald, R.M., "General Relativity" Uni. Chicago P. (1984)

SUBJECT INDEX